Die Grundlehren der mathematischen Wissenschaften

in Einzeldarstellungen
mit besonderer Berücksichtigung
der Anwendungsgebiete

Band 184

Herausgegeben von

J. L. Doob · A. Grothendieck · E. Heinz · F. Hirzebruch
E. Hopf · W. Maak · S. MacLane · W. Magnus · J. K. Moser
M. M. Postnikov · F. K. Schmidt · D. S. Scott · K. Stein

Geschäftsführende Herausgeber

B. Eckmann und B. L. van der Waerden

Murray Rosenblatt

Markov Processes. Structure and Asymptotic Behavior

Springer-Verlag Berlin Heidelberg New York 1971

Prof. Murray Rosenblatt
University of California, San Diego, La Jolla, California

AMS Subject Classifications (1970)
Primary 60J05, 60F05
Secondary 60B99, 60G10, 60G50, 60J10, 60J15, 60J20, 60G25

ISBN-13:978-3-642-65240-0 e-ISBN-13:978-3-642-65238-7
DOI: 10.1007/978-3-642-65238-7

Murray Rosenblatt

Markov Processes. Structure and Asymptotic Behavior

Springer-Verlag New York Heidelberg Berlin 1971

Prof. Murray Rosenblatt
University of California, San Diego, La Jolla, California

Geschäftsführende Herausgeber:

Prof. Dr. B. Eckmann
Eidgenössische Technische Hochschule Zürich

Prof. Dr. B. L. van der Waerden
Mathematisches Institut der Universität Zürich

AMS Subject Classifications (1970)
Primary 60J05, 60F05
Secondary 60B99, 60G10, 60G50, 60J10, 60J15, 60J20, 60G25

ISBN-13:978-3-642-65240-0 e-ISBN-13:978-3-642-65238-7
DOI: 10.1007/978-3-642-65238-7

To Ady

Preface

This book is concerned with a set of related problems in probability theory that are considered in the context of Markov processes. Some of these are natural to consider, especially for Markov processes. Other problems have a broader range of validity but are convenient to pose for Markov processes. The book can be used as the basis for an interesting course on Markov processes or stationary processes. For the most part these questions are considered for discrete parameter processes, although they are also of obvious interest for continuous time parameter processes. This allows one to avoid the delicate measure theoretic questions that might arise in the continuous parameter case. There is an attempt to motivate the material in terms of applications. Many of the topics concern general questions of structure and representation of processes that have not previously been presented in book form. A set of notes comment on the many problems that are still left open and related material in the literature. It is also hoped that the book will be useful as a reference to the reader who would like an introduction to these topics as well as to the reader interested in extending and completing results of this type.

The first chapter deals with some basic properties of Markov processes as well as a number of illustrations. The limiting behavior of Markov chains with stationary transition mechanism is dealt with. There are remarks on independent random variables and the "theory of errors" as a motivation for classical limit theorems. The simplest continuous parameter processes, the Poisson and Wiener (or Brownian motion) processes are introduced. A generalization of the classical result of Polya on recurrence for random walks on commutative countable groups is given.

A number of models in statistical mechanics, "learning" theory in psychology, and in statistical economics are discussed in the second chapter. The object is to show how Markovian-like models arise in a variety of applications and how often questions concerning collapsing of the state space can arise. The concept of ergodicity is already very important in statistical mechanics. A desire to retain at least an approximate version of the Markov property as well as interest in a central limit theorem for dependent processes are evident in the heuristic discussion of the foundations of "non-equilibrium" statistical mechanics.

The third chapter considers a number of results on functions of Markov processes. Conditions under which a function of a Markov process is still Markovian and the relationship between the Chapman-Kolmogorov equation and the Markov property are examined. Functions of Markov processes are generally not Markovian. It is of interest to ask when a finite state process is a function of a finite state Markov chain. An interesting algebraic treatment of this problem is given.

The fourth chapter deals with ergodic problems and first discusses the restriction of a Markov process to a subset of the state space. The L^1 ergodic theorem due to Chacon and Ornstein is presented. The concepts of ergodicity and mixing are introduced and illustrated. Many of the results in "ergodic theory" assume the existence of an invariant measure. Conditions of a topological character on the transition operator that insure the existence of an invariant measure are introduced. Finally, results are obtained on the asymptotic behavior of unaveraged powers of the transition probability operator. Such results can be related to a prediction problem for Markov processes.

Chapter 5 is devoted to random walks or, more properly, convolution of regular measures on compact groups and semigroups. A limit theorem of P. Lévy dealing with the circle group is introduced to motivate the development that then follows. The uniform or Haar measure on a compact group is discussed as the limit law of a convolution sequence of measures. The corresponding type of limit law for convolution sequences of measures on a compact semigroup is an idempotent measure under convolution. The so-called "Rees-Suschkewitsch" theorem describing the structure of compact semigroups is developed in order to characterize the idempotent measures on a compact semigroup. The results are illustrated in the case of semigroups of $n \times n$ (n finite) transition probability matrices. These are perhaps the simplest types of limit theorems for products of independent, identically distributed operators which may not commute.

The sixth chapter deals with nonlinear one-sided representations of Markov processes in terms of independent random variables. Such a treatment is motivated in part by prediction problems. A brief discussion of a corresponding linear representation (the Wold representation) in the linear prediction problem is given. N. Wiener dealt with such nonlinear representations in his book *Nonlinear Methods in Random Theory* where he discussed coding and decoding. Rather complete results are obtained for finite state Markov chains. Partial results are obtained for real-valued Markov processes. A relation between such representations and the isomorphism problem for stationary processes is briefly indicated.

The last chapter is concerned with conditions for the validity of central limit theorems for Markov processes. Cogburn's condition of uniform ergodicity and its relation to infinitely divisible laws as limiting laws for partial sums of stationary Markov variables are given. Uniform (or strong) mixing is also introduced. Both conditions are examined in the case of Markov processes. Finally a central limit theorem is obtained using a condition similar to uniform mixing.

A series of notes that relate the material in the text to relevant results in the literature for Markov processes and more general processes are given after each chapter. Open questions that are of interest are also discussed. It is hoped that the text will be appropriate for an audience with a general mathematical outlook as well as for those with a probabilistic (or statistical) orientation. Readers with a good strong mathematical interest and background whose primary concern is in such areas as statistical physics, mathematical economics, or learning theory should find ideas and methods that are relevant. Several of the questions examined illustrate the interplay of concepts from probability theory with other areas of mathematics. The appendices give a discussion of those ideas from other mathematical areas that bear on the material developed in the text.

The chapters are divided into numbered sections. Sections and formulas cited contain just enough information to identify them. For example, the formulas numbered 2, 1.2, 3.1.2 and referred to in section 6.3 (third section of the sixth chapter) are formula 2 of the same section, formula 2 of section 1 of the same chapter, and formula 2 of section 3.1, respectively.

I am grateful to the John Simon Guggenheim Foundation for its support in the year 1965—66 and the Office of Naval Research for funding of much of the research on which this book is based. I am indebted to Peter Bickel, D. Brillinger, J. L. Doob, John Evans, D. Rosenblatt, T. C. Sun and K. Wickwire for their helpful comments and suggestions. Lastly, my thanks go to Olive Lee and Lillian Johnson for typing and correcting the manuscript, and to Diana Marcus for help in proofreading.

La Jolla, California 1971 Murray Rosenblatt

Table of Contents

Chapter I

Basic Notions and Illustrations

Chapter II

Remarks on Some Applications

Chapter III

Functions of Markov Processes

Chapter IV

Ergodic and Prediction Problems

Chapter V

Random Walks and Convolution on Groups and Semigroups

Chapter VI

Nonlinear Representations in Terms of Independent Random Variables

Chapter VII

Mixing and the Central Limit Theorem

Chapter I

Basic Notions and Illustrations

0. Summary

The probability space for a one sided Markov process with stationary transition mechanism is set up in the discrete and continuous time parameter case under appropriate conditions in section 1. The extension (if possible) to a two-sided process is discussed, as well as the Chapman-Kolmogorov equation for first order transition probabilities. A number of illustrative examples are taken up in the following sections. The asymptotic properties of transition probabilities for Markov chains (Markov processes with a countable state space) are considered in section 2. This motivates in part the later development of an ergodic theorem (in Chapter 4 section 2) for Markov processes with a general state space. The classical example of a sequence of independent random variables is taken up in section 3. There is a brief discussion of the theory of errors and then a derivation of the Poisson approximation to the Binomial distribution and the normal approximation to the distribution of a sum of independent random variables, both with error terms. The theorems on the Poisson and normal approximation are not only of independent interest but are also used later in Chapter 7 section 1 to obtain a remarkable result of Kolmogorov on the approximation of the distribution of a sum of independent and identically distributed random variables by an infinitely divisible distribution with error term. A brief discussion of the continuous parameter Poisson and Wiener (Brownian motion) processes is given in section 4. The classical result of Polya on recurrence of one and two dimensional and nonrecurrence of three dimensional random walks is given in section 5. A generalization (due to Dudley) for random walks on countable Abelian groups is then developed.

1. Markov Processes and Transition Probability Functions

Markov processes are structurally the simplest models of dependent random behavior through time that have been dealt with. Our concern will be with Markov processes whose generating mechanism is stable through time. To avoid complications let us assume that observations of some phenomenon are made at times $n=0,1,2,\ldots$. Let Ω (the state space) be a space of points x representing the possible observations at any given fixed time. The possible events for which a probability is well defined will be the elements of a Borel field $\mathscr{A}$ of subsets of Ω. The stable generating mechanism for a Markov process is given by its *transition probability function* $P(x,A)$ which is assumed to be $\mathscr{A}$-measurable as a function of x for each set (or event) A in $\mathscr{A}$ and a probability measure on the Borel field $\mathscr{A}$ for each x in Ω. Intuitively, $P(x,A)$ represents the probability that the observation at time $n+1$ of the Markov process will fall in the set A given that, at time n, observation x was made. With an initial probability measure μ (at time 0) on the Borel field $\mathscr{A}$, a discrete time parameter Markov process with initial distribution μ and stationary transition probability function $P(\cdot,\cdot)$ can be constructed as follows. For any finite collection of sets $A_0, A_1, \ldots, A_n \in \mathscr{A}$ let

$$P_\mu(A_0 \times A_1 \times \cdots \times A_n) = P_\mu(x_0 \in A_0, \ldots, x_n \in A_n) \tag{1}$$

$$= \int_{A_0} \mu(dx_0) \int_{A_1} P(x_0, dx_1) \cdots \int_{A_{n-1}} P(x_{n-2}, dx_{n-1}) P(x_{n-1}, A_n).$$

An extension theorem of C. Ionescu-Tulcea (see Appendix 3) can be used to extend this set function to a measure P_μ on the Borel field $\mathscr{A}_\infty$ generated by sets of the form $A_0 \times A_1 \times \cdots \times A_n$ on the space Ω_∞ of points $(x_0, x_1, x_2, \ldots) = \omega$.

This probability measure P_μ describes the relative likelihood of observing the different possible trajectories $\omega = (x_0, x_1, x_2, \ldots)$ of the random system being studied through time. The observation on the system at time n is given by the nth coordinate function or random variable $X_n(\omega) = x_n$ and the random process is written $\{X_n\} = \{X_n(\omega); n = 0,1,\ldots\}$. The theorem of Ionescu-Tulcea can be applied to define a one-sided Markov process $\{X_n(\omega); n=0,1,\ldots\}$ with transition probability $P(\cdot,\cdot)$ without any additional conditions in the discrete time parameter case we are now considering. The extension theorem of Kolmogorov (see Appendix 3) can also be used to define a Markov process with transition probability function $P(\cdot,\cdot)$ if additional conditions of a mixed topological and measure theoretic character are imposed on the transition probability function $P(\cdot,\cdot)$ and an initial or marginal probability measure. However, the Kolmogorov extension theorem is especially useful in constructing two-sided Markov processes $\{X_n(\omega); n = \ldots, -1, 0, 1, \ldots,\}$ or continuous time parameter Markov processes and whenever we apply the Kolmogorov theorem we shall implicitly assume that the conditions required for its application are satisfied. The extension theorem of Ionescu-

Tulcea is not useful in the construction of continuous time parameter Markov processes.

Higher step transition probability functions $P_n(\cdot,\cdot)$ can be generated from the one-step transition probability function $P(\cdot,\cdot)$ by the following recursive procedure:

$$P_{n+1}(x,A)=\int_{\Omega} P(x,dy)P_n(y,A), \quad (n=1,2,\ldots.) \tag{2}$$

The relation

$$P_{n+m}(x,A)=\int_{\Omega} P_n(x,dy)P_m(y,A) \quad (n,m=1,2,\ldots,) \tag{3}$$

follows from (2) and is commonly called the Chapman-Kolmogorov equation.

Under certain circumstances, the study of a Markov process with stationary transition mechanism can be extended backward in time. This can be done if there exists a sequence of probability measures μ_n, $n=0, \pm 1,\ldots$, on $\mathscr{A}$ such that $\mu_0=\mu$ and

$$\int \mu_n(dx)P(x,A)=\mu_{n+1}(A), A\in\mathscr{A}.$$

Equation (1) with $\mu=\mu_m$ is used to define the probability of sets $A_m\times\cdots\times A_n$, $-\infty<m<n<\infty$. The Kolmogorov extension theorem can then be employed to set up a measure on the space of points $(\ldots,x_{-1},x_0,x_1,\ldots)=\omega$ describing the history of a system from the infinite past to the infinite future. A random process described by the probability measure is now written $\{X_n\}=\{X_n(\omega);\ n=0, \pm 1,\ldots\}$ with $X_n(\omega)=x_n$ as before. Whether the process is one-sided or two-sided, a *shift transformation* τ corresponding to a forward time shift can be introduced. In the one-sided case, $\omega=(x_0,x_1,\ldots)$ and $(\tau\omega)_n=x_{n+1}$. An inverse τ^{-1} is not always well-defined. In the two-sided case,

$$\omega=(\ldots,x_{-1},x_0,x_1,\ldots)$$

with $(\tau\omega)_n=x_{n+1}$ and the inverse τ^{-1} is always defined. In both the one-sided and two-sided situations $\mathscr{A}_\infty$ and Ω_∞ will be used to denote the Borel field and space of infinite sequences, respectively. If the probability measure μ is *invariant with respect to* $P(\cdot,\cdot)$

$$\int \mu(dx)P(x,A)=\mu(A), \tag{4}$$

then the process can clearly be extended backward in time using the Kolmogorov extension theorem. Then for any event $C\in\mathscr{A}_\infty$,

$$P_\mu(\tau C)=P_\mu(C).$$

If a process is two-sided, $\tau^{-1}C$ is defined and

$$P_\mu(\tau^{-1}C)=P_\mu(C).$$

Such processes are called *stationary Markov processes* because their probability structure is invariant with respect to time translation.

The Borel field $\mathscr{A}_\infty$ was constructed so that it is exactly the Borel field generated by the random variables $\{X_n\}$. Let $\mathscr{A}_m^n$ be the Borel field generated by the random variables $X_k, m \leqslant k \leqslant n$. Notice that $\mathscr{A}_\infty = \mathscr{A}_0^\infty$ in the case of a one-sided process and $\mathscr{A}_\infty = \mathscr{A}_{-\infty}^\infty$ for a two-sided process. $\mathscr{A}_m^n$ carries the information given by the random variables X_k, $m \leqslant k \leqslant n$. Let $\mathscr{B}_m$ be the Borel field generated by X_k, $k \leqslant m$, and $\mathscr{F}_n$ the Borel field generated by X_k, $k \geqslant n$. $\mathscr{B}_m$ and $\mathscr{F}_n$ are the backward and forward Borel fields relative to times m and n, respectively. Sometimes we will write $\mathscr{A}_n$ in place of $\mathscr{A}_n^n$.

Suppose P is some probability measure on the Borel field $\mathscr{A}_\infty$ of points ω of Ω_∞. Given any event C of $\mathscr{A}_\infty$ and a sub-Borel field $\mathscr{B} \subset \mathscr{A}_\infty$ let $P(C|\mathscr{B})(\omega)$ denote the Radon-Nikodym derivative of the measure $P(C \cap B)$ with respect to $P(B)$, $B \in \mathscr{B}$, as measures on $\mathscr{B}$. The derivative is $\mathscr{B}$-measurable and is called the conditional probability of C given the Borel field $\mathscr{B}$. Since the first measure is absolutely continuous with respect to the second,

$$P(C \cap B) = \int_B P(C|\mathscr{B})(\omega) P(d\omega), \quad (B \in \mathscr{B}.)$$

If $\mathscr{B}$ is generated by a family of random variables, then intuitively the conditioning is with respect to the family of random variables. The Markov property can now be simply given for such a general process on the space of sequences Ω_∞. Let F be any event of $\mathscr{F}_n$, $n \geqslant m$. The process is *Markovian* if

$$P(F|\mathscr{B}_m)(\omega) = P(F|\mathscr{A}_m^m)(\omega)$$

for any such event. The conditional probability $P(F|\mathscr{B}_m)(\omega)$ is $\mathscr{A}_m^m$-measurable and so depends only on the past information given at the last time that a completely specified observation is made on the process. It can be shown that the Markov property also may be written

$$P(B|\mathscr{F}_n)(\omega) = P(B|\mathscr{A}_n^n)(\omega)$$

for any event $B \in \mathscr{B}_m$, $m \leqslant n$. The Markov property is independent of time direction. Under fairly broad conditions the probability of an event $A_0 \times A_1 \times \cdots \times A_n$ can be written in terms of an initial distribution and one-step transition probabilities

$${}_kP(X_{k+1} \in A|\mathscr{A}_k^k)(\omega)$$

just as in (1) except that the transition probability function may not be stationary. In (1) it is rather curious and perhaps unesthetic that even though the process has stationary transition function in the forward direction, with time reversed the process may not have a stationary transition mechanism. Occasionally we will have to deal with derived Markov processes not having a stationary transition mechanism. The probability structure of such a process on sets of the form $A_0 \times A_1 \times \cdots \times A_n$ is then given by a sequence ${}_kP(\cdot,\cdot)$, $k=0,1,\ldots$, of one-step transition probability functions describing the transition from state x at time k into set A at time $k+1$ as follows

$$\begin{aligned} P_\mu(A_0 \times A_1 \times \cdots \times A_n) &= P_\mu(x_0 \in A_0, \ldots, x_n \in A_n) \qquad (5) \\ &= \int_{A_0} \mu(dx_0) \int_{A_1} {}_0P(x_0, dx_1) \cdots \int_{A_{n-1}} {}_{n-2}P(x_{n-2}, dx_{n-1})\, {}_{n-1}P(x_{n-1}, A_n). \end{aligned}$$

P_μ is then extended to a probability measure on the Borel field $\mathscr{A}_\infty$ generated by sets of the form $A_0 \times A_1 \times \cdots \times A_n$ just as in the case of a stationary transition probability function.

A transition probability function $P(x,A)$ induces an operator T taking probability measures μ on $\mathscr{A}$ into probability measures μT

$$\nu(A) = (\mu T)(A) = \int \mu(dx) P(x,A) \qquad (6)$$

on $\mathscr{A}$. This is also true for finite measures μ, that is, measures such that $\mu(\Omega) < \infty$. The operation (6) need not be well-defined for σ-finite measures μ. The transition probability function also induces an operator taking bounded functions into bounded functions. We again use the letter T to denote this operator. T acts on the left as an operator on measures; on the right it is to be considered as an operator on $\mathscr{A}$ measurable functions

$$(Tf)(x) = \int P(x,dy) f(y). \qquad (7)$$

The Hölder inequality implies that

$$|(Tf)(x)|^p \leqslant \int P(x,dy) |f(y)|^p = (T\,|f|^p)(x), \quad (\infty > p \geqslant 1.) \qquad (8)$$

Given a σ-finite measure μ let $L^p(d\mu)$ denote the set of $\mathscr{A}$ measurable functions f whose p^{th} $(1 \leqslant p < \infty)$ absolute mean with respect to μ is finite

$$\int |f(x)|^p \mu(dx) < \infty.$$

Inequality (8) implies that if $f \in L^p(d\nu)$, $\nu = \mu T$, then $Tf \in L^p(d\mu)$ with

$$\|Tf\|_{\mu,p} \leqslant \|f\|_{\nu,p}.$$

Here $\|f\|_{\mu,p}$ is the norm given by

$$\|f\|_{\mu,p} = \{\int |f(x)|^p \mu(dx)\}^{1/p}.$$

Further, there is equality in (8) if and only if for almost every x (with respect to μ), f is constant almost everywhere with respect to $P(x,\cdot)$.

The construction of Markov processes with continuous time parameter proceeds in a similar manner. As before, the state space Ω consists of points x representing a possible observation at any given fixed time t.

Here t may be the set of all real numbers or the set of all nonnegative real numbers. For convenience, we shall consider the set of all nonnegative real numbers. The events (at a fixed time) for which probabilities are well defined will be elements of a Borel field $\mathscr{A}$ of Ω. The stable generating mechanism for the Markov process is now given by a transition probability function $P_t(x,A)$, $0 \leqslant t < \infty$, which is $\mathscr{A}$-measurable as a function of x for each set (or event) $A \in \mathscr{A}$ and a probability measure on $\mathscr{A}$ for each x. We now have to *assume* that $P_t(x,A)$ satisfies

$$\int P_t(x,dy)P_\tau(y,A) = P_{t+\tau}(x,A),$$

for $t,\tau \geqslant 0$. This is called the Chapman-Kolmogorov equation as it was in the discrete time parameter case. It is natural to assume that

$$P_0(x,A) = \delta(x,A) = \begin{cases} 1 \text{ if } x \in A, \\ 0 \text{ otherwise} \end{cases}$$

and this will be taken for granted. Let μ be an initial probability measure (at time 0) on $\mathscr{A}$. A Markov process with continuous time parameter $\{X(t);\ 0 \leqslant t < \infty\}$ with initial distribution μ and stationary transition probability function $P_t(x,A)$ is constructed very much as it was in the discrete parameter case. Given any finite collection of events

$$A_0, A_1, \ldots, A_n \in \mathscr{A} \text{ and } t_i > 0$$

let

$$P_\mu(x(0) \in A_0, x(t_1) \in A_1, \ldots, x(t_1 + \cdots + t_n) \in A_n)$$
$$= \int_{A_0} \mu(dx_0) \int_{A_1} P_{t_1}(x_0, dx_1) \cdots \int_{A_{n-1}} P_{t_{n-1}}(x_{n-2}, dx_{n-1}) P_{t_n}(x_{n-1}, A_n).$$

The Kolmogorov extension theorem can be used under appropriate conditions on μ and $P_t(\cdot,\cdot)$ (see Appendix 3) to extend this set function to a measure P_μ on the Borel field generated by sets of the form

$$\{x(0) \in A_0, x(t_1) \in A_1, \ldots, x(t_1 + \cdots + t_n) \in A_n\},\ t_1, t_2, \ldots, t_n > 0$$

on the space of points $(x(t); 0 \leqslant t < \infty) = \omega$. Notice that the points of this space are functions, so that a measure is being constructed on a function space. Each point $\omega = (x(t); 0 \leqslant t < \infty)$ represents a possible trajectory (as a function of time) of the system whose probability structure is described by the Markov process. If there is a family of probability measures μ_t on $\mathscr{A}$, $-\infty < t < \infty$, such that

$$\mu_t(A) = \int \mu_\tau(dx)\, P_{t-\tau}(x,A)$$

for all pairs t, τ with $-\infty < \tau < t < \infty$, then one can construct a Markov process whose instantaneous distribution at time t is given by μ_t. Simply set

$$P(x(t_0) \in A_0, \ldots, x(t_n) \in A_n)$$
$$= \int_{A_0} \mu_{t_0}(dx_0) \int_{A_1} P_{t_1 - t_0}(x_0, dx_1) \cdots \int_{A_{n-1}} P_{t_{n-1} - t_{n-2}}(x_{n-2}, dx_{n-1})\, P_{t_n - t_{n-1}}(x_{n-1}, A_n)$$

for any finite collection of sets $A_0, \ldots, A_n \in \mathscr{A}$ and real numbers $t_0 < t_1 < \cdots < t_n$. Extend this set function to a measure P on the Borel field generated by the sets of the form $\{x(t_0) \in A_0, \ldots, x(t_n) \in A_n\}$, $t_0 < t_1 < \cdots < t_n$, on the space of points $(x(t); -\infty < t < \infty) = \omega$. We shall write the process as $\{X(t)\} = \{X(t)(\omega); -\infty < t < \infty\}$ with $X(t)(\omega) = x(t)$.

2. Markov Chains

The simplest interesting case of a Markov process is the Markov chain, a Markov process with a countable number of states that we shall label by the integers 1, 2, 3, ... for convenience. Let the one-step transition probability from state j to state k be $p_{j,k} = p_{j,k}^{(1)} \geqslant 0$, $\sum_k p_{j,k} = 1$. The $(n+1)$-step transition probability from state j to state k is given recursively by

$$p_{j,k}^{(n+1)} = \sum_l p_{j,l} p_{l,k}^{(n)}, \quad (n = 1, 2, \ldots) \tag{1}$$

The transition function is sometimes conveniently represented by the matrix $P^{(n)} = (p_{i,j}^{(n)}; i, j = 1, 2, \ldots) = P^n$, $n = 1, 2, \ldots$, and the Chapman-Kolmogorov equation is then simply given by

$$P^{(n+m)} = P^{n+m} = P^{(n)} P^{(m)} = P^n P^m . \tag{2}$$

The chain is said to be *irreducible* if every state can be reached from any other state with positive probability in a finite number of steps, that is, given any pair of states j and k there is an integer $n = n(j, k) > 0$ such that $p_{j,k}^{(n)} > 0$. A state j is *recurrent* if $\sum_{n=1}^{\infty} p_{j,j}^{(n)} = \infty$ and *transient* if $\sum_{n=1}^{\infty} p_{j,j}^{(n)} < \infty$. The state j is recurrent (transient) if the mean number of returns to the state j from the present to the infinite future is infinite (finite). The state j is *periodic with period* s if s is the greatest common divisor of the integers n for which $p_{j,j}^{(n)} > 0$. In the case of an irreducible chain, if one state is recurrent (periodic with period ρ) then all the states are recurrent (periodic with period ρ). Assume that j is recurrent with k any other state. Since the chain is irreducible there are integers r, s such that $p_{j,k}^{(r)}, p_{k,j}^{(s)} > 0$. Therefore

$$\sum_{n=1}^{\infty} p_{k,k}^{(n)} \geqslant \sum_m p_{k,j}^{(s)} p_{j,j}^{(m)} p_{j,k}^{(r)} = \infty$$

and k is recurrent. It immediately follows that if one state in an irreducible chain is transient, all the states are transient. Assume that state j is periodic with period ρ and that k is any other state. Again let r and s be integers such that $p_{j,k}^{(r)}, p_{k,j}^{(s)} > 0$. Since $r + s$ is divisible by ρ and

$$\begin{aligned} p_{j,j}^{(r+s+m)} &\geqslant p_{j,k}^{(r)} p_{k,k}^{(m)} p_{k,j}^{(s)}, \\ p_{k,k}^{(r+s+m)} &\leqslant p_{k,j}^{(s)} p_{j,j}^{(m)} p_{j,k}^{(r)}, \end{aligned} \tag{3}$$

it follows that k is periodic with period ρ.

Let $f_{j,k}^{(n)}=P\{X_m(\omega)\neq k, 0<m<n, X_n(\omega)=k \mid X_0(\omega)=j\}$ be the conditional probability of going from j to k for the first time in precisely n steps. The conditional probabilities $f_{j,k}^{(n)}$ and $p_{j,k}^{(n)}$ satisfy the equation

$$p_{j,k}^{(n)}=f_{j,k}^{(n)}+\sum_{m=1}^{n-1} f_{j,k}^{(m)} p_{k,k}^{(n-m)}. \tag{4}$$

Introducing the generating functions

$$F_{j,k}(s)=\sum_{n=1}^{\infty} f_{j,k}^{(n)} s^n,$$

$$G_{j,k}(s)=\delta_{j,k}+\sum_{n=1}^{\infty} p_{j,k}^{(n)} s^n$$

we see that

$$G_{j,j}(s)=(1-F_{j,j}(s))^{-1},\ G_{i,j}(s)=F_{i,j}(s)\,G_{j,j}(s),\ i\neq j. \tag{5}$$

The state j is transient if and only if $G_{jj}(1)<\infty$ or equivalently $F_{jj}(1)<1$. Let

$$\mu_j=\sum_{n=1}^{\infty} n f_{j,j}^{(n)}=F_{j,j}'(1), \tag{6}$$

$$\lim_{s\to 1-}(1-s)\,G_{j,j}(s)=\frac{1}{\mu_j}=u_j.$$

Now

$$F_{k,k}(1)\leqslant F_{k,j}(1)\,F_{j,k}(1)+(1-F_{k,j}(1)).$$

If the chain is irreducible and recurrent $F_{k,k}(1)=1$, $F_{k,j}(1)>0$. Then $F_{k,j}(1)\leqslant F_{k,j}(1)\,F_{j,k}(1)$ which implies that $F_{j,k}(1)=1$. If j is a recurrent state, the number μ_j is called the *mean recurrence time* for the state j. A recurrent state j is called *positive recurrent* if $\mu_j<\infty$ and *null recurrent* if $\mu_j=\infty$. Equation (6) and inequalities (3) imply that *an irreducible chain with one state positive (null) recurrent has all its states positive (null) recurrent.*

If a stationary probability distribution $\{v_j\}$ exists then

$$v_j=\sum_i v_i p_{ij}$$

which implies that

$$s v_j=(1-s)\,G_{jj}(s)\sum_i v_i F_{ij}(s)$$

by (5). Letting $s\to 1-$ we have

$$v_j=u_j\sum_i v_i F_{ij}(1)$$

so that $v_j \leqslant u_j$. Thus the states are positive recurrent. Conversely, if the chain is positive recurrent, then $u_j > 0$. But

$$(1-s)\sum_k G_{k,k}(s) F_{j,k}(s) = s$$

and if $s \to 1-$

$$\sum_k F_{jk}(1) u_k \leqslant 1$$

so that $\sum u_k \leqslant 1$. Using

$$p_{j,k}^{(n+1)} = \sum p_{j,i}^{(n)} p_{i,k}$$

we have

$$\frac{1}{s}[F_{j,k}(s) G_{k,k}(s) - p_{j,k} s] = \sum F_{j,i}(s) G_{i,i}(s) p_{i,k}.$$

Multiplying by $1-s$ and letting $s \to 1-$, one has

$$u_k \geqslant \sum_i u_i p_{i,k}.$$

Summing over k we have equality. The proof just given of the following result is due to Levinson [65].

Theorem 1. *There is a stationary probability distribution*

$$\{v_j\},\ v_k = \sum_j v_j p_{j,k},$$

for the transition matrix P of an irreducible chain if and only if the chain is positive recurrent. The distribution is unique with $v_k = 1/\mu_k$.

An irreducible finite state Markov chain must be recurrent since $G_{i,j}(1) = F_{i,j}(1) G_{jj}(1) > 0$ for all $i \neq j$ and $\sum_j G_{i,j}(1) = \sum_{n=1}^{\infty} 1 = \infty$. At least one state and therefore all states are recurrent. In fact a finite state irreducible Markov chain is positive recurrent because $\lim_{s \to 1-} (1-s) \sum_j G_{ij}(s) = 1$ which implies that (6) is positive for at least one j and therefore for all j.

Much more can be said in the case of an irreducible recurrent chain by virtue of a Tauberian theorem of Erdös, Feller, and Pollard. Since a proof can be found in many places (see for example [26], [9], and [93]), we shall just give a statement of the theorem here.

Theorem 2. *Let* $f^{(n)} \geq 0$, $n=1,2,\ldots$, $\sum_{n=1}^{\infty} f^{(n)}=1$, *be a probability distribution with mean* $\mu=\sum n f^{(n)}$ *(possibly infinite) such that the greatest common divisor of the integers n with* $f^{(n)}>0$ *is one. If*

$$F(s)=\sum_{n=1}^{\infty} f^{(n)} s^n$$

and

$$G(s)=\sum_{n=0}^{\infty} p^{(n)} s^n=(1-F(s))^{-1}$$

for $|s|<1$, *then*

$$p^{(n)} \to \frac{1}{\mu} \tag{7}$$

as $n \to \infty$.

This theorem implies that for any state j in an irreducible aperiodic (period one) recurrent Markov chain

$$p_{j,j}^{(n)} \to \frac{1}{\mu_j} \tag{8}$$

as $n \to \infty$ because of (7). The recurrence and irreducibility imply

$$\sum_{n=1}^{\infty} f_{j,k}^{(n)}=1$$

for all j, k. It follows from (4) and (8) that

$$p_{j,k}^{(n)} \to \frac{1}{\mu_k}$$

as $n \to \infty$ for all j and k. The case of an irreducible recurrent Markov chain with period $\rho>1$ can be reduced to the case of an aperiodic chain by considering only transitions of $n\rho$ steps, $n=1,2,\ldots$. It is then clear that for such a chain

$$p_{j,j}^{(n\rho)} \to \frac{\rho}{\mu_j}$$

as $n \to \infty$ for all j and

$$p_{j,j}^{(m)}=0$$

if m is not a multiple of ρ. The states of such a periodic chain can be partitioned into ρ disjoint classes of states $C_0, C_1, \ldots, C_{\rho-1}$ such that one can go from the states of C_i to C_{i+1}, $i=0,1,\ldots,\rho-2$, and $C_{\rho-1}$ to C_0 with positive probability only in $n\rho+1$ steps, $n=1,2,\ldots$. It then follows that

$$p_{j,k}^{(n\rho)} \to \frac{\rho}{\mu_k}$$

as $n \to \infty$ for all $j, k \in C_\alpha$, $\alpha=0,1,\ldots,\rho-1$.

Further, if $j \in C_\alpha$ and $k \in C_\beta$ then

$$p_{j,k}^{(n\rho+\beta-\alpha)} \to \frac{\rho}{\mu_k}$$

as $n \to \infty$ while $p_{j,k}^{(m)} = 0$ when $m \neq n\rho + \beta - \alpha$ for some integer $n \geqslant 0$. The asymptotic results derived thus far from the theorem of Erdös, Feller, and Pollard imply the following simpler but less detailed asymptotic statement in the general irreducible recurrent case. By taking averages one can see that

$$\frac{1}{n} \sum_{m=1}^{n} p_{j,k}^{(m)} \to \frac{1}{\mu_k}$$

as $n \to \infty$ for all states j and k.

Theorem 3. *Consider an irreducible recurrent Markov chain. If the chain is aperiodic, then*

$$p_{j,k}^{(n)} \to \frac{1}{\mu_k}$$

as $n \to \infty$ for all j, k. If the chain has period $\rho \geqslant 1$ one can in any case state that

$$\frac{1}{n} \sum_{m=1}^{n} p_{j,k}^{(m)} \to \frac{1}{\mu_k}$$

as $n \to \infty$ for all states j and k.

Of course, for a null recurrent chain, $1/\mu_k = 0$ for all k.

There may be a stationary distribution $\{u_j\}$, $0 \leqslant u_j < \infty$ for a transition matrix P

$$\sum_j u_j p_{j,k} = u_k$$

that is not a probability distribution. If we exclude the trivial case $u_j \equiv 0$, this means that $\sum_j u_j = \infty$. The following development is presented to give some insight into this possibility. For convenience and simplicity irreducible Markov chains are considered.

Lemma 1. *If $P = (p_{i,j})$ is the transition matrix of an irreducible chain, then*

$$\lim_{N \to \infty} \frac{\sum_{n=1}^{N} p_{i,j}^{(n)}}{\sum_{n=1}^{N} p_{j,j}^{(n)}}$$

exists and is finite. If the chain is recurrent, the limit is one.

Equation (4) can be written

$$p_{i,j}^{(n)} = \sum_{v=1}^{n} f_{i,j}^{(v)} p_{j,j}^{(n-v)} \tag{9}$$

if one sets $p_{j,j}^{(0)} = 1$. Sum (9) from $n = 1$ to N to obtain

$$\sum_{n=1}^{N} p_{i,j}^{(n)} = \sum_{n=1}^{N} \sum_{v=0}^{n-1} f_{i,j}^{(n-v)} p_{j,j}^{(v)} = \sum_{v=0}^{N-1} p_{j,j}^{(v)} \sum_{n=v+1}^{N} f_{i,j}^{(n-v)} = \sum_{v=0}^{N-1} p_{j,j}^{(v)} \sum_{n=1}^{N-v} f_{i,j}^{(n)}.$$

Irrespective of whether $\sum_{v=1}^{\infty} p_{j,j}^{(v)}$ is convergent or divergent, one obtains a limit on dividing by $\sum_{n=1}^{N} p_{j,j}^{(n)}$ and letting $N \to \infty$. The limit is one if $\sum_{v=1}^{\infty} p_{j,j}^{(v)}$ diverges.

Let

$$_k p_{i,j}^{(n)} = P\{X_n = j; X_v \neq k, 0 < v < n | X_0 = i\}, \quad n \geqslant 1,$$

$$_k p_{i,j}^{*} = \sum_{n=1}^{\infty} {}_k p_{i,j}^{(n)}.$$

Notice that $_j p_{i,j}^{*} = F_{i,j}(1)$. The equation

$$p_{i,j}^{(n)} = {}_k p_{i,j}^{(n)} + \sum_{v=1}^{n-1} p_{i,k}^{(v)} {}_k p_{k,j}^{(n-v)} \tag{10}$$

follows by considering the last passage through the state k before entering state j. Equation

$$\sum_{n=1}^{N} p_{i,j}^{(n)} = \sum_{n=1}^{N} {}_k p_{i,j}^{(n)} + \sum_{v=1}^{N-1} p_{i,k}^{(v)} \sum_{n=1}^{N-v} {}_k p_{k,j}^{(n)} \tag{11}$$

is obtained from (10) by summing from $n = 1$ to $n = N$ and then interchanging the order of summation on the right hand side.

Lemma 2. *If P is the transition matrix of an irreducible Markov chain, then $0 < {}_i p_{i,j}^{*} < \infty$.*

On letting $N \to \infty$ in (11) with $k = i$, we obtain

$$G_{i,j}(1) = {}_i p_{i,j}^{*}(1 + G_{i,j}(1))$$

where $G_{i,j}(1) = \sum_{n=1}^{\infty} p_{i,j}^{(n)}$. $G_{i,j}(1)$ is positive (possibly infinite) since the chain is irreducible and therefore $_i p_{i,j}^{*}$ is positive. But $_i p_{j,i}^{(m)} > 0$ for some $m \geqslant 1$ because the chain is irreducible. Since

$$_i p_{i,j}^{(n)} {}_i p_{j,i}^{(m)} \leqslant {}_i p_{i,i}^{(n+m)},$$

summing over n we find that

$$_ip^*_{i,j} \leqslant ({}_ip^{(m)}_{j,i})^{-1} \sum_{n=1}^{\infty} {}_ip^{(n+m)}_{i,i} < \infty .$$

Lemma 3. *Let P be the transition matrix of an irreducible recurrent Markov chain. Then*

$$\lim_{N\to\infty} \frac{\sum_{n=1}^{N} p^{(n)}_{i,j}}{\sum_{n=1}^{N} p^{(n)}_{i,i}} = {}_ip^*_{i,j} . \tag{12}$$

Divide (11) by $\sum_{n=1}^{N} p^{(n)}_{i,i}$ with $k=i$ and let $N\to\infty$. Since $\sum_{n=1}^{\infty} p^{(n)}_{i,i} = \infty$ we obtain (12).

Let us write

$$e_{h,i} = {}_hp^*_{h,i}$$

for h and i states of an irreducible Markov chain. Then $0 < e_{h,i} < \infty$ by Lemma 2.

Theorem 4. *Let $P=(p_{i,j})$ be the transition probability matrix of an irreducible recurrent Markov chain. The only nonnegative solution $\{u_i\}$ of*

$$u_i = \sum_k u_k p_{k,i}$$

is given by $u_i = c\,e_{h,i}$ where h is any state of the chain and c is any nonnegative constant, that is, the vectors $\{e_{h,i};\ i=1,2,\ldots\}$ are multiples of each other.

Notice that

$$\sum_{k\neq h} {}_hp^{(n)}_{h,k} p_{k,i} = {}_hp^{(n+1)}_{h,i} .$$

This implies that

$$\begin{aligned}\sum_k e_{h,k} p_{k,i} &= \sum_k \sum_{n=1}^{\infty} {}_hp^{(n)}_{h,k} p_{k,i} \\ &= \sum_{n=1}^{\infty} \{{}_hp^{(n+1)}_{h,i} + {}_hp^{(n)}_{h,h} p_{h,i}\} \\ &= e_{h,i} - {}_hp^{(1)}_{h,i} + {}_hp^*_{h,h} p_{h,i} = e_{h,i}\end{aligned}$$

since the chain is recurrent. Thus $u_i = c\,e_{h,i}$ is a solution. We now prove uniqueness. If $\{u_i\}$ is a nonnegative solution, then

$$u_i = \sum_k u_k p^{(n)}_{k,i} .$$

If one $u_i = 0$, then all $u_i = 0$. Assume therefore that all $u_i > 0$. For $n \geqslant 0$ let

$$q_{i,j}^{(n)} = \frac{u_j}{u_i} p_{j,i}^{(n)} .$$

Then $q_{i,j}^{(n)} \geqslant 0$, $\sum_j q_{i,j}^{(n)} = 1$ with

$$\sum_k q_{i,k} q_{k,j}^{(n)} = q_{i,j}^{(n+1)} .$$

$Q = (q_{i,j})$ is the transition probabilty matrix of an irreducible recurrent Markov chain since $\sum_n q_{i,i}^{(n)} = \sum_n p_{i,i}^{(n)} = \infty$. Now

$$\frac{\sum_{v=1}^{n} q_{i,h}^{(v)}}{\sum_{v=1}^{n} q_{h,h}^{(v)}} = \frac{u_h}{u_i} \frac{\sum_{v=1}^{n} p_{h,i}^{(v)}}{\sum_{v=1}^{n} p_{h,h}^{(v)}} . \tag{13}$$

Applying Lemmas 1 and 3, as $n \to \infty$ the left side of (13) tends to 1 and the right side tends to ${}_h p_{h,i}^* u_h / u_i$. Thus

$$u_i = u_h e_{h,i} .$$

Theorem 4 has an attractive intuitive interpretation since $e_{h,i}$ is the mean number of occurrences of the state i between successive occurrences of the state h as can be seen by noting that

$$\begin{aligned} e_{h,i} = {}_h p_{h,i}^* &= \sum_{n=1}^{\infty} {}_h p_{h,i}^{(n)} \sum_{m=1}^{\infty} f_{i,h}^{(m)} \\ &= \sum_{s=2}^{\infty} \sum_{n=1}^{s-1} P[X(n) = i, X(s) = h; X(v) \neq h, 0 < v < s \mid X(0) = h]. \end{aligned}$$

Notice that one cannot move with positive probability from one irreducible class of states to another irreducible class of states. Of course, there may be no irreducible class of states for a Markov chain if the number of states is infinite and all the states are transient as is the case for the chain with transition probabilities

$$p_{i,i+1} = 1, \qquad i = \ldots, -1, 0, 1, \ldots .$$

Theorem 4 indicates that if the chain has an irreducible class of recurrent states, there is an invariant vector $u = (u_i) \neq 0$ of the transition probability matrix P with nonnegative entries. If the states are positive recurrent the vector u can be taken to be a probability vector while if the states are null recurrent, the sum of the entries $\sum u_i$ is infinite. If all the states of the chain are transient, there may not be an invariant vector. However, there will still be a nontrivial subinvariant vector $u=(u_i)$, that is,

$$\sum_i u_i p_{i,j} \leqslant u_j .$$

If i is transient, then

$$\sum_{n=1}^{\infty} p_{i,i}^{(n)} < \infty .$$

If k is another state

$$\sum_{n=1}^{\infty} p_{k,i}^{(n)} = \sum_{n=1}^{\infty} \sum_{m=1}^{n} f_{k,i}^{(m)} p_{i,i}^{(n-m)}$$

$$= \sum_{m=1}^{\infty} f_{k,i}^{(n)} \sum_{n=0}^{\infty} p_{i,i}^{(n)} \leqslant \sum_{n=0}^{\infty} p_{i,i}^{(n)} < \infty .$$

The vector $u=(u_i)$ with components

$$u_i = \sum_{n=1}^{\infty} p_{k,i}^{(n)}$$

is subinvariant since

$$\sum_i u_i p_{i,j} \leqslant u_j .$$

Lemma 3 is also quite interesting in that it indicates that the limit

$$\lim_{N\to\infty} \frac{\sum_{n=1}^{N} p_{i,j}^{(n)}}{\sum_{n=1}^{N} p_{i,k}^{(n)}} \tag{14}$$

is well-defined and finite for states i, j, and k in the same irreducible class of recurrent states. However, it is also clear that (14) is well-defined and finite if i, j, and k are transient states and the denominator is not zero. In section 2 of Chapter 4 we shall consider an ergodic theorem for general state Markov processes that is a natural generalization of the ratio limit results for Markov chains of the type noted in (12).

The following random walks provide simple examples of null recurrent and transient Markov chains. Let

$$p_{i,i+1} = p, \qquad p_{i,i-1} = q = 1-p$$

with $0<p<1$ and the states of the chain $i=0, \pm 1, \ldots$. The chain is irreducible. Since there are $\binom{2n}{n}$ distinct paths leading from 0 to 0 in precisely $2n$ steps

$$p_{0,0}^{(2n)} = \binom{2n}{n} p^n q^n .$$

Now

$$G_{0,0}(s) = \sum_{n=0}^{\infty} \binom{2n}{n} p^n q^n s^{2n} = \sum_{n=0}^{\infty} \binom{-\frac{1}{2}}{n} (-1)^n (4pq)^n s^{2n} = (1-4pqs^2)^{-\frac{1}{2}} .$$

Notice that $G_{0,0}(1)=\infty$ if and only if $p=q=\frac{1}{2}$. Thus, the random walk is transient if and only if $p \neq \frac{1}{2}$. By Theorem 4 a recurrent chain has an essentially unique invariant measure. The random walks (recurrent or not) all have the obvious invariant measure $u_i \equiv 1$. In the recurrent case $p=\frac{1}{2}$, this must be the unique invariant measure. Since $\sum u_i = \sum 1 = \infty$, the chain must be null recurrent. Of course, this could have been obtained much more simply by a direct computation of the recurrence time distribution. In the transient case $u_i = \left(\frac{q}{p}\right)^i$ is another linearly independent invariant measure.

A probabilistic interpretation of finite invariant measures or distributions for a Markov chain (or process) is obvious. Invariant measures which are σ-finite but not finite arise naturally in the discussion of null recurrent and transient Markov chains. For this reason, it is of some interest to get a direct probabilistic interpretation for σ-finite invariant measures. Let $\{u_i\}$, $0<u_i<\infty$, be an initial measure (not necessarily invariant). *Assume that the numbers* $\sum_i u_i p_{i,j}^{(n)} = u_j^{(n)}$ *are finite for all j and n.* It is easy to construct specific examples in which this is not the case for proper σ-finite measures $\{u_i\}$. For example, if $p_{ij}=\beta_j>0$ for all i and j then $\sum_i u_i p_{ij} = \infty$ for all $\{u_i\}$ with $\Sigma u_i = \infty$. Let $A_i(0)$ be a random number of particles in state i at time zero. The numbers of particles $A_{i_1}(0), \ldots, A_{i_r}(0)$ in states $i_1, \ldots, i_r$ at time zero are assumed to be independent Poisson random variables with means $u_{i_1}, \ldots, u_{i_r}$ (see section 3)

$$P\{A_{i_\alpha}(0)=n_\alpha;\ \alpha=1,\ldots,r\} = \exp\left\{-\sum_{\alpha=1}^{r} u_{i_\alpha}\right\} \prod_{\alpha=1}^{r} \left(\frac{u_{i_\alpha}^{n_\alpha}}{n_\alpha!}\right).$$

Furthermore, all the particles are assumed to change position independently according to the transition mechanism given by P. Let us compute the joint probability distribution of $A_{i_1}(n), \ldots, A_{i_r}(n)$, the numbers of particles in states $i_1, \ldots, i_r$ at time n. We compute the joint characteristic function of $A_{i_1}(n), \ldots, A_{i_r}(n)$, that is, the Fourier-Stieltjes transform of their joint probability distribution. It is

$$\begin{aligned}\varphi_n(t_1, \ldots, t_r) &= E\exp\left\{i\sum_{\alpha=1}^{r} t_\alpha A_{i_\alpha}(n)\right\}\\ &= \prod_{k=0}^{\infty}\sum_{x=0}^{\infty}\{1+p_{k,i_1}^{(n)}(e^{it_1}-1)+\cdots+p_{k,i_r}^{(n)}(e^{it_r}-1)\}^x \frac{e^{-u_k}u_k^x}{x!}\\ &= \prod_{k=0}^{\infty}\exp\{u_k[p_{k,i_1}^{(n)}(e^{it_1}-1)+\cdots+p_{k,i_r}^{(n)}(e^{it_r}-1)]\}\\ &= \exp\left\{\sum_\alpha\left[\sum_k u_k p_{k,i_\alpha}(e^{it_\alpha}-1)\right]\right\}.\end{aligned} \tag{15}$$

From (15), it is apparent that $A_{i_1}(n), \ldots, A_{i_r}(n)$ are independent Poisson variables with means $\sum_k u_k p_{k,i_\alpha}^{(n)}$, $\alpha=1, \ldots, r$, respectively. Thus if $\{u_j\}$ is invariant, the joint distribution of $A_{i_1}(n), \ldots, A_{i_r}(n)$ is independent of n and is in statistical equilibrium. This interpretation of σ-finite measures can be made meaningful in a similar manner for Markov processes with general state space.

3. Independent Random Variables

A very complete and extensive study has been made of random sequences which are not statistically dependent through time. In our context, this corresponds to the situation in which the transition function $P(x, A)$ does not depend on x, that is, $P(x, A) = P(A)$. If the initial distribution μ is taken to be P, the measure (1.1) becomes a product measure and the random variables X_n, $n = 0, 1, 2, \ldots$, of the process are called independent and identically distributed random variables with common distribution P. If the random variables X_n are real-valued (the state space is the set of real numbers), then the common distribution of the random variables is effectively given by their distribution function

$$F(x) = P[X_n(\omega) \leqslant x], \qquad -\infty < x < \infty, \tag{1}$$

a right continuous nondecreasing function with $F(-\infty) = \lim_{x\to-\infty} F(x) = 0$, $F(+\infty) = \lim_{x\to+\infty} F(x) = 1$. At this point we make some very elementary remarks about the Poisson and Gaussian distributions. Simple examples of discrete (jump) and continuous distribution functions are given by the Poisson and Gaussian distribution. The Poisson distribution is given by

$$P[X(\omega) = n] = \frac{\lambda^n}{n!}e^{-\lambda}, \qquad n = 0, 1, \ldots,$$

for some $\lambda > 0$ so that the corresponding distribution function is

$$F(x) = \sum_{\substack{n \leqslant x \\ n \geqslant 0}} \frac{\lambda^n e^{-\lambda}}{n!}. \tag{2}$$

This is a distribution that is useful in describing, for example, the number of particles emitted in a given time period during radioactive decay (see [26]). The standard Gaussian (or normal) distribution function is

$$\Phi(x) = \int_{-\infty}^{x} \frac{e^{\frac{-u^2}{2}}}{\sqrt{2\pi}} du. \tag{3}$$

A random variable X with standard Gaussian distribution (3) has mean zero

$$EX = \int x P(dx) = \int x dF(x) = 0$$

and variance

$$\sigma^2(X) = EX^2 = \int x^2 P(dx) = \int x^2 dF(x) = 1.$$

A random variable Y with the same distribution as $\sigma X + m$ where σ, m are real numbers with $\sigma \geqslant 0$, $-\infty < m < \infty$, is said to have a Gaussian distribution with mean m and variance σ^2 since the mean of Y

$$E(Y) = E(\sigma X + m) = m$$

and the variance of Y is

$$\sigma^2(Y) = E(X - E(X))^2 = \sigma^2.$$

Notice that the Poisson distributions are a one-parameter family of distributions (in λ) while the Gaussian distributions are a two-parameter family of distributions (in m and σ). Given real-valued random variables $X_1, \ldots, X_n$, they are said to be independently (but not necessarily identically) distributed if their joint distribution is the product measure generated by their individual distribution functions. If X and Y are real-valued independent random variables with distribution functions F and G respectively

$$F(x) = P[X \leqslant x],$$

$$G(x) = P[Y \leqslant x],$$

one can show that the distribution function of their sum $X + Y$ is given by the convolution

$$F * G(x) = G * F(x) = \int_{-\infty}^{\infty} F(x-y) dG(y) \tag{4}$$

of F and G. The Poisson and Gaussian distributions are interesting in that they are closed under convolution.

A sequence of independent and identically distributed real-valued random variables $X_1, X_2, \ldots$ is often used as a model for a sequence of appropriately controlled measurements of a desired numerical value as, for example, the length of a standard bar or some critical temperature. The experimenter tries to control the conditions under which the measurements are made so that they are the same from one time to the next. This suggests that it is plausible to assume that the individual distributions of the X_j's are the same, if X_j is to represent the jth measurement. Furthermore, the experiments should be carried out in such a way that the results of some of the experiments will not influence those obtained in others. This motivates the assumption of independence of the random variables X_j. Assume that second moments of the X_j's exist so that the mean $m = EX$ and variance $\sigma^2 = E(X-m)^2$ of the random variables exist. Assume that the numerical value to be estimated as a result of the experiments is the mean $m = EX$ (no bias in the experiments). The error in the jth experiment is then

$$X_j - m. \tag{5}$$

There will be a nontrivial error, that is, $\sigma^2 > 0$ due to uncontrollable factors like limited resolvability in experimental equipment, reading error, etc. A reasonable estimate for the unknown mean value m is

$$\hat{m} = \frac{1}{n} \sum_{j=1}^{n} X_j .$$

The mean and variance of this estimate are

$$E\hat{m} = \frac{1}{n} \sum_{j=1}^{n} EX_j = m$$

and

$$\sigma^2(\hat{m}) = \frac{1}{n^2} n \sigma^2(X) = \frac{1}{n} \sigma^2 .$$

Let the common distribution of the X_j's be $F(x)$. Then the cumulative distribution function of $\sum_{j=1}^{n} X_j$ is $F^{(n)}(x)$, the nth convolution of F with itself. A classical result in the theory of probability, the so-called *central limit theorem* (see [93]) tells us that

$$\lim_{n\to\infty} P\left[\frac{\sqrt{n}}{\sigma}(\hat{m}-m)\leqslant x\right] = \lim_{n\to\infty} P\left[\frac{1}{\sqrt{n}\,\sigma}\sum_{j=1}^{n}(X_j-m)\leqslant x\right]$$
$$= \lim_{n\to\infty} F^{(n)}(\sqrt{n}\,\sigma\, x+m)$$
$$= \int_{-\infty}^{x} \frac{e^{\frac{-u^2}{2}}}{\sqrt{2\pi}}\,du = \Phi(x). \tag{6}$$

There is a classical variant of the central limit theorem that holds for independent random variables X_k, $k=1,2,\ldots,$ which need not be identically distributed. This version of the central limit theorem is due to Liapounov. Let m_k, σ_k^2 be the mean and variance respectively of X_k, $k=1,2,\ldots$. The central absolute moments $\gamma_k = E|X_k-m_k|^{2+\delta}$, $\delta>0$, of order $2+\delta$ of the random variables X_k are also assumed to be finite. Let

$$\begin{aligned} A_n &= \left(\sum_{k=1}^{n}\sigma_k^2\right)^{\frac{1}{2}}, \\ B_n &= \left(\sum_{k=1}^{n}\gamma_k\right)^{\frac{1}{2+\delta}}. \end{aligned} \tag{7}$$

The theorem of Liapounov states that $A_n^{-1}\sum_{k=1}^{n}(X_k-m_k)$ is asymptotically normally distributed

$$\lim_{n\to\infty} P\left(A_n^{-1}\sum_{k=1}^{n}(X_k-m_k)\leqslant x\right) = \Phi(x)$$

if

$$\lim_{n\to\infty} B_n/A_n = 0. \tag{8}$$

We shall now actually prove two limit theorems with error terms. In these theorems, the Poisson and normal distributions arise as the natural approximation. In the first, the Poisson distribution is used as an approximation to the binomial distribution while the second is a version of the central limit theorem with error term obtained by Berry and Esseen.

Let $X_1, X_2, \ldots, X_n$ be independent and identically distributed random variables with common distribution

$$P(X_i=1) = \alpha = 1-P(X_i=0), \qquad i=1,\ldots,n,$$

where $0 \leqslant \alpha \leqslant 1$. The sum $S_n = \sum_{j=1}^{n} X_j$ of the random variables has the binomial distribution (with sample size n)

$$P(S_n = k) = \binom{n}{k} \alpha^k (1-\alpha)^{n-k}. \tag{9}$$

The Poisson distribution is usually obtained in this context as the limiting distribution when $\alpha = \alpha(n) = \lambda/n$, $0 < \lambda < \infty$, as $n \to \infty$. The object is to obtain a result of Prohorov giving a bound for the difference between the binomial distribution and the Poisson approximation for all α. Of course, the approximation will only be good for small α.

Lemma 1. *Let $X_1, \ldots, X_n$ be independent binomial random variables with*

$$P(X_i = 1) = \alpha_i = 1 - P(X_i = 0), \qquad i = 1, \ldots, n,$$

and $Y_1, \ldots, Y_n$ be independent Poisson random variables with means $\alpha_1, \ldots, \alpha_n$ respectively. Then if $S = \sum X_i$ and $T = \sum Y_i$ it follows that

$$\sum_{k=0}^{\infty} |P(S=k) - P(T=k)| \leqslant 2\left\{1 - \prod_j (1-\beta_j)\right\} \leqslant 2 \sum \alpha_j^2 \tag{10}$$

with $\beta_j = \alpha_j(1 - e^{-\alpha_j})$.

We can choose the pair (X_j, Y_j) of dependent r.v.'s such that

$$P(X_j = Y_j = 0) = (1 - \alpha_j)$$

and

$$P(X_j = Y_j = 1) = \alpha_j e^{-\alpha_j}.$$

Then

$$P(X_j \neq Y_j) \leqslant 1 - (1-\alpha_j) - \alpha_j e^{-\alpha_j} \leqslant \alpha_j(1 - e^{-\alpha_j}).$$

Take the pairs (X_j, Y_j), $j = 1, \ldots, n$, independent of each other. Then

$$P(S \neq T) \leqslant 1 - \prod_j (1 - \beta_j).$$

Theorem 1. *If $0 \leqslant \alpha \leqslant 1$, then*

$$\sum_{k=0}^{\infty} |b_k(\alpha) - p_k(\alpha)| < C\alpha, \tag{11}$$

where

$$b_k(\alpha) = \begin{cases} \binom{n}{k} \alpha^k (1-\alpha)^{n-k}, & 0 \leqslant k \leqslant n, \\ 0 & \text{otherwise}, \end{cases}$$

$$p_k(\alpha) = e^{-n\alpha} \frac{(n\alpha)^k}{k!},$$

and C is an absolute constant independent of n.

According to Lemma 1 it is enough to prove the desired result for $2n\alpha > K > 0$. Set $q = 1-\alpha$ and $\varphi(v) = p_{n-v}/b_{n-v}$ or infinity according to whether $b_{n-v} > 0$ or $b_{n-v} = 0$. Now

$$\log\varphi(v) = -n + nq + (n-v)\log n - v\log q - \log n! + \log v!$$
$$= \frac{1}{2}\log\left(\frac{v}{n}\right) - v\left\{\log\frac{nq}{v} - \frac{nq}{v} + 1\right\} + A(v,n)$$

where $A(v,n) \geqslant 0$ (see the bounds on the error in Stirling's formula in Feller [26] volume 1, p. 52). $\varphi(v)$ takes on its minimal value at an integer m satisfying

$$nq - 1 < m \leqslant nq\,.$$

This implies that

$$\varphi(v) \geqslant \left(\frac{m}{n}\right)^{\frac{1}{2}} = 1-\gamma\,.$$

Let A be the set of integers for which $b_k > p_k$. Then

$$\sum_k |b_k - p_k| \leqslant 2\sum_{k\in A}(b_k - p_k) = 2\sum_{k\in A}[(1-\gamma)b_k - p_k] + 2\gamma\sum_{k\in A} b_k \leqslant 2\gamma P(A) \tag{12}$$

where $P(A)$ is the probability attributed to A by the binomial distribution (9). The expression (12) is therefore less than or equal to

$$2\left(1 - \sqrt{\frac{m}{n}}\right)P(A) \leqslant 2\left(1 - \frac{m}{n}\right)P(A)\,.$$

However

$$\frac{1+n\alpha}{n} > \frac{n-m}{n} \geqslant \alpha$$

and from this it is clear that one can find a constant C such that (11) holds. The constant C can be taken to be 4.

Before proceeding with the proof of the theorem of Berry and Esseen we shall make a few general remarks on limit theorems.

Suppose we try to consider a more general class of possible limit theorems, at least for independent and identically distributed random variables. Let $X_{n,k}$, $k = 1, 2, \ldots, n$, be n independent and identically distributed random variables. However, as n increases the common distribution of the n random variables is allowed to change. We can then ask whether there are constants A_n, B_n such that

$$\frac{1}{B_n}\sum_{k=1}^{n} X_{n,k} - A_n \tag{13}$$

has a nontrivial limiting probability distribution as $n \to \infty$. There may or may not be such sequences of constants. However, the class of possible limiting distributions (or possible limit laws) can be characterized in the following way. Let F be a possible limiting distribution function and

$$\varphi(t) = \int_{-\infty}^{\infty} e^{itx} dF(x)$$

the Fourier-Stieltjes transform of F or (in the probability literature) the characteristic function of F. Then it has been shown that F is a possible limit law if and only if for each integer $n > 1$ there is a distribution function F_n with corresponding characteristic function φ_n such that

$$\varphi_n(t)^n = \varphi(t). \tag{14}$$

Now there is a one-one correspondence between distribution functions and characteristic functions. Further, convolution of distribution functions corresponds to multiplication of characteristic functions so that if F and G are distribution functions with corresponding characteristic functions φ, ψ the characteristic function of the convolution $F * G$ is $\varphi\psi$. This means that (14) implies that for each $n > 1$ one can set up a probability space on which there are n independent random variables $X_{n,k}$, $k = 1, \ldots, n$, with common distribution function F_n such that the distribution of $\sum_{k=1}^{n} X_{n,k}$ is F. These possible limit laws are called the *infinitely divisible laws* and the property just cited is one motivation for this name. We have discussed and stated some of the properties of the infinitely divisible laws without proof. An extended discussion and derivation of their properties as well as the related limit theorems can be found in the beautiful book of Gnedenko and Kolmogorov [31].

We mention a few infinitely divisible laws. It is almost obvious that the normal and Poisson distributions are infinitely divisible. The characteristic function of a normal distribution with mean m and variance σ^2 is

$$\varphi(t) = \exp\left\{itm - \sigma^2 \frac{t^2}{2}\right\}. \tag{15}$$

It is clear that one can take

$$\varphi_n(t) = \exp\left\{\frac{itm}{n} - \sigma^2 \frac{t^2}{2n}\right\},$$

the characteristic function of a normal distribution with mean m/n and variance σ^2/n. The characteristic function of the Poisson distribution with mean $\lambda > 0$ is

$$\varphi(t) = \exp\{\lambda(e^{it} - 1)\} \tag{16}$$

and $\varphi_n(t)$ is the characteristic function of the Poisson distribution with mean λ/n. A last example is provided by the family of compound Poisson distributions. Suppose that during a given time there occurs a random number of events $N=0,1,\ldots$, governed by the Poisson distribution with mean λ. If $N=k$, we associate with each of the k events a gain or loss whose magnitude has distribution function $G(x)$. The gains and losses are assumed independent. Then the distribution of the total gain (or loss) is given by

$$F(x)=\sum_{k=0}^{\infty} e^{-\lambda}\frac{\lambda^k}{k!}G^{(k)}(x)\,. \tag{17}$$

The corresponding characteristic function is

$$\varphi(t)=\exp\{\lambda(\psi(t)-1)\} \tag{18}$$

where ψ is the characteristic function of G. F is seen to be infinitely divisible just as in the case of the Poisson distribution. It is also clear that the class of infinitely divisible laws is closed under convolution.

The proof of the Berry-Esseen theorem, which is now given, proceeds via Fourier analysis and makes use of estimates based on conveniently chosen weight functions. The characteristic function of the density function

$$v(x)=\frac{1}{\pi}\,\frac{1-\cos x}{x^2} \tag{19}$$

is

$$w(t)=\begin{cases}1-|t| & \text{if } |t|\leq 1,\\ 0 & \text{otherwise}\end{cases} \tag{20}$$

since

$$\frac{1}{2\pi}\int e^{-itx}w(t)\,dt=\frac{1}{\pi}\int_0^1 \cos tx\,(1-t)\,dt=\frac{1}{\pi}\,\frac{1-\cos x}{x^2}=v(x)\,.$$

Let F and G be two distribution functions with mean 0 and variance one. Consider the difference

$$d(x)=F(x)-G(x)$$

and the averaged weighted difference

$$d_T(t)=\int_{-\infty}^{\infty} d(t-x)\,T v(Tx)\,dx, \qquad T>0\,.$$

The following Lemma allows us to get a bound for

$$\eta = \sup_x |d(x)|$$

in terms of

$$\eta_T = \sup_x |d_T(x)| .$$

Lemma 2. *Let F and G be distribution functions with* $G'(x) \leqslant m < \infty$. *Then*

$$\eta_T \geqslant \frac{\eta}{2} - \frac{12m}{\pi T}. \tag{21}$$

There is a value x_0 such that either $|d(x_0+)|$ or $|d(x_0-)| = \eta$. We can just as well assume that $d(x_0) = \eta$. Then

$$d(x_0+s) \geqslant \eta - ms \quad \text{for } s > 0 . \tag{22}$$

If one sets

$$h = \frac{\eta}{2m}, \quad t = x_0 + h, \quad x = h - s ,$$

then (22) implies that

$$d(t-x) \geqslant \eta - m(h-x) = \frac{\eta}{2} + mx \tag{23}$$

for $|x| \leqslant h$. Use inequality (23) and the fact that $d(t-x) \geqslant -\eta$ for $|x| > h$. Since the mass given to $|x| > h$ by the $Tv(Tx)$ is less than or equal to $4/(\pi T h)$ it follows that

$$d_T(x_0) \geqslant \frac{\eta}{2}\left[1 - \frac{4}{\pi T h}\right] - \eta \frac{4}{\pi T h} = \frac{\eta}{2} - \frac{6\eta}{\pi T h} = \frac{\eta}{2} - \frac{12m}{\pi T} .$$

The proof of the Lemma is essentially complete.

A change of variable indicates that $d_T(x)$ is continuously differentiable and its derivative is given by

$$d_T'(t) = \int \frac{d}{du}\{Tv(T(t-u))\}\, d(u)\, du = \int Tv(T(t-u))\, d\{F(u) - G(u)\}$$

$$= \frac{1}{2\pi} \int_{-T}^{T} e^{-itx} w\left(\frac{x}{T}\right) [\varphi(x) - \psi(x)]\, dx \tag{24}$$

where φ and ψ are the characteristic functions of F and G respectively. Assume that F and G both have mean zero. Since $\varphi(0) = \psi(0)$ and $\varphi'(0) = \psi'(0) = 0$

$$\frac{\varphi(x) - \psi(x)}{-ix}$$

is a bounded continuous function vanishing at zero. It is natural to conjecture that

$$d(x) = \frac{1}{2\pi} \int_{-T}^{T} e^{-ixt} \frac{\varphi(t) - \psi(t)}{-it} w\left(\frac{t}{T}\right) dt. \tag{25}$$

The right side of (25) tends to zero as $|t| \to \infty$ by the Riemann-Lebesgue theorem. Since the continuous derivative of the right side is (24), it follows that the conjecture is valid. Using the Lemma 2 and (21) it follows that

$$|F(x) - G(x)| \leqslant \frac{1}{\pi} \int_{-T}^{T} \left| \frac{\varphi(t) - \psi(t)}{t} \right| dt + \frac{24m}{\pi T}. \tag{26}$$

It is this inequality that is crucial in obtaining the proof of the following Theorem.

Theorem 2 *(Berry-Esseen). Let $X_1, X_2, \ldots, X_n$ be independent random variables with means zero, variances σ_k^2 and finite third moments $E(|X_k|^3) = \rho_k$ satisfying*

$$\frac{1}{\sigma_k^2} \rho_k < \lambda < \infty, \qquad k = 1, 2, \ldots, n. \tag{27}$$

Let $F_n(x)$ be the distribution function of

$$\frac{1}{s_n} \sum_{k=1}^{n} X_k$$

where $s_n^2 = \sum_{k=1}^{n} \sigma_k^2$. Then

$$|F_n(x) - \Phi(x)| < \frac{33}{4} \frac{\lambda}{s_n}. \tag{28}$$

Let ψ_k be the characteristic function of X_k. If we let $F = F_n$ and $G = \Phi$ in relation (26) the following inequality is obtained

$$|F_n(x) - \Phi(x)| \leqslant \frac{1}{\pi} \int_{-T}^{T} \left| \prod_{k=1}^{n} \psi_k\left(\frac{t}{s_n}\right) - e^{-\frac{1}{2}t^2} \right| \frac{dt}{|t|} + \frac{24}{\pi T \sqrt{2\pi}}. \tag{29}$$

In a sufficiently small interval about the origin free of zeros of $\psi_k(t)$

$$-\log \psi_k(t) - \tfrac{1}{2}\sigma_k^2 t^2 = -\log\{1 - [1 - \psi_k(t)]\} - \tfrac{1}{2}\sigma_k^2 t^2$$

$$= (1 - \psi_k(t) - \tfrac{1}{2}\sigma_k^2 t^2) + \sum_{k=2}^{\infty} \frac{1}{k} (1 - \psi_k(t))^k. \tag{30}$$

The Hölder inequality implies that $\sigma_k^6 \leqslant \rho_k^2$. Now

$$1-\psi_k(t) = -\tfrac{1}{2}\psi_k''(\theta t)\, t^2 \tag{31}$$

for some θ, $0 \leqslant \theta \leqslant 1$. However

$$\psi_k''(t) = \int (it)^2 e^{itx}\, dH_k(x)$$

where H_k is the distribution function of X_k so that

$$|\psi_k''(t)| \leqslant \sigma_k^2 \tag{32}$$

for all t. Relations (31) and (32) imply that

$$|1-\psi_k(t)| \leqslant \tfrac{1}{2}\sigma_k^2 t^2 \leqslant \tfrac{1}{2} \tag{33}$$

for $|t| \leqslant \sigma_k^2/\rho_k$ and therefore for $|t| \leqslant 1/\lambda$. Similarly, since

$$1-\psi_k(t) - \tfrac{1}{2}\sigma_k^2 t^2 = -\frac{it^3}{6}\psi_k'''(\theta t)$$

for some θ, $0 \leqslant \theta \leqslant 1$, and

$$|\psi_k'''(t)| = \left|\int (it)^3 e^{itx}\, dH_k(x)\right| \leqslant \rho_k,$$

it follows that

$$|1-\psi_k(t) - \tfrac{1}{2}\sigma_k t^2| \leqslant \rho_k |t|^3. \tag{34}$$

Thus, relation (30) and inequalities (33) and (34) indicate that

$$|\log \psi_k(t) + \tfrac{1}{2}\sigma_k^2 t^2| \leqslant \tfrac{1}{6}\rho_k |t|^3 + \tfrac{1}{4}\sigma_k^4 t^4 \leqslant \tfrac{5}{12}\rho_k |t|^3 \tag{35}$$

for $|t| \leqslant 1/\lambda$. Now

$$|e^t - 1| \leqslant |t|\, e^{|t|}. \tag{36}$$

Let $T = s_n/\lambda$ in (29). Both (35) and (36) indicate that the integrand in (29) is less than or equal to

$$\frac{5}{12}\frac{R_n}{s_n^3}|t|^3 \exp\left(-\frac{1}{2}t^2 + \frac{5}{12}\frac{R_n}{s_n^3}|t|^3\right) \leqslant \frac{5}{12}\frac{R_n}{s_n^3} t^2 \exp\left(-\frac{1}{12}t^2\right),$$

$$R_n = \sum_{k=1}^{n} \rho_k,$$

for $|t| < s_n/\lambda$. The integral of $t^2 \exp(-\frac{1}{12}t^2)$ is $6\sqrt{12\pi}$ so that

$$\frac{s_n^3}{R_n}|F_n(x) - \Phi(x)| \leqslant \frac{1}{\sqrt{2\pi}}\left(5\sqrt{6} + \frac{24}{\pi}\right) < \frac{12.5+8}{\sqrt{2\pi}} < 8.2.$$

The proof is complete.

It has already been noted that if $X_{n,k}$, $k=1,\dots,n$, are independent random variables with a common distribution there may or may not be constants A_n, B_n such that the centered and normalized sums (13) have a limiting distribution as $n\to\infty$. However, there is a remarkable theorem due to Kolmogorov stating that *there is an infinitely divisible distribution function* $H_n(x)$ *such that*

$$\sup_x \left| P\left(\sum_{k=1}^{n} X_{n,k}\leqslant x\right) - H_n(x)\right| \leqslant C n^{-\frac{1}{3}}$$

where C *is an absolute constant independent of the distribution function of the* $X_{n,k}$'s. This is surprising since it says that the distribution of $\sum_{k=1}^{n} X_{n,k}$ is increasingly well approximated by appropriately chosen infinitely divisible laws as $n\to\infty$ even if there is no nontrivial limiting distribution for any centered and normalized sums (13) as $n\to\infty$.

The theorems of Prohorov and Berry-Esseen are of great interest in themselves. We shall also use them in Chapter 7, section 1 to prove this theorem of Kolmogorov.

4. Some Continuous Parameter Markov Processes

There are two simple but especially important continuous parameter Markov processes that can be defined in terms of the Poisson and Gaussian distributions introduced in the discussion of independent random variables. The first process is commonly called the *Poisson process.* In discussing this process, it will be convenient to introduce the cumulative distribution function of the exponential distribution

$$F(x) = \begin{cases} 1-e^{-\lambda x}, & x\geqslant 0,\\ 0 \text{ if} & x<0 \end{cases} \tag{1}$$

for some $\lambda>0$. Let $X_1, X_2, \dots$ be independent and identically distributed random variables with common distribution given by (1). Let S_n denote the partial sum

$$S_n = \sum_{j=1}^{n} X_j.$$

We now think of a process starting at zero at time zero and jumping from state j to state $j+1$ at time S_{j+1}. If the cumulative distribution function of S_j is denoted by $F_j(x)$, it is clear that $F_j(x)=F^{(j)}(x)$. Since F has the density

$$f(x) = \begin{cases} \lambda e^{-\lambda x} & \text{if } x>0,\\ 0 & \text{otherwise} \end{cases}$$

with respect to Lebesgue measure, F_j has a density f_j and

$$f_{j+1}(x)=\int_0^x f_j(u)\,f(x-u)\,du$$

if $x>0$ and is zero if $x\leqslant 0$. Recursively one can easily verify that

$$f_{j+1}(x)=\begin{cases}\dfrac{\lambda(\lambda x)^j}{j!}\,\mathrm{e}^{-\lambda x} & \text{if } x>0,\\ 0 & \text{otherwise}\end{cases}$$

But then

$$F_{j+1}(x)=\int_0^x f_{j+1}(u)\,du=\int_0^x \frac{\lambda(\lambda u)^j}{j!}\,e^{-\lambda u}\,du$$

$$=-\left.\frac{(\lambda u)^j e^{-\lambda u}}{j!}\right|_0^x+\int_0^x \frac{\lambda(\lambda u)^{j-1}}{(j-1)!}\,e^{-\lambda u}\,du$$

$$=1-e^{-\lambda x}\left(1+\frac{\lambda x}{1!}+\cdots+\frac{(\lambda x)^j}{j!}\right)$$

for $x>0$. The states of the process are the nonnegative integers $j=0,1,2,\ldots$. If we call the process $\{N(t);\, t\geqslant 0\}$ then

$$P[N(t)=j]=P[S_j\leqslant t,\, S_{j+1}>1]=F_j(t)-F_{j+1}(t)=\frac{(\lambda t)^j}{j!}\,e^{-\lambda t},$$

the Poisson distribution. The exponential distribution has the interesting property that

$$P[X>\tau+t\,|\,X>\tau]=\frac{P[X>\tau+t]}{P[X>\tau]}=\frac{e^{-\lambda(t+\tau)}}{e^{-\lambda\tau}}=e^{-\lambda t}=P[X>t]$$

if X is exponentially distributed. One can show (see Doob [18]) that the number of jumps of $\{N(t)\}$ in disjoint intervals are independent so that

$$\begin{aligned}P[N(\tau)=j,\, N(t)=k]&=P[N(\tau)=j,\, N(t)-N(\tau)=k-j]\\&=P[N(\tau)=j]\,P[N(t)-N(\tau)=k-j]\end{aligned}$$

if $t>\tau>0$. Therefore the conditional probability

$$\begin{aligned}P[N(t)=k\,|\,N(\tau)=j]&=P[N(t)-N(\tau)=k-j]\\&=P[N(t-\tau)=k-j]=\frac{\{\lambda(t-\tau)\}^{k-j}e^{-\lambda(t-\tau)}}{(k-j)!}\end{aligned}$$

if j, k are nonnegative integers with $k \geqslant j$ and $t > \tau > 0$. The reader can verify that the transition probability function

$$P_t(j,k) = \begin{cases} \dfrac{(\lambda t)^{(k-j)} e^{-\lambda t}}{(k-j)!} & \text{if } k \geqslant j, \\ 0 & \text{otherwise} \end{cases} \tag{2}$$

defined for nonnegative integers j, k and real $t \geqslant 0$ satisfies the Chapman-Kolmogorov equation and that the Poisson process $N(t)$ is the Markov process with $N(0) \equiv 0$ and the transition probability function (2). The Poisson process is a jump process and as already suggested, it has been used to describe the emission of particles from radioactive material.

The second process $B(t)$, $t \geqslant 0$, that we shall construct is usually called the Brownian motion process (or occasionally the Wiener process) because it has been used to describe the horizontal excursion of a particle in Brownian motion. Let $B(0) \equiv 0$. Then if $B(t)$ is thought of as the location of the particle (in linear Brownian motion) at time t, the difference $B(t) - B(\tau)$, $t > \tau > 0$, is the distance the particle has been displaced in the time interval $(\tau, t]$. The process is assumed to be a Gaussian process, that is, all linear combinations of finite numbers of random variables $B(t_i)$, $t_i \geqslant 0$, are assumed to have Gaussian distributions. The mean drift is assumed to be zero so that

$$E\,B(t) \equiv 0.$$

The displacements in position over disjoint time intervals are assumed to be independent and the variance of the displacement $B(t) - B(\tau)$ over the interval $(\tau, t]$, $0 \leqslant \tau \leqslant t$, is

$$E(B(t) - B(\tau))^2 = t - \tau.$$

This implies that the Brownian motion $B(t)$, $t \geqslant 0$, is a Gaussian Markov process with initial location $B(0) \equiv 0$ and transition probability function

$$P_t(x, A) = P[B(t+\tau) \in A \mid B(\tau) = x] = \int_A \{2\pi t\}^{-\frac{1}{2}} \exp\left\{-\frac{(y-x)^2}{2t}\right\} dy \tag{3}$$

defined for x real, A a linear Borel set and $t, \tau \geqslant 0$. One can again verify that (3) satisfies the Chapman-Kolmogorov equation. In some problems, it is convenient to define a Brownian motion process for all time t, $-\infty < t < \infty$. One still assumes that process is Gaussian, with displacements over disjoint intervals independent and with mean

$$E\,B(t) \equiv 0, \quad -\infty < t < \infty,$$

and variance

$$E|B(t)-B(\tau)|^2=|t-\tau|, \quad -\infty<t,\tau<\infty.$$

The process is then completely specified by assuming that

$$B(0)\equiv 0.$$

5. Random Walks on Countable Commutative Groups

An interesting class of Markov chains is given by random walks on countable commutative groups. If the group elements are labeled $g_i \in G$, $i=1,2,\ldots$, then the transition probabilities are given by

$$p_{g_i,g_j}=p_{g_j-g_i}\geqslant 0 \tag{1}$$

with

$$\sum_{g\in G} p_g=1. \tag{2}$$

Here the group operation is given by addition. The simplest example of such a random walk is one on the integers $i=0,\pm 1,\ldots$ with

$$p_{i,i+1}=p, \qquad p_{i,i-1}=q$$

where $0<p<1$ and $q=1-p$. This example was discussed earlier in section 2. It has already been noted that this random walk is recurrent (actually null recurrent) if and only if $p=q=\frac{1}{2}$. This suggests that for random walks on the integers, recurrence ought to hold if and only if the mean (assuming it exists) of the distribution p_i is zero. An argument showing that this is actually the case can be found in Spitzer's book [105] on random walks.

Another simple random walk is the one on the two-dimensional lattice points (i,j), $i,j=0,\pm 1,\ldots$, with transition probabilities

$$p_{(i,j),(i\pm 1,j)}=p_{(i,j),(i,j\pm 1)}=\tfrac{1}{4}.$$

The probability of returning to the point $(0,0)$ at the $(2n)^{\text{th}}$ step given that one started there initially is

$$\binom{2n}{n}^2\frac{1}{4^{2n}}. \tag{3}$$

The state $(0,0)$ is recurrent if and only if the sum of these probabilities diverges. If Stirling's formula is used as an approximation for the factorials, it is seen that (3) behaves asymptotically as $n\to\infty$ like

$$\frac{1}{\pi n}. \tag{4}$$

The sum of the probabilities (4) diverges so that (0,0) is recurrent. That it is null recurrent can be seen by projecting the walk on the line $i+j=0$ in which case one is reduced to the one-dimensional symmetric random walk discussed in the first paragraph.

The remarks just made indicate that there are recurrent random walks on the lattice points on the line and in two-space in which each state is reached. We can ask whether there are any such walks on the lattice points in 3-space or more generally whether there are such walks on any specific commutative group with a countable number of elements. We shall first show that there aren't any recurrent walks on the lattice points in 3-space and then derive a neat result of Dudley that indicates what happens on a general countable commutative group.

Let $\bar{n}=(n_i; i=1,\dots,d)$ be a vector of integers $n_i=0,\pm 1,\dots$ and $\theta=(\theta_i; i=1,\dots,d)$ a vector of real numbers. The transition probabilities of a random walk on the d-dimensional lattice points are given by

$$p_{\bar{n}-\bar{n}'}\geqslant 0 \tag{5}$$

with

$$\sum_{\bar{n}} p_{\bar{n}}=1\,. \tag{6}$$

Consider the trigonometric function

$$\varphi(\theta)=\sum_{\bar{n}} p_{\bar{n}}\, e^{i\bar{n}\cdot\theta} \tag{7}$$

where $\bar{n}\cdot\theta$ is the inner product of the vectors $\bar{n}$ and θ. Then if $\bar{0}=(0,\dots,0)$

$$P(X_r=\bar{0})=(2\pi)^{-d}\int_{-\pi}^{\pi}\cdots\int[\varphi(\theta)]^n\,d\theta$$

if $X_0=\bar{0}$. However, this implies that

$$\sum_{r=0}^{\infty} s^r P(X_r=\bar{0})=\mathrm{Re}(2\pi)^{-d}\int_{-\pi}^{\pi}\cdots\int\frac{1}{1-s\varphi(\theta)}\,d\theta,\qquad 0<s<1\,.$$

Since

$$\mathrm{Re}\frac{1}{1-s\varphi(\theta)}=\frac{1-s\,\mathrm{Re}\,\varphi(\theta)}{|1-s\varphi(\theta)|^2}\leqslant\frac{1}{1-s\,\mathrm{Re}\,\varphi(\theta)}\leqslant\frac{1}{1-\mathrm{Re}\,\varphi(\theta)},$$

it follows that

$$\sum_{r=0}^{\infty} P(X_r=\bar{0})\leqslant(2\pi)^{-d}\int_{-\pi}^{\pi}\cdots\int\frac{1}{1-\mathrm{Re}\,\varphi(\theta)}\,d\theta\,.$$

We have the following simple Lemma.

Lemma 1. *In a random walk on the d-dimensional lattice points, if*

$$\int_{-\pi}^{\pi} \cdots \int \frac{1}{1-\operatorname{Re}\varphi(\theta)}\, d\theta \tag{8}$$

is finite, then $\overline{0}$ is a transient state.

Let

$$[\varphi(\theta)]^m = \sum_{\bar{n}} p_{\bar{n}}^{(m)} e^{i\bar{n}\cdot\theta}, \qquad m \geqslant 1, \tag{9}$$

so that

$$p_{\bar{n}-\bar{n}'}^{(m)}$$

is the transition probability of going from $\bar{n}'$ to $\bar{n}$ in precisely m steps. If the random walk is irreducible and aperiodic then for each lattice point $\bar{n}$ it follows that

$$p_{\bar{n}}^{(m)} > 0 \tag{10}$$

for $m \geqslant m(\bar{n})$ where $m(\bar{n})$ is a positive integer depending on $\bar{n}$. Suppose that θ is such that

$$\varphi(\theta) = 1 .$$

Then (9) and (10) imply that $(2\pi)^{-1}\bar{n}\cdot\theta$ is an integer for each lattice point $\bar{n}$. However, this can only be the case if each component of $(2\pi)^{-1}\theta$ is an integer.

Lemma 2. *For an irreducible aperiodic random walk on the d-dimensional lattice points*

$$\varphi(\theta) = 1$$

implies that the coordinates of θ must be integer multiples of 2π.

Lemma 2 is now used to get a useful bound on $1-\operatorname{Re}\varphi(\theta)$.

Lemma 3. *Consider an irreducible aperiodic random walk on the d-dimensional lattice points. Then there is a $\lambda > 0$ such that*

$$1-\operatorname{Re}\varphi(\theta) \geqslant \lambda|\theta|^2 \tag{11}$$

for all θ with $|\theta_i| \leqslant \pi$, $i=1,\ldots,d$.

Since the random walk is irreducible there are d linearly independent vectors $\bar{n}_1,\ldots,\bar{n}_d$ in the set $\{\bar{n} : p_{\bar{n}} > 0\}$. Let $L = \max|\bar{n}_i|$. The quadratic form

$$Q(\theta) = \sum_{|\bar{n}| \leqslant L} (\bar{n}\cdot\theta)^2 p_{\bar{n}}$$

is positive definite since

$$Q(\theta) \geqslant \sum_{i=1}^{d} (\bar{n}_i\cdot\theta)^2 p_{\bar{n}_i} . \tag{12}$$

The right hand side of (12) cannot vanish unless θ is orthogonal to the $\bar{n}_i$, $i=1,\dots,d$, and this is impossible since the $\bar{n}_i$ are linearly independent. Now

$$\operatorname{Re}[1-\varphi(\theta)] = \sum_{\bar{n}} [1-\cos\bar{n}\cdot\theta] p_{\bar{n}} \geqslant 2 \sum_{|\bar{n}|\leqslant L} \sin^2\left(\frac{\bar{n}\cdot\theta}{2}\right) p_{\bar{n}}.$$

However,

$$\left|\sin\left(\frac{\bar{n}\cdot\theta}{2}\right)\right| \geqslant \pi^{-1}|\bar{n}\cdot\theta|$$

when $|\bar{n}\cdot\theta|\leqslant\pi$ and $|\bar{n}\cdot\theta|\leqslant\pi$ if $|\theta|\leqslant\pi L^{-1}$.

Thus

$$\operatorname{Re}[1-\varphi(\theta)] \geqslant \frac{2}{\pi^2} \sum_{|\bar{n}|\leqslant L} (\bar{n}\cdot\theta)^2 p_{\bar{n}} = \frac{2}{\pi^2} Q(\theta)$$

when $|\theta|\leqslant\pi L^{-1}$. This implies that

$$1-\operatorname{Re}\varphi(\theta) \geqslant \frac{2}{\pi^2}\lambda_1|\theta|^2$$

for $|\theta|\leqslant\pi L^{-1}$ where $\lambda_1>0$ is the smallest eigenvalue of $Q(\theta)$. Lemma 2 tells us that $1-\operatorname{Re}\varphi(\theta)>0$ when $|\theta|\geqslant\pi L^{-1}$ and $|\theta_i|\leqslant\pi$, $i=1,\dots,d$. Since $1-\operatorname{Re}\varphi(\theta)$ is continuous, there is *a* $\lambda>0$ such that (11) is satisfied.

We can now prove the following theorem.

Theorem 1. *All irreducible random walks on the d-dimensional lattice points with $d\geqslant 3$ are transient.*

We need only consider aperiodic random walks. For if the period of the random walk is $\gamma>1$, an aperiodic walk is obtained by considering the chain only at times which are integer multiples of γ. Lemma 3 gives the estimate

$$1-\operatorname{Re}\varphi(\theta)\geqslant\lambda|\theta|^2$$

for some $\lambda>0$ when $|\theta_i|\leqslant\pi$, $i=1,\dots,d$. But this implies that

$$\int_{-\pi}^{\pi}\cdots\int \frac{1}{1-\operatorname{Re}\varphi(\theta)}\,d\theta$$

is finite when $d\geqslant 3$. By Lemma 1 $\bar{0}$ is transient and thus the random walk is transient.

Before stating and proving Dudley's result, we shall have to introduce certain ideas from the theory of commutative (or Abelian) groups. For a detailed discussion of these ideas the reader can refer to Kurosh [62]. The binary group operation will be given with additive notation. A commutative group G is a collection of elements with a binary operation "+" defined for all elements that is associative:

$$g+(g'+g'')=(g+g')+g'',$$

commutative:

$$g+g'=g'+g,$$

has an identity 0 under addition:

$$g+0=g,$$

and with an inverse $-g$ for each element g:

$$g+(-g)=0.$$

The *cyclic* Abelian groups are the simplest. They are the groups with the property that all elements of the group are generated by one element a of the group, that is, all elements of the group are of the form

$$na=\underbrace{a+\cdots+a}_{n}, \qquad n=0,\pm 1,\ldots.$$

If $na=0$ for some positive integer n, the group is finite cyclic. Actually, if n is the smallest positive integer with this property the group is isomorphic (in a $1-1$ correspondence preserving the group operation) to the integers modulo n under addition. If there is no such integer n then the cyclic group is infinite cyclic and is isomorphic to the full set Z of integers under addition. If M is a set of elements of G then $\{M\}$ denotes the intersection of all groups (the smallest group) containing M. If $\{M\}=G$ the elements of M are called *generators* of G. If no proper subset of M generates the group G, M is an irreducible set of generators. As already indicated the element 1 generates a finite or infinite cyclic group of integers. However, some Abelian groups do not have an irreducible set of generators. For example, in the case of the group of rational numbers under addition, the set $\{1/n!, n=1,2,\ldots\}$ is a set of generators but there is no irreducible set of generators. An Abelian group is said to be *finitely generated* if there is a finite set of generators. The cyclic groups are finitely generated but the group of rationals under addition is not.

An Abelian group G is called the *direct sum of a finite number of subgroups* $H_1, H_2, \ldots, H_n$.

$$G=H_1\oplus H_2\oplus\cdots\oplus H_n$$

if each element $g\in G$ has a unique representation as a sum of elements $h_i\in H_i$, $i=1,\ldots,n$,

$$g=\sum_{i=1}^{n} h_i.$$

G is the *restricted direct sum* of an infinite number of subgroups $H_1, H_2, \ldots$ if each element $g \in G$ has a unique representation as a sum of elements $h_i \in H_i$, $i = 1, 2, \ldots$

$$g = \sum_{i=1}^{\infty} h_i$$

where all but a finite number of summands are the identity element 0. It is easily seen that the set of d-dimensional lattice points is the direct sum of d copies of the group Z of integers under addition. Another example is given by the group of rationals under addition which is the direct sum of Z and the group of rationals modulo one under addition.

The idea of the *rank* of an Abelian group is now introduced. A finite set of elements $g_1, \ldots, g_k$ of an Abelian group is called *linearly dependent* if there are integers $a_1, \ldots, a_k$ not all zero such that

$$a_1 g_1 + a_2 g_2 + \cdots + a_k g_k = 0.$$

A finite set of elements which don't have this property is called *linearly independent.* An element g is said to be linearly dependent on $g_1, \ldots, g_k$ if there is an integer $a \neq 0$ and integers $a_1, \ldots, a_k$ such that

$$a g = a_1 g_1 + \cdots + a_k g_k.$$

Two finite sets of elements of G are said to be *equivalent* if each element of the first set is linearly dependent on the second set and similarly each element of the second set is linearly dependent on the first set. One can show that if a group G *has finite maximal linearly independent sets of elements, then all these sets are equivalent and have the same number of elements.* This number is called the *rank* of the Abelian group G and G is said to be a group of *finite rank.* If G does not have an element g such that $ng \neq 0$ for all integers $n \neq 0$, G is said to have rank zero. Notice that all finite Abelian groups and the additive group of rationals modulo one have rank zero. The group of d-dimensional lattice points has rank d while the additive group of rationals has rank one.

For convenience we can assume the random walk on the countable Abelian group G starts at zero at time zero. The random walk is called *symmetric* if

$$p_g = p_{-g}$$

for each $g \in G$. Let $\{X_n, n = 0, 1, \ldots\}$ be the random variables of the random walk (with state space G) whose transition probabilities are given by (1). Set

$$p_g^{(n+1)} = \sum_{h \in G} p_{g-h} p_h^{(n)}, \qquad n = 1, 2, \ldots.$$

Dudley's result is given in the following theorem.

Theorem 2. *A countable Abelian group G has an irreducible recurrent random walk if and only if its rank is at most two.*

In proving this theorem we shall need a number of Lemmas. There will also occasionally be a result on Abelian groups that will be required. Such results cannot be derived in the text. We hope they will be intuitively plausible. The interested reader should refer to Kurosh [62] for a derivation.

Lemma 4. *If a countable Abelian group G has an irreducible recurrent random walk, then so does any subgroup H of G.*

Let

$$q_h = p_h + \sum_{g_1 \notin H} p_{g_1} p_{h-g_1} + \sum_{g_i \notin H} p_{g_1} p_{g_2} p_{h-g_1-g_2} + \cdots .$$

Notice that q_h is the probability, given that it started at 0 at time 0, of the random walk hitting H for the first time thereafter at $h \in H$. Because of the recurrence of the random walk on G with probabilities p_g, it is clear that

$$\sum_{h \in H} q_h = 1 .$$

Let $q_{h-h'}$ be the transition probabilities of a random walk on H. In this new random walk on H, a visit by time n to a fixed element $h \in H$ must be more likely than in the original random walk on G. Thus the new random walk is irreducible and recurrent on H.

Lemma 5. *A symmetric irreducible random walk on the two-dimensional lattice points with $p_{\bar{n}} = 0$ except for finitely many $\bar{n} \neq (0,0)$ is recurrent.*

Let $p_{\bar{n}_j} = \alpha_j$, $j = 1, \ldots, N$ with $\sum_{j=1}^{N} 2\alpha_j = 1$ and $\bar{n}_j \neq -\bar{n}_k$ for all j and k. Now

$$\sum_{k=0}^{\infty} P(X_k = \bar{0}) = \sum_{k=0}^{\infty} \int_{-\pi}^{\pi}\!\!\int [\varphi(\theta)]^k \, d\theta = \lim_{k \to \infty} \int_{-\pi}^{\pi}\!\!\int \frac{1 - [\varphi(\theta)]^{k+1}}{1 - \varphi(\theta)} \, d\theta , \quad (13)$$

where

$$\varphi(\theta) = \sum_{j=1}^{N} 2\alpha_j \cos(\bar{n}_j \cdot \theta) .$$

As $k \to \infty$ through the odd integers, the integrands are nondecreasing so that by the monotone convergence theorem, (13) is equal to

$$\int_{-\pi}^{\pi}\!\!\int \frac{1}{1 - \varphi(\theta)} \, d\theta = \int_{-\pi}^{\pi}\!\!\int \frac{d\theta_1 \, d\theta_2}{\alpha\theta_1^2 + \beta\theta_1\theta_2 + \gamma\theta_2^2 + o(|\theta|^2)} = +\infty .$$

The zero element $\overline{0}=(0,0)$ is recurrent and this implies that the random walk is recurrent.

Lemma 4 implies that the rank of a discrete Abelian group with an irreducible recurrent random walk is at most two since there is no irreducible recurrent random walk on the three-dimensional lattice points. The converse is established by constructing a sequence of recurrent walks on finitely generated subgroups which converge sufficiently rapidly to the desired recurrent random walk on the full group. Assume that G is infinite since otherwise the converse is trivial.

If G has rank one, let g_1 be any element of infinite order, that is, such that $ng_1 \neq 0$ for all integral $n \neq 0$. If G has rank two, let g_1, g_2 be two linearly independent elements. If G has rank zero and is nontrivial (not the group of one element), take g_1 any element not 0. Further, the g_i's are to be chosen so that $g_1, g_2, \ldots$ generate G and yet g_{n+1} is not in the subgroup G_n generated by $g_1, \ldots, g_n$. A sequence of probability distributions $\{{}_np\}$ determining symmetric random walks is specified with $\{{}_np\}$ concentrated on G_n. Let ${}_1p_{g_1}={}_1p_{-g_1}=\frac{1}{2}$ (unless $2g_1=0$ in which case set ${}_1p_{g_1}=1$) and ${}_2p_{g_1}={}_2p_{-g_1}={}_2p_{g_2}={}_2p_{-g_2}=\frac{1}{4}$ (unless $2g_1=0$ and/or $2g_2=0$ in which case set ${}_2p_{g_1}$ and/or ${}_2p_{g_2}=\frac{1}{2}$). The $\{{}_np\}$ are to be constructed so that they determine an irreducible recurrent random walk on G_n for each n. This is true for $n=2$ since Lemma 5 implies that a symmetric random walk on an Abelian group with two generators is recurrent.

If ${}_np$ determines a symmetric irreducible recurrent walk on G_n let

$$ {}_{n+1}p_g=(1-q_{n+1})\,{}_np_g $$

if $g \in G_n$, ${}_{n+1}p_{g_{n+1}}={}_{n+1}p_{-g_{n+1}}=q_{n+1}/2$ (${}_{n+1}p_{g_{n+1}}=q_{n+1}$ if $2g_{n+1}=0$) and ${}_{n+1}p_g=0$ otherwise, where q_{n+1} is a small positive number to be determined.

A basic result on finitely generated Abelian groups (see Kurosh [62]) implies that G_n, which is at most of rank two, must be of the form $Z \oplus Z \oplus H$ or $Z \oplus H$ with H a finite group if it has positive rank. Lemma 4 implies that the walk returns to H with probability one and this in turn implies that the probability of returning to 0 is one. Thus ${}_np$ determines an irreducible recurrent walk on G_n.

The q_n's will now be specified. Let $X_k^{(n)}$, $k=1,2,\ldots$, be the random walk starting at 0 on G_n determined by ${}_np$. Now $\sum_{k=1}^{\infty} P(X_k^{(n)}=0)=\infty$ for each n. Let N_1 be such that

$$ \sum_{k=1}^{N_1} P(X_k^{(1)}=0)>1 $$

and choose $Q_{11}, Q_{12}, \ldots > 0$ so that

$$\prod_{j=1}^{\infty} (1 - Q_{1j})^{N_1} \sum_{k=1}^{N_1} P(X_k^{(1)} = 0) > 1 .$$

With $N_1, \ldots, N_{n-1}, Q_{11}, \ldots, Q_{n-1,1}, \ldots$ already determined let N_n be such that

$$\sum_{k=N_{n-1}+1}^{N_n} P(X_k^{(n)} = 0) > 1$$

and $Q_{n1}, Q_{n2}, \ldots > 0$ with

$$\prod_{j=1}^{\infty} (1 - Q_{nj})^{N_n} \sum_{k=N_{n-1}+1}^{N_n} P(X_k^{(n)} = 0) > 1. \tag{14}$$

Set $q_n = \min(Q_{1n}, \ldots, Q_{nn}, r_n)$ where $0 < r_n < 1$ and $\prod_{n=1}^{\infty} (1 - r_n) > 0$. Let

$$p_g = \prod_{n=m+1}^{\infty} (1 - q_n)\, {}_m p_g \quad \text{for } g \in G_m.$$

Then p is defined on all of G since $G = \bigcup G_n$. Furthermore, since

$$\prod_{k=n}^{\infty} (1 - r_k) = \prod_{k=n}^{\infty} (1 - r_k) \sum_{g \in G_n} {}_n p_g \leqslant \sum_{g \in G_n} p_g < 1,$$

it follows that $\sum_{g \in G} p_g = 1$. Let p determine the random walk on all of G. Call X_k, $k = 1, 2, \ldots$, the random walk on G starting at 0 with transition probabilities given by p. For each positive integer n

$$P(X_k \in G_n, k = 1, \ldots, m) = \prod_{k=n+1}^{\infty} (1 - q_k)^m .$$

The conditional probability of returning to 0 on the jth step given that the walk has remained on G_n is exactly $P(X_j^{(n)} = 0)$. Therefore

$$\begin{aligned}
\sum_{j=N_{n-1}+1}^{N_n} P(X_j = 0) &\geqslant \sum_{j=N_{n-1}+1}^{N_n} \prod_{k=n+1}^{\infty} (1 - q_k)^j P(X_j^{(n)} = 0) \\
&\geqslant \prod_{k=n+1}^{\infty} (1 - q_k)^{N_n} \sum_{j=N_{n-1}+1}^{N_n} P(X_j^{(n)} = 0) \\
&\geqslant \prod_{k=n+1}^{\infty} (1 - Q_{nk})^{N_n} \sum_{j=N_{n-1}+1}^{N_n} P(X_j^{(n)} = 0) \\
&\geqslant 1
\end{aligned}$$

by (14). Thus $\sum_{j=1}^{\infty} P(X_j=0)=\infty$ and we have shown that the random walk is irreducible recurrent.

Notice that Dudley's result implies that the restricted direct sum of a countable number of copies of the rationals modulo one will support an irreducible recurrent random walk.

Notes

1.1 The Markov property is given as

$$P(F\,|\,\mathscr{B}_m)(w)=P(F\,|\,\mathscr{A}_m^m)(w)$$

for any event F of the forward field $F_m=\mathscr{B}(X_k, k\geq m)$ in section 1.1. An equivalent more symmetrical definition is the following. Let F and B be any two events of the forward and backward fields $\mathscr{F}_m=\mathscr{B}(X_k, k\geq m)$ and $\mathscr{B}_m=\mathscr{B}(X_k, k\leq m)$, respectively. The process is Markovian if

$$P(F\cap B\,|\,\mathscr{A}_m^m)(\omega)=P(F\,|\,\mathscr{A}_m^m)(\omega)\,P(B\,|\,\mathscr{A}_m^m)(\omega)$$

for any two such events. This is conditional independence of the past and the future given precise knowledge of the present (the time m being regarded as the present). The second formulation of the Markov property indicates that if a process is Markovian with time increasing, then it is Markovian with time reversed. A discussion of the equivalence of these two descriptions of the Markov property can be found in section 2.6 of Doob [18]. The equivalence is easy to verify in the case of a countable state process.

The Markovian property as stated above is formulated in terms of processes with a linearly ordered index set. For this reason, it cannot be extended naturally to processes with an index set not linearly ordered. However, one can formulate a related property which can be extended to processes with, say, a multidimensional index set. For convenience, we consider only processes with index set the lattice points in k-space (k a fixed positive integer) with integer components. Given any set S of index points, let S^c denote the complement of S and H the boundary of S consisting of lattice points in S^c but at a distance of one from S. Let $\mathscr{B}_S$, $\mathscr{B}_{S^c}$, and $\mathscr{B}_H$ be the Borel fields of sets generated by the random variables with indices in S, S^c and H respectively. The property is then that

$$P(B\,|\,\mathscr{B}_{S^c})(\omega)=P(B\,|\,\mathscr{B}_H)(\omega)$$

for any event $B\in\mathscr{B}_S$. Loosely, one could speak of this as conditional independence of the interior and exterior of S given precise knowledge on the boundary of S. Of course one wants this to hold for every set S of lattice points. The analogue of this property for continuous time parameter processes has to be formulated more carefully. A discussion of this property for discrete time parameter processes can be found in Dobrushin [16]. Dobrushin's motivation in looking at such processes with multidimensional index set appears to have been certain problems in statistical mechanics.

We have already noted that even though a Markov process has a stationary transition function in the forward direction, with time reversed it will generally not have a stationary transition function. We shall briefly sketch some ideas of Hunt [43] that indicate there is a formulation (not as intuitive as that ordinarily given) within which there is greater symmetry with respect to time reversal. Think of the initially given state space as a countable set of proper states (say the integers) with two additional states a, b corresponding to the specification of the process (or chain) before entering and after leaving the proper states respectively. Let $(\Lambda, \mathscr{B}, m)$ be a measure space with measure m. Λ is to be thought of as the space of possible trajectories ω of the chain (a generalized chain) with $\mathscr{B}$ a Borel field of subsets of Λ on which m is defined. Let α and β be measurable functions on Λ with the values of α integers or $-\infty$ and the values of β integers or $+\infty$. Further let $\alpha \leqslant \beta$. The functions α and β are to be the initial and final times of a sojourn through the integers. Let $X_n(\omega)$, $\omega \in \Lambda$, be an integer for almost all ω and all integers n with $\alpha(\omega) \leqslant n \leqslant \beta(\omega)$. Extend the definition of X_n by setting $X_n(\omega) = a$ for $n < \alpha(\omega)$ and $X_n(\omega) = b$ for $n > \beta(\omega)$. Assume that the extended X_n is measurable over $\mathscr{B}$ for each n. Of course, X_n represents the state of the generalized chain (and m is allowed to be an infinite measure). The triple (X, α, β) is called a *random chain* on R if $A_{n,r} = \{\omega : \alpha(\omega) \leqslant n \leqslant \beta(\omega), X_n(\omega) = r\}$ has finite m measure for all integers n and r. The measures of the sets on which X_n takes the values a or b may be infinite. The random chain (X, α, β) is called a Markov chain if past and future are conditionally independent given precise knowledge of the present when the process takes on integer values as states. The Markov chain (X, α, β) is said to have P as stationary probability matrix *(is a P-chain)* if

$$m(A_{n,r} \cap A_{n+1,s}) = m(A_{n,r}) p_{r,s}$$

for all integers r, s, n with $P = (p_{j,k})$. Consider a random chain (X, α, β) and a $\mathscr{B}$ measurable function having $\pm\infty$ or integers as values. Let $\Lambda' = \{\omega : \sigma(\omega)$ is finite, $\alpha(\omega) \leqslant \sigma(\omega) \leqslant \beta(\omega)\}$ with

$$\gamma(\omega) = \beta(\omega) - \sigma(\omega), \; Y_n(\omega) = X_{n+\sigma(\omega)}(\omega).$$

Thus $(Y, 0, \gamma)$ is a random chain obtained by starting to observe X_n at time $\sigma(\omega)$ (relabeled as time 0) and ending at time β. The random time σ is said to reduce (X, α, β) to a P-chain if $(Y, 0, \gamma)$ is a P-chain defined over Λ'. (X, α, β) is said to be an *approximate P-chain* if there is a sequence of random times α_n whose values are $+\infty$ or the integers such that the α_n decrease to α almost surely and each α_n reduces (X, α, β) to a P-chain. If (X, α, β) is a random chain, then (X', α', β') defined over the same measure space with $\alpha' = -\beta$, $\beta' = -\alpha$, and $X'_n(\omega) = X_{-n}(\omega)$ is also a random chain, the *reversed chain*. Given a set of integers S let I_S be the indicator function of S. Consider the measure η on the integers given by

$$\eta(S) = \int_\Lambda \sum_{\alpha \leqslant n \leqslant \beta} I_S(X_n(\omega)) m(d\omega).$$

Hunt proves the following interesting result. *Let (X, α, β) be an approximate P-chain for which the measure η is finite on finite sets. Then the reversed chain (X', α', β') is an approximate Q-chain with $Q = (q_{j,k})$ given by*

$$q_{j,k} = \frac{\eta(k) p_{k,j}}{\eta(j)}$$

if $\eta(j) > 0$. A discussion of a slightly modified version of some of Hunt's results can be found in Chapter 10 of [54].

1.2 An extensive discussion of Markov chains can be found in Chung's book [9] on the subject. Notice that class properties like transience, recurrence, or persistence are of especial interest. If one state in an irreducible class of states has this property, then all the states in that irreducible class have the same property.

These properties are determined in terms of the recurrence time distribution of the state. Let $m_i^{(p)}$ be the pth moment $(p \geqslant 0)$ of the recurrence time distribution of the state i (assuming that the state is recurrent). If $m_i^{(p)} < \infty$ then $m_j^{(p)} < \infty$ for every state j in the same irreducible class of states as i. Thus finiteness of the pth recurrence time moments is also a class property. A derivation of this result can be found in section 11 of the first part of Chung [9].

1.3 Treatments of the infinitely divisible laws can be found in Doob [18], Loève [70], or Feller [27]. However, the most extended and complete development of infinitely divisible laws and limit theorems for sums of independent random variables is still to be found in the book of Gnedenko and Kolmogorov [31]. Only a few of the infinitely divisible laws can be neatly written out. The most famous of these few laws are the Poisson and Gaussian distributions. However, characteristic functions (Fourier-Stieltjes transform) $\varphi(t)$ of the general infinitely divisible law can be written in terms of a nondecreasing bounded function G:

$$\log \varphi(t) = i\gamma t + \int \left\{ e^{itu} - 1 - \frac{itu}{1+u^2} \right\} \frac{1+u^2}{u^2} \, dG(u) . \tag{1}$$

The integrand in the integral on the right is defined by continuity at $t=0$. The characteristic function of hX where X is a Poisson random variable with mean λ is

$$\exp\{\lambda(e^{ith} - 1)\} .$$

This suggests that on replacing the integral in (1) by an approximating finite Riemann-Stieltjes sum one would get an approximating characteristic function, the Fourier-Stieltjes transform of a weighted sum of independent Poisson variables. The basic character of the Poisson distribution in the class of infinitely divisible laws is suggested by this heuristics. Such an argument is given in rigorous form in [31].

There is still discussion of the best constants that can be attained in the Berry-Esseen theorem (see Feller [27] pp. 516) and Prohorov's result (see LeCam [63]). Related investigations have been carried out by Meshalkin [76].

1.4 From the discussion in Appendix 3 it is clear that in one method of setting up the measure space for a continuous time parameter process, the basic space is a function space. The derivation of some of the properties of the Poisson process as given in this section suggests that the measure space can be taken to be the space of nondecreasing right continuous step functions whose jumps are all of magnitude one (see Doob [18] p. 398 for a more extended discussion). One can also show that the measure space in the case of the one-dimensional Brownian motion process can be set up so that it consists of continuous functions (see Doob [18] p. 392). Thus the possible paths of a particle in such a Brownian motion can be taken to be continuous. Both Poisson and Brownian motion processes are Markovian. Those continuous time Markov processes with continuous paths (the measure space consisting of continuous functions) are called *diffusion processes*. The one-dimensional diffusion processes have been analyzed in great detail (see Dynkin [23] and Ito and Mckean [49]). S. Bernstein [3] and K. Ito [47] dealt with a way of deriving a large class of one-dimensional diffusion processes in terms of the one-dimensional Brownian motion. Let $X(t)$ be a one-dimensional diffusion with local mean and variance $m(t,x)$ and $\sigma^2(t,x)$ respectively, that is,

$$E[X(t+\Delta) \mid X(t)=x] = m(t,x)\Delta + o(\Delta)$$
$$E[(X(t+\Delta) - X(t))^2 \mid X(t)=x] = \sigma^2(t,x)\Delta + o(\Delta) .$$

They suggested solving the equation

$$X(t)-X(\tau)=\int_{\tau}^{t} m(s,X(s))\,ds+\int_{\tau}^{t} \sigma(s,X(s))\,dB(s) \tag{2}$$

to obtain a diffusion $X(t)$ with the desired properties in terms of the Brownian motion process. This would mean that the process $X(t)$ is defined on the measure space of the Brownian motion process. The definition of the integrals in equation (1) and the class of functions for which they are defined can be found in Doob [18] p. 277. Under appropriate bounds and regularity conditions on the functions m and σ, it is shown in [18] that the equation (2) has a unique solution $X(t)$ that is a diffusion process.

1.5 A very detailed discussion of random walks on the group of lattice points in k-space for $k=1,2,3$ can be found in Spitzer's book on random walk [105]. The result of Dudley developed in this section can be found in [19]. Chung and Fuchs dealt with random walks in Euclidean k-space, $k=1,2,3$, in their paper [8]. They call a value b recurrent if for every $\varepsilon>0$

$$P\{|X_n-b|<\varepsilon \text{ for infinitely many integers } n\}=1$$

where X_n is the random walk considered. Let F be the distribution function of one step in the random walk, that is, of $X_{n+1}-X_n$ for $n=1,2,\ldots$, and let φ be the characteristic function of F. They showed that if

$$\int_V \frac{du}{1-\varphi(u)}=\infty$$

for some compact neighborhood V of 0 in k-space, then there are recurrent values.

Kesten and Spitzer [56] obtained a corresponding result for a random walk on a countably infinite locally compact Abelian group G. Let μ be the probability measure on G corresponding to one step in the random walk. Assume that the support of μ is not contained in a proper subgroup of G. Let $\hat{\mu}$ be the Fourier transform of μ (see Rudin [101] for a discussion of harmonic analysis on Abelian topological groups). Then $\hat{\mu}$ is defined on the character group Γ of G. They show that the random walk is recurrent if and only if

$$\int_\Gamma \operatorname{Re}\frac{1}{1-\hat{\mu}(\lambda)}\,d\lambda=\infty$$

where the integration of the real part of $(1-\hat{\mu}(\lambda))^{-1}$ is carried out on Γ with respect to the Haar (or uniform) measure on Γ. This result of Kesten and Spitzer implicitly contains Dudley's result but also allows one to determine whether a specific random walk is recurrent or not (if one can carry out the computations indicated).

The result of Kesten and Spitzer has recently been extended by Port and Stone [85] to the case of a general locally compact Abelian group. The random walk is said to be recurrent if

$$P(X_n\in N \text{ infinitely often})=1$$

for some compact neighborhood of 0 in G. Let μ be a regular probability measure generating the random walk (corresponding to one step in the random walk) whose support is not contained in a closed subgroup of G. Then the random walk is recurrent if and only if

$$\int_V \mathrm{Re}\,\frac{1}{1-\hat{\mu}(\lambda)}\,d\lambda=\infty$$

where $\hat{\mu}$ is the Fourier-Stieltjes transform of μ and the integral is taken over a compact neighborhood V of 0 in the character group Γ of G with respect to the Haar measure on Γ. Kesten discusses some related questions for noncommutative groups in [57].

Chapter II

Remarks on Some Applications

0. Summary

We now briefly sketch a few rather specific applied problems that suggest or even motivate a number of the questions taken up in this book. The aim will not be to give a full development, which would lie outside the scope and intent of the book, but rather to indicate how the problems arise and the mathematical questions suggested. References are given that will allow the intersted reader to look into the the details more extensively on his own. At times the description we give will be heuristic since there are still wide gaps in the mathematical development. This is especially so in an example from statistical mechanics. There is a brief discussion of a class of models in so-called "learning theory" in mathematical psychology. The last example is a resource flow model which is not probabilistic as usually given but obviously could be considered with a probabilistic interpretation. It is suggested formally by the Quesnay-Leontief models which have been of some interest in mathematical and statistical economics.

1. A Model in Statistical Mechanics

The first problem we consider is one in statistical mechanics. It involves a molecular model of certain gross features of matter (the laws of thermodynamics). Our concern will be with the model rather than actual derivation of the laws of thermodynamics. Assume that one has N molecules whose motion is governed by the laws of classical mechanics. Typically the molecules are thought of as being contained in a box of finite extent. The system (the N molecules and the box) is assumed to be conservative, that is, there is no dissipation of energy. The equations describing the system of N molecules are set up in terms of a Hamiltonian

$$H = T + V \tag{1}$$

where the kinetic energy T has the form

$$T=\sum_{i=1}^{N}\frac{\bar{p}_i^2}{2m} \tag{2}$$

and the potential energy V the form

$$V=\sum_{i=1}^{N}U(\bar{r}_i)+\sum_{i<j}\varphi(|\bar{r}_i-\bar{r}_j|). \tag{3}$$

Here $\bar{p}_i$ and $\bar{r}_i$ are the momentum and position of the ith molecule and $|\bar{r}_i-\bar{r}_j|$ is the distance between the ith and j^{th} molecule. All the molecules are assumed to have the same mass m. The term $U(\bar{r}_i)$ can be thought of as the potential of exterior forces acting on the ith molecule (including forces exerted by the walls of the box to keep the molecule inside). The function φ describes the interaction potential between molecules. The interaction potential is often taken with an attractive region extending roughly from r_0 to r_1 $(0<r_0<r_1)$ and a steep repulsive part corresponding to a hard core of radius r_0. Of course, such a Hamiltonian is highly simplified even in the classical (nonquantum mechanical) context we are considering.

The state of this mechanical system of N molecules (or particles) is represented by a point in $2n$-dimensional phase space (momentum-position space) with $n=3N$. The coordinates of the point are $\bar{p}_1,\ldots,\bar{p}_N,\bar{r}_1,\ldots,\bar{r}_N$ the momenta and positions of the particles, respectively. The Hamiltonian H of the system is given by (1) and the trajectory of the point in phase space (its path as a function of time) is governed by the Hamilton equations, a system of ordinary differential equations

$$\left.\begin{aligned}\frac{dq_i}{dt}&=\frac{\partial H}{\partial p_i}\\ \frac{dp_i}{dt}&=-\frac{\partial H}{\partial q_i}\end{aligned}\right\}\quad i=1,\ldots,n. \tag{4}$$

The q_i's and p_i's are the generalized coordinates and momenta of the system, which may coincide with the components of the initial coordinates and initial momenta, respectively (see Goldstein [30] for an extended discussion of the Hamilton equations). If they do not coincide, they are usually introduced so as to incorporate some restraints on the initially given Euclidean coordinates or momenta that are part of the problem or to take advantage of some natural symmetry in the problem. In order for the Hamilton equations to make sense, the Hamiltonian must be smooth. Let us *assume that the Hamiltonian $H(p,q)$ is a continuously differentiable function of the p's and q's*. Notice that the Hamiltonian does not depend explicitly on time. This means that H is an integral of the equations (4), that is,

$$\frac{dH}{dt}=\sum_{i=1}^{n}\left(\frac{\partial H}{\partial p_i}\frac{dp_i}{dt}+\frac{\partial H}{\partial q_i}\frac{dq_i}{dt}\right)=0\,.$$

This is only a trivial verification of the law of conservation of energy for the system, which says that all points on any solution curve or trajectory of (4) have the same energy H. For this reason we can consider the solutions of the system (4) in the domain

$$M=\{(p,q)\colon -\infty<E_1\leqslant H(p,q)\leqslant E_2<\infty\}\,.$$

The domain M in (p,q) space is compact if the potential energy V in (3) is continuous and bounded below. We can then use standard results on solutions of ordinary differential equations to guarantee the existence of solutions of the system for all time. For each point $(p(0),q(0))$ of the domain M there is a neighborhood U and a number $\varepsilon>0$ such that (4) has a unique smooth solution $(p(t),q(t))$ for $|t|<\varepsilon$ with

$$(p(0),q(0))=(p,q)\in U$$

(see [10] for the standard existence and uniqueness theorem). The compact set M can be covered by a finite number of these neighborhoods. Let ε_0 be the smallest of the ε's corresponding to these neighborhoods. By piecing together the solutions from the neighborhoods of the finite covering we conclude that the system (4) has a unique solution $(p(t),q(t))$ for $|t|<\varepsilon_0$ with $(p(0),q(0))=(p,q)$ for all $(p,q)\in M$. The solution $(p(t),q(t))$ is smooth as a function of t and the initial condition $(p(0),q(0))=(p,q)$. Let

$$(p(t),q(t))=S_t(p,q)\,.$$

Then it is clear that

$$S_t\cdot S_\tau=S_{t+\tau}$$

as long as $|t|$, $|\tau|$, $|t+\tau|<\varepsilon_0$. For each $|t|<\varepsilon_0$ it follows that S_t is a continuous differentiable mapping of M onto itself. We now have to define S_t for $|t|\geqslant\varepsilon_0$. It is enough to do this for $t>0$ since a similar argument will define S_t for all $t<0$. For $t=k(\varepsilon_0/2)+r$ with k a non-negative integer and $|r|<\varepsilon_0/2$ *let*

$$S_t=(S_{\varepsilon_0/2})^k S_r\,.$$

One can then readily show that S_t, $-\infty<t<\infty$, is a continuous group $(S_t\cdot S_\tau=S_{t+\tau})$ of continuously differentiable mappings of the domain M onto itself with $S_0=I$, the identity map. The domain M is often called an "energy shell". It is now clear that under the assumptions made on the potential energy we can extend the domain (and range) of the group $\{S_t\}$ so that it maps the admissible part of the phase space

$$\{(p,q): H(p,q) < \infty\}$$

onto itself.

As a very simple illustrative example consider a two-dimensional linear oscillation with the Hamiltonian

$$H = \frac{\omega_1}{2}(p_1^2 + q_1^2) + \frac{\omega_2}{2}(p_2^2 + q_2^2).$$

The kinetic and potential energy are given by

$$T = \frac{\omega_1}{2} p_1^2 + \frac{\omega_2}{2} p_2^2$$

and

$$V = \frac{\omega_1}{2} q_1^2 + \frac{\omega_2}{2} q_2^2$$

respectively. The Hamilton equations in this case become

$$\frac{dq_i}{dt} = \omega_i p_i, \quad \frac{dp_i}{dt} = -\omega_i q_i, \quad i=1,2.$$

These equations reduce to

$$\frac{d^2 q_i}{dt^2} = -\omega_i^2 q_i, \quad i=1,2,$$

the equations for two uncoupled linear oscillators.

Let us now show that there is a natural measure (ordinary Euclidean volume in phase space) that is left invariant by the group of transformations S_t. *Assume* that *the Hamiltonian H has continuous second order derivatives.* For convenience relabel the coordinates and momenta

$$x_1 = q_1, \ldots, x_n = q_n$$
$$x_{n+1} = p_1, \ldots, x_{2n} = p_n.$$

Let A be a subset of the energy shell M with finite volume. The transformation S_t takes the set A into the set $A_t = S_t A$ with volume

$$\int_{A_t} dx_1 \ldots dx_{2n} = \int_A J\, dy_1 \ldots dy_{2n},$$

where

$$S_t(y_1, \ldots, y_{2n}) = (x_1, \ldots, x_{2n})$$

and J is the corresponding Jacobian

$$J=\frac{\partial(x_1,\ldots,x_{2n})}{\partial(y_1,\ldots,y_{2n})}.$$

Notice that

$$\frac{dJ}{dt}=\sum_{k=1}^{2n}J_k$$

with

$$J_k=\frac{\partial(x_1,\ldots,x_{k-1},\dot{x},x_{k+1},\ldots,x_{2n})}{\partial(y_1,\ldots,y_{2n})}$$

and $\dot{x}=dx_k/dt$. However,

$$J_k=\sum_{v=1}^{2n}\frac{\partial\dot{x}}{\partial x_v}\frac{\partial(x_1,\ldots,x_{k-1},x_v,x_{k+1},\ldots,x_{2n})}{\partial(y_1,\ldots,y_{2n})}=\frac{\partial\dot{x}}{\partial x_k}J$$

so that

$$\frac{dJ}{dt}=J\sum_{k=1}^{n}\frac{\partial^2 H}{\partial p_k\partial q_k}-J\sum_{k=1}^{n}\frac{\partial^2 H}{\partial q_k\partial p_k}=0.$$

Thus J does not depend on t so that $J\equiv 1$. The volume is left invariant by S_t. This interesting result is called *Liouville's theorem*. Since energy is conserved, it seems reasonable to restrict oneself to a surface of constant energy, that is $H(p,q)=E$. If the length of the gradient of H is bounded away from zero, Liouville's theorem suggests that the measure on the surface of constant energy given by

$$d\sigma(\Omega)=\frac{\text{const.}\,d\Omega}{|\text{grad}\,H|} \tag{5}$$

is preserved by the group of transformations S_t. This can be seen by choosing a local coordinate frame with one axis normal to the surface (parallel to the gradient) and the other axes in the tangent plane. Assume that the measure (5) is normalized so as to be a probability measure on the surface. If the system of N particles is initially given the invariant measure (5) on the surface of constant energy, the stochastic process generated on the surface is stationary. The process is trivially Markovian since it is given by a group of point transformations. Suppose we consider any observable $f(p,q)$ integrable with respect to the measure (5). We might hope that the time average

$$\frac{1}{T}\int_0^T f(p(t),q(t))\,dt$$

will tend to the space average

$$\int_{H(p,q)=E} f(p,q)\,d\sigma$$

in some reasonable sense as $T \to \infty$ for every such observable. This is the well-known ergodic hypothesis. Ergodicity has only recently been established for some simple systems of interest in statistical mechanics, for example, the case of hard spheres with elastic collisions (see Sinai [103]). The system considered on a surface of constant energy E can be regarded as an isolated thermally insulated system with total energy E. A system with the invariant distribution described on a surface of constant energy is often referred to as a *microcanonical ensemble.* Of course, in the potential energy (3), the interaction potential might not be the same for all particles if there were different types of particles in the system. However, the discussion and the development given would hold under appropriate conditions even for more complicated systems of this type. The object is to consider systems of this type with a very large number N of particles and to hope that an analysis of the mean behavior of quantities of thermodynamic interest will then be possible.

We shall now make some remarks of a heuristic character that may motivate some of the problems we discuss later on. In statistical physics one would like to have an extension of equilibrium thermodynamics that will cover situations close to equilibrium, that is, to study "non-equilibrium thermodynamics". In setting up the foundations for such a study, it is usual to consider a microcanonical ensemble. It is assumed that the system can be described macroscopically by m variables $\alpha_1(\overline{r},\overline{p}), \ldots, \alpha_m(\overline{r},\overline{p})$, where m is much smaller than the number N of molecules in the system. Here $\overline{r}$ and $\overline{p}$ denote the positions and momenta of the molecules of the system. The variables α might be the energies of smaller subsystems. Each of these subsystems is assumed to contain many molecules. The microcanonical ensemble itself is obviously Markovian since its evolution is deterministic. In order to study the macroscopic variables $\alpha_1, \ldots, \alpha_m$ it is assumed that the process

$$\alpha(\overline{r}(t),\overline{p}(t)) = \big(\alpha_1(\overline{r}(t),\overline{p}(t)), \ldots, \alpha_m(\overline{r}(t),\overline{p}(t))\big)$$

is Markovian. Of course, this will not be literally true for all time, but the hope is that it would be satisfied on some time scale between the microscopic collision time and some macroscopic relaxation time characteristic of the system. In any case these considerations provide a motivation for the study of the following idealized problem. *When is a function of a Markovian system itself Markovian?*

We will assume that the process $\alpha(\overline{r},\overline{p})$ is approximately Gaussian. This assumption is usually motivated in the literature by an appeal to the central limit theorem though it is not clear what version of such a result is referred to. In any event this assumption provides another one of the many reasons for studying the central limit problem.

Let us discuss the process $\alpha(\bar{r},\bar{p})$ under the further assumption that it is Markovian. The variables α are often assumed to be *invariant under time reversal*, that is,

$$\alpha_i(\bar{r},\bar{p}) = \alpha_i(\bar{r}, -\bar{p}) \tag{6}$$

for all i. This would be true, for example, if the α's were functions of the momenta. Since the microscopic equations of motion are time reversible ($H(\bar{r},\bar{p}) = H(\bar{r}, -\bar{p})$), it then follows that (6) is valid. The process $\alpha(\bar{r}(t),\bar{p}(t))$ is stationary since the instantaneous distribution for the microcanonical ensemble is invariant under the action of the group S_t. The stationarity of $\alpha(\bar{r}(t),\bar{p}(t))$ coupled with the invariance of α under time reversal implies that

$$\begin{aligned} &P\big(\alpha(\bar{r}(t),\bar{p}(t))\in A, \alpha(\bar{r}(t+\tau),\bar{p}(t+\tau))\in B\big) \\ &= P\big(\alpha(\bar{r}(t),\bar{p}(t))\in B, \alpha(\bar{r}(t+\tau),\bar{p}(t+\tau))\in A\big) \end{aligned}$$

for any two Borel sets A and B (in n-space). The process $\alpha(\bar{r}(t),\bar{p}(t))$ is thus a reversible Markov process. This is commonly called the *principle of detailed balance.*

The Gaussian Markov character of the α process allows one to use linear relationships to describe the approach to equilibrium. This provides a basis for the derivation of the Onsager reciprocal relations which relate macroscopic fluxes (such as flows of heat, matter, and charge) to conjugate thermodynamic forces (temperature difference, concentration difference and electrical potential difference). We shall not pursue these ideas any further. The interested reader can find a detailed discussion and derivation of the Onsager relations using the approach described above in de Groot and Mazur's book "Nonequilibrium Thermodynamics" [32].

2. Some Models in Learning Theory

A very large number of experiments have been made to examine the modification of behavior in animals or man. Descriptions of such experiments can be found, for example, in the psychological literature (see [5], [80], [81]). Mathematical models have been constructed in an attempt to describe the response of a subject to a sequence of experiments. The models are usually referred to as "learning models". Before describing a format into which some of these models fit, let us briefly describe a few of the experiments to which one has tried to apply them.

One simple case is that of T-maze learning. Suppose that on each trial an animal is placed at the beginning of a T-maze. At the fork in the T-maze the animal (usually a rat or guinea pig) makes a choice of the left or right arm of the maze. There the animal may or may not be rewarded by food. Receiving food will reinforce the choice the animal has just made. The choice of the right arm (or left arm) of the maze can be thought of as an event in this trial. The reinforcement (or lack of it) coupled with the animal's choice can be regarded as the outcome of the trial. Suppose one has a sequence of trials. At the beginning of the nth trial there is a probability p_n that the animal will choose the left arm of the maze. The probability p_{n+1} that the animal will choose the left arm on the next trial will depend on p_n, the choice of a direction on the nth trial and the reinforcement or lack of it. Such a model would be Markov or a function of a Markovian scheme. Notice that the reinforcement mechanism could be of several different types. Suppose that there is a probability of reinforcing one event (right or left turn) on the $(n+1)^{\text{st}}$ trial given a particular prior event on the nth trial (right or left turn). These reinforcement probabilities may or may not depend on the occurrence of specific prior events.

Another experiment with many possible responses is the following. A subject is seated before a large disc. At each trial a spot of light appears somewhere on the rim of the disc. The subject tries to predict where the light will appear. Suppes [108] has proposed a model for a sequence of trials of such an experiment.

Let us now consider the formulation for the framework for a class of models in "learning theory" as given in a paper of Norman [80]. Consider a state space S and an event space E. For convenience, assume that the event space E is finite. To fit the second example into this context we are forced to think of the disc as having only a finite number of locations where the light to be predicted could appear. If the number of these locations is large enough, the approximation will be good. Corresponding to each event $e \in E$ there is a mapping $f_e(\cdot)$ of S into S. If the subject is in state S_n at trial n, an event E_n is determined (corresponding to the reinforcement or lack of reinforcement if one thinks of the T-maze experiment) and S_{n+1} is given by

$$S_{n+1} = f_{E_n}(S_n) .$$

Assume that S can be taken to be a compact metric space with metric d and the functions $f_e(\cdot)$ to be distance-diminishing operators, that is,

$$d(f_e(s), f_e(s')) \leqslant d(s, s')$$

for points $s, s' \in S$. A function $\varphi_e(s)$ on $E \times S$ is given which determines the conditional probability that $e \in E$ occurs, namely

$$P(E_1 = e | S_1 = s) = \varphi_e(s)$$

and

$$P(E_{n+1} = e_{n+1} | E_j = e_j, 1 \leqslant j \leqslant n, S_1 = s) = \varphi_{e_{n+1}}(f_{e_1 \dots e_n}(s)),$$

where

$$f_{e_1 \dots e_n}(s) = f_{e_n}(f_{e_{n-1}}(\dots(f_{e_1}(s)))).$$

Then

$$\begin{aligned} \varphi_{e_1 \dots e_n}(s) &= P(E_j = e_j, 1 \leqslant j \leqslant n | S_1 = s) \\ &= \varphi_{e_1}(s) \varphi_{e_2}(f_{e_1}(s)) \dots \varphi_{e_n}(f_{e_1 \dots e_{n-1}}(s)). \end{aligned}$$

The sequence of states $S_1, S_2, \dots$ forms a Markov process with transition function

$$P(s, A) = \sum_{e: f_e(s) \in A} \varphi_e(s).$$

A very simple illustration of this set up can be found in a model posed to describe an experiment carried out by Solomon and Wynne (see [5]). In it a dog was able to avoid an intense shock by jumping a barrier within ten seconds after a signal was given or escape further discomfort by jumping while the shock was being applied. The probability of avoiding the shock on trial n is denoted by p_n. If avoidance occurs on trial n, operator Q_1 is applied to p_n to obtain p_{n+1}

$$Q_1 p = \alpha_1 p + (1 - \alpha_1), \quad 0 < \alpha_1 < 1,$$

while if nonavoidance (shock) occurs operator Q_2 is applied to p_n to obtain p_{n+1}

$$Q_2 p = \alpha_2 p + (1 - \alpha_2), \quad 0 < \alpha_2 < 1.$$

In this simple model, it is clear as $n \to \infty$ that $p_n \to 1$ with probability one. However, in applications the primary interest would be to estimate α_1 and α_2 from the data obtained from the experiments. A discussion of such questions can be found in [5].

A qualitative idea as to what happens asymptotically in these learning theory models can be found in the discussion of random walks on semigroups, even though there is strictly speaking a limited overlap of these two classes of random models. The book of Iosifescu and Theodorescu [46] deals with an interesting class of related models. The reader should refer to recent issues of the Journal of Mathematical Psychology for insight into recent work.

3. A Resource Flow Model

The following descriptive model is not probabilistic as usually presented. However, it can be put into a probabilistic framework as a finite state stationary Markov chain in a reasonably natural way. We discuss the model because it suggests problems of aggregation or pooling data with a view toward deriving a system that would be more tractable from a computational point of view. Some of these problems of aggregation are related to the questions on Markovian functions of Markov processes that are treated in Chapter 3.

Let us consider a so-called input-output table for a resource flow. This would be a matrix

$$M=(m_{i,j};\ i,j=1,\dots,n)$$

of nonnegative elements $m_{i,j}\geqslant 0$, where $m_{i,j}$ represents the value of the input to industry or activity i from industry or activity j. The sum

$$\sum_{h=1}^{n} m_{h,i}=m_i>0$$

represents the value of the output of industry i; in M, the ith row sum equals the ith column sum for $i=1,\dots,n$, that is, "marginal balance" obtains. Let

$$a_{i,j}=\frac{m_{i,j}}{m_i}$$

be thought of as the normalized input to activity i from activity j. The matrix $A=(a_{i,j})$ is a stochastic matrix, that is,

$$a_{i,j}\geqslant 0$$

and

$$\sum_j a_{i,j}=1\,.$$

The equation

$$x=xA \tag{1}$$

can be viewed as representing a closed input-output model with a nonnegative solution vector x representing an activity level. If B is a proper principal submatrix of A, it will be substochastic, that is,

$$b_{i,j}\geqslant 0$$

$$\sum_j b_{i,j}\leqslant 1\,.$$

An equation

$$x(I-B)=w \tag{2}$$

with x, w nonnegative vectors can be thought of as representing an open input-output model with w representing a bill of goods for those activities represented in A but not in B as a single aggregate and x as a corresponding activity level. Such representations are often referred to as Leontief models because they have been extensively discussed as approximations (or descriptions) of economic activity by him.

If the tableau (matrix) A is thought of as giving the normalized representation of an economy, it is clear that the number of sectors taken into account will often be exceedingly large. A question that typically arises is whether there exists an aggregation or consolidation (pooling) of sectors that will leave the form of the equations (1) and (2) the same for the aggregated model. A typical aggregation matrix C represents a function mapping the n sectors into $r<n$ sectors. The matrix C will be an $n \times r$ matrix with exactly one entry in each row equal to one and all the rest zero.

Let us now return to equation (2). If B has no principal stochastic submatrix, the inverse $(I-B)^{-1}$ exists and is represented by the power series

$$(I-B)^{-1} = \sum_{j=0}^{\infty} B^j .$$

One can then simply solve for x by

$$x = w(I-B)^{-1} .$$

The aggregated solution vector is

$$y = xC = w(I-B)^{-1} C \tag{3}$$

and the aggregated bill of goods vector is

$$wC .$$

Suppose we can express y directly in terms of the aggregated bill of goods vector

$$y = wCH^{-1} \tag{4}$$

by means of an invertible $r \times r$ matrix H. If this is to hold for every admissible vector w we shall require that

$$(I-B)^{-1} C = CH^{-1}$$

by (3) and (4) or

$$CH = (I-B)C . \tag{5}$$

But then we must have

$$H = (C'C)^{-1} C'(I-B)C . \tag{6}$$

Coupling (5) with (6) we find that

$$BC = C(C'C)^{-1}C'BC. \tag{7}$$

Condition (7) on the matrix B is very similar to a condition we will obtain in section 1 of Chapter 3 where we investigate conditions under which a function of a Markov process is still Markovian. In much of our brief discussion on aggregation we have used an approach of D. Rosenblatt [89], [90]. As already indicated, our discussion is primarily focused on a Leontief model of resource (or mass) flow. For a discussion of the role of such a model in a simple theoretical construction of a price mechanism see the book of Schwartz [102].

Notes

2.1 A discussion of some of the questions that arise in setting up the foundations of statistical mechanics can be found in the books of Kac [52] and Uhlenbeck and Ford [110]. Recently a set of lectures of Minlos [75] on statistical physics has been published. Minlos considers the basic concepts of statistical physics and developes mathematically a number of interesting results.

2.2 There is a survey of theoretical and experimental work on a variety of stochastic learning models in the book of Bush and Mosteller [5].

Chapter III

Functions of Markov Processes

0. Summary

This chapter is concerned with many-to-one functions of Markov processes. Neither the Markov property nor the Chapman-Kolmogorov equation are generally satisfied by the derived processes determined by such functions. The first section considers special circumstances under which the Chapman-Kolmogorov equation is still satisfied by the derived process. These involve conditions relating the structure of the function to the initially given Markov process. The case in which the Markov process is stationary and the transition function is self-adjoint or compact is discussed in some detail. The second section deals with preservation of the Markov property whatever the initial distribution. This involves the retention of a stronger property by the derived process. Finally, in the last section finite state non-Markovian processes are considered. The object is to determine which processes are instantaneous functions of finite state Markov chains. Algebraic concepts and tools are used to obtain a result of Heller.

1. Collapsing of States and the Chapman-Kolmogorov Equation

There are a variety of problems, particularly in applications in physics and economics (see Chapter 2), in which one is led to consider a function of a Markov process. It is natural to inquire under what circumstances (in terms of the function and the original process) the derived process will retain some aspect of the Markov property. The following two problems will be considered: 1. When will the first order transition probabilities of the derived process still satisfy the Chapman-Kolmogorov equation? 2. When will the derived process be itself Markovian? One is naturally led to consider the relationship between the Chapman-Kolmogorov equation and the full Markov property. The first order transition probabilities of a Markov process always satisfy the Chapman-Kolmogorov equation. However, there are many examples of non-Markovian processes whose first order transition probabilities also satisfy the Chapman-Kolmogorov equation. For example, consider a process $Y(n)$ with m states ($m>3$) and second order transition probability

$$P(Y(n+2)=u_2 \mid Y(n+1)=u_1,\, Y(n)=u_0)$$
$$=\frac{1}{m}\left\{1-\cos\left[\frac{2\pi}{m}(2u_2-u_1-u_0)\right]\right\} \quad u_0,u_1,u_2=0,1,\ldots,m-1.$$

Let the process $X(n)=(Y(n), Y(n-1))$ be Markov with a uniform distribution on the states (u_0,u_1). The first order transition probabilities of $Y(n)$ satisfy the Chapman-Kolmogorov equation but $Y(n)$ is not Markovian. This example is obtained by a simple modification of a construction due to P. Lévy [68].

Let $\mathscr{A}$ be a Borel field of subsets of the state space Ω. Take $X_n(\omega)$, $n=0,1,2,\ldots$, to be a Markov process with initial distribution μ, transition probability function $P(\cdot,\cdot)$ and state space Ω constructed as in section 1.1. Consider the new process $Y_n(\omega)=\varphi(X_n(\omega))$, $n=0,1,2,\ldots$, obtained by an instantaneous measurable point transformation φ mapping Ω onto Ω' where Ω' is a space that carries a Borel field $\mathscr{A}'$. Measurability of φ is just the requirement that

$$\varphi^{-1}B'=\{x:\varphi(x)\in B'\}\in\mathscr{A} \quad \text{if} \quad B'\in\mathscr{A}'.$$

The case of interest is that of a many-to-one transformation φ. For convenience we assume that the sets consisting of single points of Ω are elements of $\mathscr{A}$. A measure P'_μ is induced by φ and P_μ (see section 1.1) on the Borel field $\mathscr{A}'_\infty$ generated by sets of the form $A'_0\times A'_1\times\cdots\times A'_n$ with $A'_i\in\mathscr{A}'$. Questions concerning the Chapman-Kolmogorov property and retention of the Markov property are questions about the induced measure P'_μ. One can consider these problems in terms of the original space. This will often be the convenient way to analyze some of them. Then we consider the sub-Borel field $\mathscr{C}=\varphi^{-1}\mathscr{A}'$ of $\mathscr{A}$ and the Borel field $\mathscr{C}_\infty$ generated by sets of the form $C_0\times C_1\times\cdots\times C_n$ with $C_i\in\mathscr{C}$. Questions concerning the validity of the Chapman-Kolmogorov equation and retention of the Markov property relate to the structure of P_μ on $\mathscr{C}_\infty$.

Although the measure P_μ has been set up in terms of an initial probability measure μ, it will occasionally be useful to let μ be a positive σ-finite measure with $\mu(\Omega)=\infty$. P_μ can be constructed as before with the obvious modifications where necessary. We shall do this only when μ is an invariant measure with respect to $P(\cdot,\cdot)$, that is

$$\int\mu(dx)\,P(x,A)=\mu(A).$$

It is convenient to discuss the Chapman-Kolmogorov equation (and occasionally even questions about the preservation of the Markov property) in terms of the operator T induced by $P(\cdot,\cdot)$ on appropriate Hilbert spaces of real-valued functions. Given the σ-finite measure μ on $\mathscr{A}$ let $(f,g)_\mu$ denote the inner product

$$(f,g)_\mu=\int f(x)g(x)\mu(dx)=\mu f g$$

for $\mathscr{A}$-measurable functions $f,g\in L^2(d\mu)$. Let $\mathscr{C}$ be the sub-Borel field of $\mathscr{A}$ induced by φ. Assume that the measure μ is still σ-finite on $\mathscr{C}$. Let $f\in L^1(d\mu)$. By the Radon-Nikodym theorem the set function

$$\int_C f\,d\mu, \qquad C\in\mathscr{C},$$

has a well-defined derivative with respect to μ on $\mathscr{C}$, the conditional expectation $E_\mu(f|\mathscr{C})$ of f with respect to $\mathscr{C}$. This is the uniquely determined (up to sets of μ measure zero) $\mathscr{C}$ measurable function satisfying (see Appendix 1)

$$\int_C f\,d\mu=\int_C E_\mu(f|\mathscr{C})\,d\mu \tag{1}$$

for all $C\in\mathscr{C}$. The conditional probability $P(A|\mathscr{C})$ discussed in Appendix 1 is obtained as a special case by taking f to be the set indicator function of A. Let $f\in L^2(d\mu)$. The conditional expectation $E_\mu(f|\mathscr{C})$ is still well-defined and uniquely determined (up to a set of μ measure zero) by using the defining relation (1) for all sets C of finite μ measure. Relation (1) can be alternatively rewritten as

$$\int_C f g\,d\mu=\int_C E_\mu(f|\mathscr{C})g\,d\mu \tag{2}$$

for any function $f\in L^2(d\mu)$ and any $\mathscr{C}$ measurable function g square integrable on any set $C\in\mathscr{C}$ with $\mu(C)<\infty$. On taking $g=E_\mu(f|\mathscr{C})$ and making use of the Schwarz inequality

$$\int_C |E_\mu(f|\mathscr{C})|^2\,d\mu\leqslant\int_C f^2\,d\mu$$

follows. It is clear that $E_\mu(f|\mathscr{C})\in L^2(d\mu)$. The defining relation (2) can be reformulated as

$$\int E_\mu(f|\mathscr{C})g\,d\mu=\int E_\mu(f|\mathscr{C})E_\mu(g|\mathscr{C})\,d\mu=\int f E_\mu(g|\mathscr{C})\,d\mu$$

for any two functions $f,g\in L^2(d\mu)$. The operator S_μ taking f into the conditional expectation $E_\mu(f|\mathscr{C})$ is the unique self-adjoint projection mapping the functions of $L^2(d\mu)$ onto the $\mathscr{C}$ measurable functions in $L^2(d\mu)$. The operator is subscripted by μ since its form depends on the measure μ.

Consider a Markov process $\{X_n(\omega)\}$ with initial distribution μ and transition probability function $P(\cdot,\cdot)$. The derived process

$$\{Y_n(\omega)\}=\{\varphi(X_n(\omega))\}$$

is completely determined by the behavior of P_μ on $\mathscr{C}_\infty$ on the original measure space. The operator

$$S_{\mu T^n}\, T^j\, S_{\mu T^{n+j}} \qquad n,j=0,1,\ldots$$

mapping the $\mathscr{C}$ measurable functions of $L^2(d\mu\, T^{n+j})$ into the $\mathscr{C}$ measurable functions of $L^2(d\mu\, T^n)$ describes the action of the first order transition probability function of the process $\{Y_n(\omega)\}$ from time n to time $n+j$ on the original space. The first order transition probability functions of $\{Y_n(\omega)\}$ satisfy the Chapman-Kolmogorov equation if and only if

$$S_{\mu T^n}\, T^j\, S_{\mu T^{n+j}} S_{\mu T^{n+j}}\, T^k\, S_{\mu T^{n+j+k}} = S_{\mu T^n}\, T^{j+k}\, S_{\mu T^{n+j+k}} \tag{3}$$

for $n,j,k=0,1,\ldots$. It is understood that $T^0=I$, the identity transformation.

Two types of questions seem to be natural in regard to retention of the Chapman-Kolmogorov equation. If the first order transition probabilities of the derived process $\{Y_n(\omega)\}$ are to still satisfy the Chapman-Kolmogorov equation for a specific initial distribution μ, it seems reasonable to require that μ be an invariant, possibly σ-finite measure with respect to T (if such an invariant measure exists). Condition (3) then becomes

$$(S_\mu\, T^j\, S_\mu)(S_\mu\, T^k\, S_\mu)=S_\mu\, T^{j+k}\, S_\mu$$

or equivalently,

$$S_\mu\, T^j\, S_\mu=(S_\mu\, T S_\mu)^j, \qquad j=1,2,\ldots. \tag{4}$$

In terms of some current terminology in operator theory $S_\mu T S_\mu$ is a contraction of the operator T (T is a dilation of $S_\mu T S_\mu$) with projection S_μ the conditional expectation operator with respect to Borel field $\mathscr{C}$ and μ the measure on $\mathscr{A}$.

Alternatively, one could require the Chapman-Kolmogorov equation to be satisfied for any initial distribution on $\mathscr{A}$, that is, condition (4) should be satisfied for any probability measure μ on $\mathscr{A}$.

Let us consider the first problem and see what can be said about it in a simple situation. Let μ be an invariant σ-finite measure with respect to T. The operator T^* adjoint to T on $L^2(d\mu)$ corresponds to the transition probability function of the process going backward in time. It is natural to call the Markov process *reversible* if $T^*=T$ on $L^2(d\mu)$, that is, if T is self-adjoint on $L^2(d\mu)$. Notice that T^* really depends on the measure μ. The following result appears to be related to a result of Kadison cited in [107].

Theorem 1. *Let μ be an invariant σ-finite measure with respect to T. If T is self-adjoint on $L^2(d\mu)$, the process induced by the measurable transformation φ (with associated sub-Borel field $\mathscr{C} \subset \mathscr{A}$ on Ω) will have first order transition probabilities satisfying the Chapman-Kolmogorov equation if and only if*

$$TS_\mu = S_\mu TS_\mu = S_\mu T \tag{5}$$

where S_μ is the self-adjoint projection of $L^2(d\mu)$ on the subspace of $\mathscr{C}$ measurable functions.

The transition probability operators on $\mathscr{C}$-measurable functions induced by φ are

$$T^{(n)} = S_\mu T^n S_\mu \qquad n = 1, 2, \ldots.$$

If $T^{(1)}$ satisfies the Chapman-Kolmogorov equation, we must have

$$(T^{(1)})^2 = S_\mu T S_\mu^2 T S_\mu = S_\mu T S_\mu T S_\mu = S_\mu T^2 S_\mu = T^{(2)}.$$

Then

$$S_\mu T(I - S_\mu) T S_\mu = 0.$$

Since $I - S_\mu$ is positive semidefinite it has positive semidefinite square root M

$$M^2 = I - S_\mu.$$

Then

$$S_\mu T M^2 T S_\mu = 0$$

and since T is self-adjoint

$$M T S_\mu = 0.$$

It follows that

$$M^2 T S_\mu = (I - S_\mu) T S_\mu = 0$$

and similarly that

$$S_\mu T(I - S_\mu) = 0.$$

On the other hand, if

$$T S_\mu = S_\mu T S_\mu,$$

then

$$(T^{(1)})^n = (S_\mu T S_\mu)^n = S_\mu T^n S_\mu = T^{(n)}$$

and the Chapman-Kolmogorov equation is satisfied. It will later be seen that this condition not only implies that the Chapman-Kolmogorov equation is satisfied but also that the new process is itself Markovian.

Condition (5) can be conveniently reinterpreted. T acts on the space $L^2(d\mu)$ of $\mathscr{A}$ measurable functions square integrable with respect to μ.

Let $L^2(d\mu;\mathscr{C})$ be the space of $\mathscr{C}$-measurable functions square integrable with respect to μ. Condition (5) states that $L^2(d\mu;\mathscr{C})$ reduces the operator T since T is self-adjoint and

$$(S_\mu g, T(I-S_\mu)f)=((I-S_\mu)g, TS_\mu f)=0$$

for any $f,g\in L_2(d\mu)$. Therefore *the spectrum of* $T'=S_\mu TS_\mu$ *is a subset of the spectrum of* T (see [71]). The condition obtained in Theorem 1 is meaningful even in the case of a transition operator T that is not self-adjoint. Corollary 1 follows immediately.

Corollary 1. *If T takes bounded $\mathscr{C}$-measurable functions into $\mathscr{C}$-measurable functions, the first order transition probabilities of the derived process $\{Y_n(\omega)\}$ satisfy the Chapman-Kolmogorov equation whatever the initial distribution μ. If μ is an invariant σ-finite measure with respect to T, we can introduce the adjoint operator T^* on $L^2(d\mu)$ and the condition*

$$S_\mu T^* S_\mu = T^* S_\mu \tag{6}$$

implies that the first order transition probabilities of $\{Y_n(\omega)\}$ satisfy the Chapman-Kolmogorov equation.

If T takes bounded $\mathscr{C}$-measurable functions into $\mathscr{C}$-measurable functions it is clear that $S_\mu TS_{\mu T}=TS_{\mu T}$ for any measure μ so that (3) is satisfied. It is easily seen that the condition (6) implies

$$S_\mu T^{*j} S_\mu=(S_\mu T^* S_\mu)^j$$

which is equivalent to (4). Although (6) is equivalent to (5) in the case of self-adjoint T, it is generally not equivalent to (5). This explains the interest in (6) as a condition distinct from (5).

Theorem 1 treats preservation of the Chapman-Kolmogorov equation in the case of a self-adjoint operator T. The case of a nonselfadjoint operator seems to be much more complicated, even if the state space of the Markov process is finite (the case of a finite state Markov chain). Let us consider this special case. The operator T is now an $n\times n$ transition probability matrix (n finite) with μ an invariant probability vector for T. The preservation of the Chapman-Kolmogorov equation (4) implies that there is a homomorphism $h(M)=S_\mu MS_\mu$ of the ring of matrices generated by the identity I and the matrix T onto the ring of matrices generated by S_μ and $S_\mu TS_\mu$ with $h(T)=S_\mu TS_\mu$. Specifically, if $r(T)=\sum_{j=0}^{m}\alpha_j T^j$ is the minimal polynomial of T, then

$$\rho(S_\mu TS_\mu)=\alpha_0 S_\mu+\Sigma\,\alpha_j(S_\mu TS_\mu)^j$$

is an annihilating polynomial of $S_\mu TS_\mu$ acting on the linear space $V(\mathscr{C})$ of $\mathscr{C}$ measurable vectors. Let $\nu(\lambda)$ be the index of the eigenvalue λ of T

and let $E(\lambda)$ be the projection corresponding to λ in the spectral resolution of T. $\sigma(T)$ denotes the spectrum of T. Then

$$E(\lambda)E(\lambda')=\delta_{\lambda,\lambda'}E(\lambda)$$

and

$$I=\sum_{\lambda\in\sigma(T)}E(\lambda).$$

Set $N_\lambda^n=\{x:(T-\lambda I)^n x=0\}$. If $\lambda\in\sigma(T)$

$$E(\lambda)V=N_\lambda^{\nu(\lambda)}=\{x:(T-\lambda I)^{\nu(\lambda)}x=0\}$$

where V is the linear space of n-vectors. For a function f analytic on an open set containing $\sigma(T)$

$$f(T)=\sum_{\lambda\in\sigma(T)}\sum_{i=0}^{\nu(\lambda)-1}\frac{(T-\lambda I)^i}{i!}f^{(i)}(\lambda)E(\lambda)$$

where $f^{(i)}$ is the ith derivative of f. Given the homomorphism

$$h(M)=S_\mu M S_\mu,$$

it follows that if

$$e(\lambda)=S_\mu E(\lambda)S_\mu,$$

then

$$e(\lambda)e(\lambda')=\delta_{\lambda,\lambda'}e(\lambda)$$

and

$$S_\mu=\sum_{\lambda\in\sigma(T)}e(\lambda)$$

with

$$f(S_\mu T S_\mu)=\sum_{\lambda\in\sigma(T)}\sum_{i=0}^{\nu(\lambda)-1}\frac{(S_\mu T S_\mu-\lambda S_\mu)^i}{i!}f^{(i)}(\lambda)e(\lambda).$$

For many values $\lambda\in\sigma(T)$ the operator $e(\lambda)$ will be a trivial projection, the null operator. S_μ is the identity operator when acting on the space of $\mathscr{C}$ measurable vectors $V(\mathscr{C})$. We therefore have a resolution of the identity for $S_\mu T S_\mu$ as an operator on $V(\mathscr{C})$. It then follows that $S_\mu T S_\mu$ as an operator on $V(\mathscr{C})$ has a spectrum $\sigma(S_\mu T S_\mu)\subset\sigma(T)$.

Theorem 2. *Let T be an $n\times n$ transition probability matrix (n finite) with μ an invariant probability vector. If the process induced by the transformation φ satisfies the Chapman-Kolmogorov equation, the operators $e(\lambda)=S_\mu E(\lambda)S_\mu$ provide a resolution of the identity S_μ for the matrix $S_\mu T S_\mu$ acting on $\mathscr{C}$ measurable vectors. The spectrum of $S_\mu T S_\mu$ acting on this space of vectors is a subset of $\sigma(T)$.*

Notice that the condition that the operators $e(\lambda)$ provide a resolution of the identity S_μ (for the matrix $S_\mu T S_\mu$) is necessary and sufficient for the Chapman-Kolmogorov equation to be satisfied by the induced process if T is diagonalizable. Making use of these simple ideas we now construct another example of a non-Markovian process that is a simple function of a Markov chain, and yet whose first order transition probabilities satisfy the Chapman-Kolmogorov equation. Let the stationary Markov chain $\{X_k\}$ have the states $1,2,3,4$ with transition matrices

$$P^n=\tfrac{1}{4}\begin{pmatrix}1&1&1&1\\1&1&1&1\\1&1&1&1\\1&1&1&1\end{pmatrix}+\lambda^n\begin{pmatrix}0&1&-1&0\\0&0&0&0\\0&-1&1&0\\0&0&0&0\end{pmatrix}$$

$$+\left(\frac{\lambda'}{\sqrt{2}}\right)^n\frac{1}{\sqrt{2}}\begin{pmatrix}1&-1&1&-1\\0&0&0&0\\0&0&0&0\\-1&1&-1&1\end{pmatrix}$$

$n=1,2,\ldots$, with λ and λ' sufficiently small in absolute value. The invariant probability vector $\mu=(\tfrac{1}{4},\tfrac{1}{4},\tfrac{1}{4},\tfrac{1}{4})$. The new process $Y_n=\varphi(X_n)$, $n=0,1,\ldots$, with $\varphi(1)=\varphi(2)=1$, $\varphi(3)=\varphi(4)=2$. The transition matrices of the process $\{Y_n\}$ are

$$Q^n=\begin{pmatrix}\tfrac{1}{2}&\tfrac{1}{2}\\\tfrac{1}{2}&\tfrac{1}{2}\end{pmatrix}+\lambda^n\begin{pmatrix}\tfrac{1}{2}&-\tfrac{1}{2}\\-\tfrac{1}{2}&\tfrac{1}{2}\end{pmatrix}$$

so that the Chapman-Kolmogorov equation is satisfied. However,

$$P\{Y_0=Y_1=Y_2=1\}=\tfrac{1}{8}(1+2\lambda+\lambda\lambda')$$

so that $\{Y_n\}$ is not Markovian when $\lambda\neq\lambda'$.

The following corollary is obtained by an argument similar to the derivation of Theorem 2.

Corollary 2. *Suppose there is a positive integer n such that T^n is compact. If the process induced by the transformation φ satisfies the Chapman-Kolmogorov equation, then the spectrum of $S_\mu T S_\mu$ (μ an invariant measure with respect to T) acting on the space $L^2(d\mu;\mathscr{C})$ is a subset of $\sigma(T)$.*

The condition $TS_\mu=S_\mu TS_\mu$ arose in many of the results obtained above as a consequence of the assumption that the Chapman-Kolmogorov equation held for the process induced by φ. In section 3.2 we shall see that this also implies the full Markov property for the induced process. However, there are many examples (even when the original Markov process is a finite state Markov chain) where this condition does not arise as a consequence of the Chapman-Kolmogorov equation. Such an example (due to P. Lévy) was given at the beginning of this section.

In section 3.2 we shall examine a condition weaker than (5) which also implies the full Markov property for the induced process.

The conclusion that the spectrum $\sigma(S_\mu T S_\mu)$ of the first order transition probability operator $S_\mu T S_\mu$ of the process induced by φ is a subset of $\sigma(T)$ was obtained under a variety of assumptions. Specifically this is the case if the Chapman-Kolmogorov equation is satisfied by the induced process and the operator T is self-adjoint or compact. If T is merely assumed to be normal, this needn't be so. We construct an example by making use of an irreducible finite state stationary Markov chain with no periodic states. Assume that the chain has more than one state. Consider the stationary process with state space consisting of the infinite vectors with coordinates the states of the Markov chain from the infinite past to the infinite future. This process is a Markov process with transition probability operator T equal to the shift operator. T is a unitary operator and the transition probability matrix of the chain is obtained by projecting onto the coordinate of the infinite vector representing the present. The spectrum of T consists of complex numbers of absolute value one while the spectrum of the transition probability matrix $S_\mu T S_\mu$ aside from the eigenvalue one is inside the unit circle of the complex plane. This simple example shows that, in general, one cannot even expect the point spectrum of $S_\mu T S_\mu$ to be a subset of the point spectrum of T.

2. Markovian Functions of Markov Processes

Let us consider a discrete or continuous time parameter Markov process $\{X(t)\}$ with transition probability function $P_t(x,A)$. The state space is Ω and we assume that $P_t(\cdot\,,\cdot)$ satisfies the measurability conditions specified in Chapter 1 relative to Ω and the Borel $\mathscr{A}$ on Ω. In this section we are interested in conditions that insure that a derived process $\varphi(X(t))$ obtained from the Markov process $X(t)$ with initial measure μ and transition function $P_t(\cdot\,,\cdot)$ be Markovian whatever the initial distribution μ may be. The operator S_μ will denote as before the self-adjoint projection of the functions in $L^2(d\mu)$ onto the $\mathscr{C}$ measurable functions in $L^2(d\mu)$ where $\mathscr{C}$ is the Borel subfield of $\mathscr{A}$ induced by φ.

The condition that $\varphi(X(t))$ be Markovian for any initial probability distribution μ will be written as follows. Take $0<t_1,t_2,\ldots,t_n<\infty$ to be any n-tuple of positive numbers and $h_0,h_1,\ldots,h_n$ any $(n+1)$-tuple of bounded $\mathscr{C}$-measurable functions. The Markovian property for $\varphi(X(t))$ for arbitrary initial distribution μ amounts to the condition

$$\begin{aligned}&(h_0,T(t_1)h_1\ldots T(t_n)h_n)_\mu\\&=(h_0,T(t_1)h_1\,S_{\mu T(t_1)}\,T(t_2)h_2\ldots S_{\mu T(t_1+\cdots+t_{n-1})}\,T(t_n)h_n)_\mu.\end{aligned}\tag{1}$$

One can now easily see that the condition (1.5) referred to in Theorem 1 of the preceding section implies that $\varphi(X(t))$ is Markovian whatever the initial distribution. This is actually a restricted version of the Markov property since we don't insist that the conditional probability of the collapsed process be a probability measure for almost every initial state of the collapsed process (see Appendix 1 and reference [51]). This allows us to evade the questions involved in setting up such a conditional probability function. We have the following

Corollary 1. *Let $X(t)$ be a Markovian process with initial distribution μ and transition probability function $P_t(\cdot,\cdot)$. If the operators $T(t)$, $t\geqslant 0$, take bounded $\mathscr{C}$-measurable functions into $\mathscr{C}$-measurable functions, then $\varphi(X(t))$ is Markovian whatever the initial distribution μ of $X(t)$.*

This follows easily since for bounded $\mathscr{C}$-measurable h, $T(t)h=S_\mu T(t)h$ for any measure μ and therefore (1) is satisfied. Of course, this is rather obvious and is referred to in Dynkin [23].

As an example illustrating this corollary, consider a random walk on a locally compact topological group G (see Appendices 2 and 5). Let v be a regular measure on the Borel field $\mathscr{A}$ generated by the closed sets on G. A random walk on G determined by v is a Markov process with state space G and transition probability function

$$P(g,A)=v(g^{-1}A), \qquad A\in\mathscr{A}.$$

Let H be a closed subgroup of G. Consider the homogeneous space G/H and the canonical map π of G onto G/H, that is $\pi(g)=gH$. The elements of G/H are the left cosets $gH, g\in G$. A subset K of G/H is open in G/H if and only if the inverse image $\pi^{-1}(K)=\{g:\pi(g)\in K\}$ is open. Let us call the Borel field induced by the open (or closed) sets of G/H the Borel field $\mathscr{A}'$ (see Appendix 5). It is then clear that

$$v(g^{-1}\pi^{-1}(A')), \qquad A'\in\mathscr{A}',$$

depends only on the left coset in which g is located. The conditions of Corollary 1 are satisfied and therefore the process induced by the mapping π on the homogeneous space G/H is Markovian whatever the initial distribution of the original Markov process (random walk) on G. Notice that if H is a normal subgroup of G, the homogeneous space G/H is itself a locally compact topological group and the induced process is a random walk on G/H. It is easy to give an example of a group G with H a subgroup that is not normal. The induced process on G/H (which is not a group) is then not a random walk. As an example consider the group G of matrices

$$\begin{pmatrix} a & b \\ c & d \end{pmatrix}$$

under multiplication with a, b, c, d integers and $ad-bc=1$. The subgroup H of matrices of the form

$$\begin{pmatrix} 1 & 0 \\ c & 1 \end{pmatrix}$$

is isomorphic to the additive group of integers and is not a normal subgroup of G.

We now wish to show that $\varphi(X(t))$ is Markovian whatever the initial distribution μ of the Markov process if and only if

$$(h_0, T(t_1)h_1\, T(t_2)h_2)_\mu = (h_0, T(t_1)h_1\, S_{\mu T(t_1)}\, T(t_2)h_2)_\mu, \tag{2}$$

$t_1, t_2 > 0$, for all bounded $\mathscr{C}$ measurable h_i and probability measures μ. This is clearly a special case of (1) and the difficult part is to show that (2) implies (1). We shall make some preliminary remarks concerning assumptions that are required. *Assume that the Borel field $\mathscr{C}$ induced by φ is separable.* By this we mean that $\mathscr{C}$ is generated by a countable subclass. There is then a countable subfield $\mathscr{I}$ of $\mathscr{C}$ such that for each set $A \in \mathscr{C}$ there exists a subsequence of sets $A_n \in \mathscr{I}$ such that $\mu(A_n \wedge A) \to 0$ as $n \to \infty$. Here $A \wedge B$ denotes the symmetric difference $(A-B) \cup (B-A)$. Let Φ be the empty set. The measure μ_A is the measure μ restricted to subsets of A, that is, $\mu_A(B) = \mu(A \cap B)$. Consider any set $N \in \mathscr{C}$ such that $(\mu_A T)_N, (\mu_B T)_N$ are mutually singular as measures on $\mathscr{C}$ for each pair of sets $A, B \in \mathscr{C}$ with $A \cap B = \Phi$. Then of course $(\mu_A T)_N, (\mu_B T)_N$ are mutually singular for each pair A, B from the subfield $\mathscr{I}$ with $A \cap B = \Phi$. We would like to show that the converse implication is valid. Let $N \in \mathscr{C}$ be a set such that for any $A, B \in \mathscr{I}$ with $A \cap B = \Phi$, $(\mu_A T)_N$ and $(\mu_B T)_N$ are mutually singular on $\mathscr{C}$. Consider any two sets $A, B \in \mathscr{C}$ with $A \cap B = \Phi$. There are then sequences $A_n, B_n \in \mathscr{I}$ with $A_n \cap B_n = \Phi$ such that $\mu(A \wedge A_n)$, $\mu(B \wedge B_n) \to 0$ as $n \to \infty$. This implies that

$$\|\mu_{A_n} T - \mu_A T\| \to 0, \qquad \|\mu_{B_n} T - \mu_B T\| \to 0$$

as $n \to \infty$ where $\|v\|$ denotes the total variation of the measure v. However, $(\mu_{A_n} T)_N, (\mu_{B_n} T)_N$ are mutually singular as measures on $\mathscr{C}$. If μ_n, v_n are mutually singular measures such that $\|\mu_n - \mu\|, \|v_n - v\| \to 0$ as $n \to \infty$ it follows that μ, v are mutually singular. Therefore $(\mu_A T)_N$ and $(\mu_B T)_N$ are mutually singular as measures on $\mathscr{C}$.

Let the sets of $\mathscr{I}$ be enumerated by A_n. For any two disjoint $A_n, A_m \in \mathscr{I}$ let $N_{n,m} \in \mathscr{C}$ be the set on which $\mu_{A_n} T, \mu_{A_m} T$ are mutually singular as measures on $\mathscr{C}$. Let $N = \bigcap_{n,m} N_{n,m}$ be the intersection of the sets $N_{n,m}$ over all pairs of disjoint sets $A_n, A_m \in \mathscr{I}$. The set N constructed is the maximal set $N \in \mathscr{C}$ such that $(\mu_A T)_N, (\mu_B T)_N$ are mutually singular on $\mathscr{C}$ for any $A, B \in \mathscr{C}$ with $A \cap B = \Phi$. N depends on the measure μ and the operator T. Since we are dealing with the semigroup of operators $T(\cdot)$ we shall *call the set N determined by the operator $T=T(t)\,(t>0)$ and the measure μ the single entry set $N_{\mu,t}$ corresponding to $T(t)$ and μ.*

If two probability measures ν,μ are mutually absolutely continuous we shall call them equivalent and write $\nu\sim\mu$. The complement of a set B will be denoted by B^c.

Lemma 1. *Let $T(\cdot)$ be a semigroup of transition probability operators and let the Borel subfield $\mathscr{C}\subset\mathscr{A}$ induced by the function φ be separable. Then if ν,μ are equivalent probability measures, the single entry sets $N_{\mu,t}$ and $N_{\nu,t}$ are the same up to a set of $\mu T(t)\sim\nu T(t)$ measure zero. Condition (2) implies that*

$$S_{\nu T(t)}\,T(\tau)h=S_{\mu T(t)}\,T(\tau)h \tag{3}$$

almost everywhere on $N^c_{\nu,t}=N^c_{\mu,t}$ with respect to the measure $\mu T(t)\sim\nu T(t)$ for every bounded $\mathscr{C}$-measurable function h and each $\tau>0$.

The equivalence of ν and μ implies that $\nu_A T(t)\sim\mu_A T(t)$ for all $A\in\mathscr{C}$. It follows from the definition of single entry sets that $N_{\mu,t}=N_{\nu,t}$ up to an exceptional set of $\mu T(t)\sim\nu T(t)$ measure zero. Let us now assume condition (2), that is,

$$(h_0,T(t_1)h_1\,S_{\mu T(t_1)}\,T(t_2)h_2)_\mu=(h_0,T(t_1)h_1\,T(t_2)h_2)_\mu$$

$t_1,t_2>0$ for all probability measures μ and all bounded nonnegative $\mathscr{C}$-measurable functions h_0,h_1,h_2. Consider two measures η_1,η_2 mutually singular on $\mathscr{C}$. There is then a set $A\in\mathscr{C}$ such that $\eta_1(A)=0,\eta_2(A^c)=0$. Set $\eta=\alpha\eta_1+\beta\eta_2$ for some $\alpha,\beta>0$. Let h_0 be the indicator function of A^c and h_1 the indicator function of some set $B\in\mathscr{C}$. Then

$$\begin{aligned}(h_0,T(t_1)h_1\,S_{\eta T(t_1)}\,T(t_2)h_2)_\eta&=\alpha\int_B(\eta_1\,T(t_1))(dx)\,S_{(\alpha\eta_1+\beta\eta_2)T(t_1)}\,T(t_2)h_2\\&=\alpha\int_B(\eta_1\,T(t_1))(dx)\,T(t_2)h_2\end{aligned}$$

so that

$$\int_B(\eta_1\,T(t_1))(dx)\,S_{(\alpha\eta_1+\beta\eta_2)T(t_1)}\,T(t_2)h_2=\int_B(\eta_1\,T(t_1))(dx)\,S_{\eta_1T(t_1)}\,T(t_2)h_2$$

for any $\alpha,\beta>0$. This implies that

$$S_{(\alpha\eta_1+\beta\eta_2)T(t_1)}\,T(t_2)h_2=S_{\eta_1T(t_1)}\,\mathrm{T}\,(t_2)h_2$$

almost everywhere with respect to $(\eta_1+\eta_2)\,T(t_1)$ on the set $C\in\mathscr{C}$ on which $\eta_1\,T(t_1)$ and $\eta_2\,T(t_2)$ are mutually absolutely continuous. The case $\alpha=\beta=1$ is of especial interest. On interchanging η_1 and η_2 we can conclude that

$$S_{\eta_1T(t_1)}\,T(t_2)h_2=S_{\eta_2T(t_1)}\,T(t_2)h_2 \tag{4}$$

almost everywhere on C with respect to $(\eta_1+\eta_2)\,T(t_2)$.

The single entry set $N_{\mu,t}$ defined earlier could also be obtained as the intersection $\cap N_n$ of the sets N_n on which $\mu_{A_n} T(t)$ and $\mu_{A_n^c} T(t)$ are mutually singular as measures on $\mathscr{C}$, where A_n is any set of $\mathscr{I}$. Then $N_{\mu,t}^c = \cup N_n^c$. Let $\eta_1 = \mu_{A_n}, \eta_2 = \mu_{A_n^c}$ in (4). On setting $\alpha = \beta = 1$ we find that

$$S_{\mu T(t_1)} T(t_2) h_2 = S_{\mu_{A_n} T(t_1)} T(t_2) h_2 \tag{5}$$

almost everywhere on N_n^c with respect to $\mu T(t_1)$, where h_2 is any bounded $\mathscr{C}$ measurable function and $t_2 > 0$. Similarly by letting $\eta_1 = \nu_{A_n^c}, \eta_2 = \nu_{A_n}$ one finds that

$$S_{\nu T(t_1)} T(t_2) h_2 = S_{\nu_{A_n^c} T(t_1)} T(t_2) h_2 \tag{6}$$

almost everywhere on N_n^c with respect to $\nu T(t_1) \sim \mu T(t_1)$, where h_2 is any bounded $\mathscr{C}$ measurable function and $t_2 > 0$. N_n^c is the relevant set in both (5) and (6) since $\nu_{A_n} T(t_1) \sim \mu_{A_n} T(t_1)$, $\nu_{A_n^c} T(t_1) \sim \mu_{A_n^c} T(t_1)$. Set $\eta_1 = \mu_{A_n}, \eta_2 = \nu_{A_n^c}$ in (4) to obtain

$$S_{\mu_{A_n} T(t_1)} T(t_2) h_2 = S_{\nu_{A_n^c} T(t_1)} T(t_2) h_2$$

almost everywhere on N_n^c with respect to

$$\mu_{A_n} T(t_1) + \nu_{A_n^c} T(t_1) \sim \nu T(t_1) \sim \mu T(t_1).$$

But then

$$S_{\mu T(t_1)} T(t_2) h_2 = S_{\nu T(t_1)} T(t_2) h_2$$

almost everywhere on N_n^c with respect to $\nu T(t_1) \sim \mu T(t_1)$ for any bounded $\mathscr{C}$ measurable h_2 and $t_2 > 0$. The proof of the Lemma is complete since $N_{\mu,t}^c = \bigcup_n N_n^c$.

Lemma 2. *Let h be a given bounded $\mathscr{C}$ measurable function and g any bounded measurable function that is zero on $N_{\mu,t}^c$, There is then a bounded $\mathscr{C}$ measurable function h' corresponding to h such that*

$$\mu h T(t) g = \mu T(t) h' g. \tag{7}$$

Here μh denotes $\int \mu(dx) h(x)$. Let $A \in \mathscr{C}$. Set $\eta = \mu T(t)$ and $\eta' = \mu_A T(t)$ on $N_{\mu,t}$. η' is absolutely continuous with respect to η. Let $\lambda_t(A) \in \mathscr{C}$ be the subset of $N_{\mu,t}$ on which the derivative $d\eta'/d\eta$ (computed on $\mathscr{C}$) is positive. Since $N_{\mu,t}$ is a single entry set, $d\eta'/d\eta = 1$ on $\lambda_t(A)$ almost everywhere with respect to $\mu T(t)$. Let

$$B_{k,n} = \left\{ x : \frac{k}{2^n} \leqslant h(x) < \frac{(k+1)}{2^n} \right\}.$$

Set

$$h^{(n)}(x)=\sum_k \frac{k}{2^n} I_{B_{k,n}}(x)$$

$$h'^{(n)}(x)=\sum_k \frac{k}{2^n} I_{\lambda_t(B_{k,n})}(x)$$

where I_B denotes the indicator function of B. Then

$$\begin{aligned}\mu h^{(n)} T(t)g &= \sum_k \frac{k}{2^n} \int_{B_{k,n}} \mu(dx)(T(t)g)(x)\\ &= \sum_k \frac{k}{2^n} \int_{\lambda_t(B_{k,n})} (\mu T(t))(dx)g(x) = \mu T(t) h'^{(n)} g\,.\end{aligned}$$

As $n\to\infty$ we know that $h^{(n)}\to h$ and $h'^{(n)}$ converges to a bounded $\mathscr{C}$-measurable function h'. The desired result (7) follows.

Theorem 1. *Let $T(\cdot)$ be a semigroup of transition probability operators and let the Borel subfield $\mathscr{C}\subset\mathscr{A}$ induced by the function φ be separable. Let $X(t)$ be the Markov process determined by $T(\cdot)$ and the initial probability measure μ. The process $\varphi(X(t))$ is then Markovian whatever the initial distribution μ of $X(t)$ if and only if*

$$(h_0, T(t_1)h_1\, T(t_2)h_2)_\mu = (h_0, T(t_1)h_1\, S_{\mu T(t_1)}\, T(t_2)h_2)_\mu\,,$$

$t_1, t_2>0$ *for all bounded $\mathscr{C}$ measurable h_i and all probability measures μ.*

It is enough to show that condition (2) implies condition (1). This need only be done for $\mathscr{C}$ measurable bounded positive functions h_i that are bounded away from zero, since by means of simple linear operations one can then extend the validity of (2) to the larger class of bounded $\mathscr{C}$ measurable functions. We shall prove that

$$\begin{aligned}&(h_0, T(t_1)h_1 \dots h_{n-2}\, T(t_{n-1})h_{n-1}\, T(t_n)h_n)_\mu\\ &= (h_0, T(t_1)h_1 \dots h_{n-2}\, T(t_{n-1})h_{n-1}\, S_{\mu T(t_1+\cdots+t_{n-1})}\, T(t_n)h_n)_\mu\end{aligned} \tag{8}$$

for any bounded positive $\mathscr{C}$ measurable h_i's bounded away from zero and any probability measure μ, since repeated application of (8) will yield (1). Let N_j be the single entry set corresponding to the operator $T(t_{n-1}+\cdots+t_{n-j})$ and the measure $\mu T(t_1+\cdots+t_{n-j-1})$, $j=1,2,\dots,n-2$. N_{n-1} is understood to be the single entry set corresponding to the operator $T(t_{n-1}+\cdots+t_1)$ and the measure μ. Set

$$M_1 = \bigcap_{j=1}^{n-1} N_j, \qquad M_n = N_1^c$$

and

$$M_j = N_{n-j+1}^c \bigcap_{1}^{n-j} N_k, \qquad j=2, \dots, n-1\,.$$

Let I_{M_j} be the indicator function of the set M_j. Then

$$\begin{aligned}&(h_0, T(t_1)h_1 \dots T(t_{n-1})h_{n-1})\,T(t_n)h_n)_\mu \\ &= \sum_{j=1}^{n} (h_0, T(t_1)h_1 \dots T(t_{n-1})h_{n-1})\,I_{M_j}\,T(t_n)h_n)_\mu.\end{aligned} \tag{9}$$

First consider the term with $j=n$ in the sum of formula (9). Since the h_i's are positive and bounded away from zero

$$v = \mu h_0\, T(t_1)h_1 \dots T(t_{n-2}) \sim \mu\, T(t_1 + \dots + t_{n-2}).$$

Therefore

$$\begin{aligned}&(h_0, T(t_1)h_1 \dots T(t_{n-1})h_{n-1}\, I_{M_n}\, T(t_n)h_n)_\mu \\ &\quad= (h_{n-2}, T(t_{n-1})h_{n-1}\, I_{M_n}\, T(t_n)h_n)_v \\ &\quad= (h_{n-2}, T(t_{n-1})h_{n-1}\, I_{M_n}\, S_{vT(t_{n-1})}\, T(t_n)h_n)_v \\ &\quad= (h_{n-2}, T(t_{n-1})h_{n-1}\, I_{M_n}\, S_{\mu T(t_1 + \dots + t_{n-1})}\, T(t_n)h_n)_v \\ &\quad= (h_0, T(t_1)h_1 \dots T(t_{n-1})h_{n-1}\, I_{M_n}\, S_{\mu T(t_1 + \dots + t_{n-1})}\, T(t_n)h_n)_\mu\end{aligned}$$

by Lemma 1. A term in (9) with $j<n$ can be taken care of by a small modification of this argument using Lemmas 1 and 2. For, if $j<n$, then

$$\begin{aligned}&(h_0, T(t_1)h_1 \dots T(t_{n-1})h_{n-1}\, I_{M_j}\, T(t_n)h_n)_\mu \\ &\quad= \mu h_0\, T(t_1)h_1 \dots T(t_{n-1})h_{n-1}\, I_{M_j}\, T(t_n)h_n \\ &\quad= \mu h_0\, T(t_1)h_1 \dots h_{n-3}\, T(t_{n-1} + t_{n-2})h'_{n-2}\, h_{n-1}\, I_{M_j}\, T(t_n)h_n \\ &\quad= \cdots \\ &\quad= \mu h_0\, T(t_1)h_1 \dots h_{j-2}\, T(t_{n-1} + \dots + t_{j-1})h'_{j-1} \dots h'_{n-2}\, h_{n-1}\, I_{M_j}\, T(t_n)h_n\end{aligned} \tag{10}$$

by a repeated application of Lemma 2 where h'_k is the function corresponding to h_k and the operator $T(t_{n-1} + \dots + t_{k+1})$ in Lemma 2. Now one can apply Lemma 1 since $M_j \subset N_{n-j+1}^c$ if $j \neq 1$ and see that (10) is equal to

$$\begin{aligned}&\mu h_0\, T(t_1)h_1 \dots h_{j-2}\, T(t_{n-1} + \dots + t_{j-1}) \\ &\times h'_{j-1} \dots h'_{n-2}\, h_{n-1}\, I_{M_j}\, S_{\mu T(t_1 + \dots + t_{n-1})}\, T(t_n)h_n\,.\end{aligned} \tag{11}$$

On applying Lemma 2 repeatedly to (11) one obtains

$$\mu h_0\, T(t_1)h_1 \dots h_{n-2}\, T(t_{n-1})h_{n-1}\, I_{M_j}\, S_{\mu T(t_1 + \dots + t_{n-1})}\, T(t_n)h_n\,.$$

The corresponding identity for $j=1$ is trivial. By summing over j we obtain the conclusion of the theorem.

3. Functions of Finite State Markov Chains

In this section we shall characterize those finite state processes that can be obtained as functions of finite state Markov chains. The presentation that we will give is algebraic in character and is due to A. Heller [38].

Let $\{X_k; k=0,1,2,\ldots\}$ be a stochastic process with finite state space S. The probability structure of the process is completely specified by the probabilities $p(x_1,\ldots,x_n)$ of finite sequences $x_1,\ldots,x_n$ of elements of S. It will be convenient to introduce some algebraic concepts in describing such finite state processes. Think of a succession of elements $x_1,\ldots,x_n$ of S as being given by a formal product $x_1 \ldots x_n$. If we take linear combinations of real multiples of these formal products we are led to the notion of the free associative R-algebra (algebra over the reals) A_S generated by S. These linear combinations of products are polynomials. Write $p(x_1 \ldots x_n)$ instead of $p(x_1,\ldots,x_n)$ and set $p(1)=1$ (1 is the identity of the algebra). Since p is defined on a basis of A_S we can extend p linearly to all of A_S in the obvious manner

$$p(\alpha u+\beta v)=\alpha p(u)+\beta p(v)$$

where $u, v \in A_S$ and α, β are reals. Let P_S be the *positive cone* of A_S consisting of polynomials with nonnegative coefficients. Set $\sigma=\sum_{x\in S} x$. Then an R-linear function p on A_S into the reals is a stochastic process if and only if

(0) $p(1)=1$

(1) $p(P_S)\subset[0,\infty]$

(2) for all $\xi\in A_S$, $\quad p(\xi\sigma)=p(\xi)$.

Let N be the left ideal $\{\xi: p(A_S\xi)=0\}\subset A_S$. N is obviously a left ideal since for any $\eta\in A_S$, $\xi\in N$, clearly $\eta\xi\in N$. Let L' be the quotient A_S/N with $\lambda: A_S\to L'$ the canonical map of A_S onto L'. Since p vanishes on N there is a unique linear $q: L'\to R$ with $p=q\lambda$. Set $l_0=\lambda(1)$. Then L' is a left A_S-module, that is, for each $\xi\in A_S$ and $l\in L'$ there is a product $\xi l\in L'$ with the following properties:

for all $\xi,\xi'\in A_S$, $l,l'\in L'$, $\alpha\in R$

$$\xi(l+l')=\xi l+\xi l', \quad (\xi+\xi')l=\xi l+\xi' l$$

$$(\xi\xi')l=\xi(\xi' l), \quad 1l=l$$

$$(\alpha\xi)l=\alpha(\xi l)=\xi(\alpha l).$$

The product ξl is given by

$$\xi l=\lambda(\xi\xi')$$

where ξ' is any element of $\lambda^{-1}(l)$. L' is like a vector space with left multiplication by "scalars" in A_S. Actually a vector space is a two-sided module over a field (see [72] for a more detailed discussion of the basic elementary notions connected with rings and modules).

Consider any left A_S-module L with an element $l_0 \in L$ and a linear $q: L \to R$. Then $p(\xi) = q(\xi l_0)$ defines a linear $p: A_S \to R$.

Lemma 1. *p is a stochastic process if and only if*

(i) $q(l_0) = 1$

(ii) $q(P_S l_0) \subset [0, \infty)$

(iii) for all $\xi \in A_S$, $q(\xi(\sigma - 1) l_0) = 0$.

We call the triple (L, q, l_0), for which the conditions of the Lemma are satisfied, a *stochastic S-module* (sS module) and the stochastic process p is said to be *associated with* (L, q, l_0). The sS-module (L, q, l_0) is *reduced* if (i) $A_S l_0 = L$ (L cyclic with generator l_0) (ii) L has no nonzero submodules L'' with $q(L'') = 0$. The A_S-module $L = A_S/N$ constructed above is reduced. All reduced sS-modules arise like A_S/N with $\xi \to \xi l_0$ taking the role of the map λ.

A *morphism* of stochastic S-modules (L, q, l_0), (L', q', l'_0) is a homomorphism $\varphi: L \to L'$ of left A_S-modules such that $q' \varphi = q$, $\varphi(l_0) = l'_0$. φ is a homomorphism if

$$\varphi(l_1 + l_2) = \varphi(l_1) + \varphi(l_2), \qquad \varphi(\xi l) = \xi \varphi(l)$$

for $l_i \in L$, $\xi \in A_S$. An *isomorphism* is an invertible morphism.

Suppose (L, q, l_0) is any S-module. The reduced sS-module of the associated stochastic process is obtained by replacing L by the cyclic submodule $A_S l_0$ and then dividing by the largest submodule on which q vanishes. If (L, q, l_0) is a reduced sS-module then $\sigma l_0 = l_0$ since $A_S(\sigma - 1) l_0$ is a submodule in the kernel of q.

A state x of a stochastic process $p: A_S \to R$ is *prohibited* if any sequence of states containing x has probability zero.

Lemma 2. *Let (L, q, l_0) be a reduced sS-module with $x \in S$. The following statements are then equivalent:* (i) $x l_0 = 0$; (ii) x is *a prohibited state of the associated stochastic process;* (iii) $x L = 0$.

Statement (i) implies (ii) since

$$0 \leqslant p(x_1 \ldots x_n x x'_1 \ldots x'_m) \leqslant p(x_1 \ldots x_n x \sigma^m) = q(x_1 \ldots x_n x l_0) = 0.$$

The remainder of the Lemma is obvious.

Consider now a function $f: S \to S'$ mapping the state space S onto the state space S'. The case of real interest is that in which the number of states in S' is less than that in S. The process $\{Y_k = f(X_k); k = 0, 1, 2, \ldots\}$

obtained from $\{X_k;\ k=0,1,2,\ldots\}$ has probability structure given by $p': A_{S'} \to R$ where

$$p'(y_1 \ldots y_n) = \sum_{fx_1=y_1} \cdots \sum_{fx_n=y_n} p(x_1 \ldots x_n).$$

Some convenient notation is now introduced. If $\varphi: A \to B$ is a ring homomorphism and L is a left B-module let ${}_{[\varphi]}L$ be the left A-module whose underlying commutative group is that of L with $\alpha l = \varphi(\alpha) l$. Define $f^*: A_{S'} \to A_S$ to be the homomorphism on free generators $y \in S'$ of $A_{S'}$ by $f^* y = \sum_{fx=y} x$. We then immediately have the following Lemma.

Lemma 3. *Let* (L, q, l_0) *be an* sS'*-module with* $f: S \to S'$. *Then* $({}_{[f^*]}L, q, l_0)$ *is an* sS'*-module where the stochastic process associated with* $({}_{[f^*]}L, q, l_0)$ *is induced by* f *from that associated with* (L, q, l_0).

Call a stochastic process *finitary* if its reduced sS-module has finite dimension (over R). It is clear that a stochastic process induced from a finitary process by a function f is itself finitary.

The stochastic process $p: A_S \to R$ is a Markov chain if there is a map $t: S \times S \to R$ (the transition probability matrix) such that for any sequence $x_1, \ldots, x_n \in S$

$$p(x_1 \ldots x_n) = p(x_1) t(x_1, x_2) \ldots t(x_{n-1}, x_n). \tag{1}$$

Of course, the transition matrix is taken to be nonnegative and stochastic, that is, $\sum_y t(x, y) = 1$. Formula (1) can be written

$$p(\xi(xy - t(x,y)x)) = 0 \tag{2}$$

for all $\xi \in A_S$, $x, y \in S$.

Lemma 4. *If* (L, q, l_0) *is a reduced* sS*-module the corresponding stochastic process is a Markov chain if and only if for each* $x \in S$, xL *has dimension less than or equal to one.*

If p is a Markov chain, equation (2) states that q vanishes on the submodule $A_S(xy - t(x,y)x)l_0$ for any $x, y \in L$. Since (L, q, l_0) is reduced, $xyl_0 = t(x,y)xl_0$. But $A_S l_0 = L$ and therefore $xL \subset Rxl_0$.

Conversely if $x, y \in S$ then $xyl_0 = t(x,y)xl_0$ for some $t(x,y) \in R$. Then for $\xi \in A_S$, $q(\xi(xy - t(x,y)x)l_0) = 0$ which is just (2).

This implies that if (L, q, l_0) is an sS-module and for each $x \in S$, $\dim xL \leqslant 1$, then the associated stochastic process is a Markov chain. Call such modules Markovian. Suppose $f: S \to S'$ where $\mu(y)$ for $y \in S'$ is the number of x's for which $f(x) = y$.

Lemma 5. *If* (L', q', l'_0) *is the reduced* S'*-module of a stochastic process induced by* f *from a Markov chain on* S *then for each* $y' \in S'$, $\dim y'L' \leqslant \mu(y)$.

Let $x^{(\alpha)} \in S$ be those elements for which $f(x^{(\alpha)}) = y'$. Then $\lambda(x^{(\alpha)})$ will serve as a basis for $y'L'$.

We now define the *regular module* of a Markov chain $p: A_S \to R$ with transition matrix t. Let L be a vector space with basis $\{[x] | x \in S\}$ which corresponds one-to-one to S. Set up the operation of A_S on L by $x[y] = t(x,y)[x]$ for $x, y \in S$. Let $l_0 = \sum_x [x]$ and define $q: L \to R$ by taking $q([x]) = p(x)$. Then (L, q, l_0) is the regular module of p. The stochastic process associated with (L, q, l_0) is p since

$$x_1 \dots x_n l_0 = t(x_1, x_2) \dots t(x_{n-1}, x_n)[x_1]$$

and therefore

$$q(x_1 \dots x_n l_0) = p(x_1) t(x_1, x_2) \dots t(x_{n-1}, x_n).$$

A *cone* in a real vector space V is a union of rays from the origin. The cone C is *strongly convex* if $x, -x \in C$ imply $x = 0$. A convex cone C is *polyhedral* if it is the convex hull of finitely many rays. A subspace $W \subset V$ intersects a cone $C \subset V$ *extremally* if $x, y \in C$, $x + y \in W$ imply $x, y \in W$.

The following remarks will be useful:

1. If $W \subset V$ is a subspace and $C \subset V$ a strongly convex cone, then $W \cap C$ is a strongly convex cone.

2. If $W \subset V$ is a subspace and $C \subset V$ a polyhedral cone, then so is $W \cap C$.

3. If $W, W' \subset V$ are subspaces and W' intersects the cone $C \subset V$ extremally, then $W \cap W'$ intersects $W \cap C$ extremally.

4. If $W \subset V$ is a subspace, $\eta: V \to V/W$ the canonical map and $C \subset V$ is a polyhedral cone, then so is $\eta C \subset V/W$.

5. If $W \subset V$ is a subspace, $\eta: V \to V/W$ the canonical map, $C \subset V$ a strongly convex cone and W intersects C extremally, then ηC is strongly convex.

Statements 1., 3., and 4. are obvious. 2. can be obtained by an induction argument on the number n of rays generating the cone. If $W \cap C$ doesn't intersect any of the $(n-1)$-faces of the cone other than 0, the demonstration is finished since $W \cap C$ is then generated by one ray. If it intersects some of the $(n-1)$-faces, by induction there will be a finite number of rays generating each intersection. The union of all such rays for all the $(n-1)$-faces will generate $W \cap C$. The argument for 5. is simple. If ηC is not strongly convex, there are elements $x, -x + w \in C$ with $x \notin W$, $w \in W$. Since W intersects C extremally, $x \in W$ and we have a contradiction.

Theorem 1. *Let* (L,q,l_0) be a reduced sS*-module. The associated stochastic process is induced from a Markov chain by some function f if and only if there is a cone* $C\subset L$ *such that* (i) $l_0\in C$; (ii) $q(C)\subset[0,\infty)$; (iii) $P_S C\subset C$; (iv) *C is strongly convex and polyhedral.*

Let the process be induced by the function $f:S'\to S$ from a Markov chain with regular sS'-module (L',q',l'_0). The nonnegative orthant $C'\subset L'$ has properties (i)–(iv) with respect to S'. Set $L_1=A_S l'_0$ in the A_S-module ${}_{[f^*]}L$. Since $f^*(P_S)\subset P_{S'}$, the cone $L_1\cap C'$ has by (1) and (2) properties (i)–(iv) with respect to S.

If $N_1\subset L_1$ is the largest submodule on which q' vanishes, then L can be identified with L_1/N_1 with q',l'_0 going into q,l_0. N_1 intersects $L_1\cap C'$ extremally. For if $u,v\in L_1\cap C', u+v\in N_1$, then for any $\xi\in P_S$ one has $\xi u\in L_1\cap C'$ and $\xi(u+v)\in N_1$ so that

$$0\leqslant q'(\xi u)\leqslant q'(\xi u+\xi v)=0.$$

Thus u and likewise v are in N_1. The image of $L_1\cap C'$ in $L=L_1/N_1$ satisfies conditions (i)–(iv) by 4., 5.

Conversely, let $C\subset L$ be a cone satisfying (i)–(iv) with $U\subset C$ a finite subset such that C is the convex hull of rays through elements of U. Then

$$l_0=\sum_{u\in U}\lambda_u u \qquad \lambda_u\geqslant 0$$

$$xu=\sum_{v\in U}\alpha_{xuv}v \qquad x\in S,\ u\in U,\ \alpha_{xuv}\geqslant 0.$$

Set $S'=S\times U$ with L' a vector space with basis $\{[x,u]\}$ in one-one correspondence with S'. Make L' a left $A_{S'}$-module by setting

$$(x,u)[y,v]=\alpha_{xvu}[x,u] \qquad x,y\in S,\ u,v\in U.$$

If n is the number of elements in S, let

$$l'_0=n^{-1}\sum_{x\in S,u\in U}\lambda_u[x,u]$$

and $q':L'\to R$ be given by $q'([x,u])=q(u)$. By Lemma 4 (L',q',l'_0) is an sS'-module whose associated stochastic process is a Markov chain.

Let $f([x,u])=x$. The process induced by f from the one defined above is associated with (L,q,l_0). To see this define $g:L'\to L$ as the linear extension of $g([x,u])=u$. Then $gl'_0=l_0$ and $qg=q'$. Also $g:{}_{[f^*]}L'\to L$ is a homomorphism of A_S-modules since if $x\in S$

$$\begin{aligned} g\{(f^*x)[y,v]\}&=g\Big\{\sum_u (x,u)[y,v]\Big\}\\ &=g\Big\{\sum_u \alpha_{xvu}[x,u]\Big\}=\sum_u \alpha_{xvu}u=xv=xg([y,v])\end{aligned}$$

so that g is a morphism of sS-modules.

Notice that the proof of the theorem indicates that the number of states in the Markov chain constructed can be taken to be no more than the number of states in S multiplied by the number of elements of U.

The theorem is applied to construct a finitary stochastic process that is not induced from a Markov chain. Let L be Euclidean 3-space with orthonormal basis $\{e_0, e_1, e_2\}$. Set $l_0 = e_0 + \varepsilon e_1$ with ε small and positive, where q is defined to be the inner product with $|l_0|^{-2} l_0$ so that $q_0(l_0) = 1$. For S take the set of two elements x, y and make L into an A_S-module as follows: x acts on L as a contraction in length together with a rotation about e_0 through a small angle θ which is an irrational multiple of π, while y acts on L as $\alpha\rho$ where ρ is the orthogonal projection on $l_0 - x l_0$ and $\alpha \in R$ is chosen so that $\sigma l_0 = (x+y) l_0 = l_0$.

The closure of $P_S l_0$ is the right circular cone with axis e_0 and element $l_0 - x l_0$. If ε is small enough $q(P_S l_0) \subset [0, \infty)$. Thus (L, q, l_0) is a stochastic S-module which is irreducible. The only P_S-invariant cones in L are the right circular cones with axis e_0. If these are to be strongly convex, they cannot be polyhedral.

Notes

3.1 W. Feller also gave an example of a non-Markovian process whose first order transition probabilities satisfy the Chapman-Kolmogorov equation in [25].

A more detailed treatment of aspects of spectral theory than that given in Appendix 4 can be found in Lorch's little book on spectral theory [71] or in Dunford and Schwartz [20], [21]. Contractions and dilations of operators are dealt with at some length in Sz-Nagy's Appendix [107] to Riesz and Nagy's book on functional analysis [88]. A version of Theorem 1 can be found in [33].

3.2 We have avoided the difficulties associated with obtaining a "proper" version of conditional probabilities by taking a restricted version of the Markov property in this section. These difficulties are considered, for example, in the papers of Jirina [51] and Pfangazl [84]. The results of this section are due to M. Rosenblatt: Functions of Markov processes. Z. Wahrscheinlichkeitstheorie und Verw. Gebiete 5, 232–243 (1966).

3.3 Conditions guaranteeing that a finite state stationary process

$$\{Y_n,\ n = \ldots, -1, 0, 1, \ldots\}$$

is an instantaneous function $Y_n = f(X_n)$ of a finite state Markov chain

$$\{X_n,\ n = \ldots, -1, 0, 1, \ldots\}$$

have been considered by a number of people. We mention in particular the papers of D. Blackwell and L. Koopmans [4], E.J. Gilbert [29], and S.W. Dharmadhikari [15]. Let I be the state space of the Markov chain $\{X_n\}$ and J the instantaneous state space of $\{Y_n\}$. Consider ε a state of J and s, t finite sequences of states of J. If $s = (\varepsilon_1, \ldots, \varepsilon_n)$ let $p(s) = P[(Y_1, \ldots, Y_n) = s]$. For each ε in J let $n(\varepsilon)$ be the largest integer n for which there are sequences s_i, t_i, $i = 1, \ldots, n$, such that the matrix $\{p(s_i \varepsilon t_j);\ i, j = 1, \ldots, n\}$ is nonsingular. Gilbert [29] showed that if f takes $N(\varepsilon)$

states of I into the state ε of J then $n(\varepsilon) \leqslant N(\varepsilon)$. Thus $\Sigma n(\varepsilon) < \infty$ is a necessary condition for Y to be a function of a finite state Markov chain. Dharmidhikari's work [15] indicated the geometric character of the problem. Heller's necessary and sufficient condition for $\{Y_n\}$ to be a function of a finite state Markov chain has this geometric flavor. Some interesting recent work on functions of Markov chains can be found in the paper of Erickson [24].

Arbib [1] used Heller's ideas to give conditions for a stochastic system to be induced by a probabilistic finite automaton. *A probabilistic finite automaton* is given by a quadruple (X, Y, Q, P) with X a finite set of inputs, Y a finite set of outputs, Q the finite set of states of the automaton and $P(q', y|q, x)$ the conditional probability that the automaton in state $q \in Q$ with input $x \in X$ has output $y \in Y$ and enters state $q' \in Q$. Given the finite sets X, Y let $(X \times Y)^*$ be the set of finite sequences of pairs (x, y) with $x \in X$, $y \in Y$. $\mathscr{P}(X, Y)$ is the set of all functions p on $(X \times Y)^*$ into the real numbers. If $p \in \mathscr{P}(X, Y)$ let us write

$$p(y_1, \dots, y_n | x_1, \dots, x_n) \quad \text{for } p((x_1, y_1), \dots, (x_n, y_n)).$$

A function $p \in \mathscr{P}(X, Y)$ is called a *stochastic system* if 1. $p((X \times Y)^*) \subseteq [0,1]$ 2. $\sum_{y \in Y} p(y|x) = 1$ 3. $\sum_{y \in Y} p(y_1, \dots, y_n, y | x_1, \dots, x_n, x) = p(y_1, \dots, y_n | x_1, \dots, x_n)$ for all $x \in X$. Consider the probabilistic finite automaton (X, Y, Q, P). Let ρ be a probability distribution on Q. A stochastic system p is said to be induced by the probabilistic finite automaton if there is a ρ such that

$$\begin{aligned} p(y_1, \dots, y_n | x_1, \dots, x_n) &= p_\rho(y_1, \dots, y_n | x_1, \dots, x_n) \\ &= \sum_{q_k \in Q} \rho_{q_1} \prod_{k=1}^{n} P(q_{k+1}, y_k | q_k, x_k). \end{aligned}$$

Chapter IV

Ergodic and Prediction Problems

0. Summary

In this chapter we will be concerned with the asymptotic behavior of Markov processes with stationary transition probability function $P(\cdot,\cdot)$. It is occasionally useful to look at the process only when it falls within a proper measurable subset A of the state space Ω. In the first section, conditions under which a well-defined "process on A" can be defined will be established. Sometimes it is useful to look at such derived processes when trying to establish the existence of *subinvariant* or *invariant* measures μ, that is, σ-finite measures for which

$$\int \mu(dx) P(x, A) \leqslant \mu(A)$$

or

$$\int \mu(dx) P(x, A) = \mu(A)$$

hold for each set $A \in \mathscr{A}$, respectively. The primary object of this chapter is the study of the asymptotic behavior of partial sums

$$\sum_{j=1}^{n} P_j(\cdot\,,\cdot) \tag{1}$$

or of

$$P_n(\cdot\,,\cdot) \tag{2}$$

itself as $n \to \infty$. Problems dealing with the partial sums (1) are related to ergodic theory while those concerned with the transition probability function (2) itself are more closely related to prediction. In discussions of ergodic and prediction problems, the existence of an invariant (or subinvariant) measure is usually required. Some conditions under which such a measure exists will be derived in section 3 under assumptions of a topological character. However, in discussions of ergodic and prediction problems the existence of such a measure will be taken for granted. In section 2 an ergodic theorem due to Chacon and Ornstein will be given with the classic Birkhoff ergodic theorem for stationary processes as an almost immediate corollary. Some aspects of a least squares version of the (nonlinear) prediction problem for a Markov process are sketched in section 4. It is quite clear that they are directly related to the asymptotic behavior of the transition function or equivalently to the asymptotic behavior of powers of the induced operator.

1. A Markov Process Restricted to a Set A

Consider a Markov process starting from the point $x \in \Omega$ with stationary transition probability function $P(\cdot,\cdot)$. Let $P_A(x,B)$ be the probability of first hitting the set $B \subseteq A$ thereafter without entering A before where B and A are sets of the Borel field $\mathscr{A}$. It is clear that

$$\begin{aligned} P_A(x,B) = P(x,B) + \int_{A^c} P(x,dx_1)P(x_1,B) \\ + \int_{A^c}\int_{A^c} P(x,dx_1)P(x_1,dx_2)P(x_2,B) + \cdots . \end{aligned} \tag{1}$$

Notice that $P_A(x,B)$ is measurable with respect to $\mathscr{A}$ in x for each $B(\subseteq A) \in \mathscr{A}$ and that for each $x \in \Omega$ it is a finite measure (though not necessarily a probability measure) on $B(\subseteq A) \in \mathscr{A}$. Actually, $P_A(x,B) \leqslant 1$ for all $x \in \Omega$ and $B(\subseteq A) \in \mathscr{A}$.

Suppose we now consider $P_A(x,B)$ for all $B \in \mathscr{A}$, that is, B is not necessarily contained in A. Then $P_A(x,B)$ may well be infinite for some sets B. However, we shall show that $P_A(x,\cdot)$ *is a sigma-finite measure for all* $x \in \Omega$ *if for each such* x *there is an integer* $n = n(x) > 0$ *such that*

$$P_n(x,A) > 0 .$$

That is we assume that one can hit A with positive probability from any point $x \in \Omega$. From the definition of $P_A(\cdot,\cdot)$ as given in (1) it is clear that

$$P_A(x,B) = P(x,B) + \int_{A^c} P_A(x,dy)P(y,B) . \tag{2}$$

If $B \subseteq A^c$, then $P_A(x,B)$ can be interpreted as the mean number of visits to the set B before a visit to A if the starting point is x and the transition probability function of the Markov process is $P(\cdot,\cdot)$. Let

$$S_{i,j} = \left\{ x : x \in A^c,\ P_i(x,A) > \frac{1}{j} \right\}, \qquad i,j = 1,2,\ldots .$$

Then

$$
\begin{aligned}
P_A(x, S_{i,j}) &\leqslant j \int_{S_{i,j}} P_A(x, dy) P_i(y, A) \\
&\leqslant j \Big\{ \int_{A^c} P_A(x, dy) P(y, A) + \cdots \\
&\quad + \int_{A^c} P_A(x, dy) \int_{A^c} P(y, dy_1) \dots \int_{A^c} P(y_{i-2}, dy_{i-1}) P(y_{i-1}, A) \Big\} \\
&\leqslant j\, i \int_{A^c} P_A(x, dy) P(y, A) \leqslant j\, i
\end{aligned}
\tag{3}
$$

follows from (2) and the fact that the probability of hitting A is less than or equal to one. Since $A^c = \bigcup_{i,j} S_{i,j}$ it is clear from (3) that $P_A(x, \cdot)$ is sigma-finite for each $x \in \Omega$.

Lemma 1. *If there is positive probability of hitting the set A from every point $x \in \Omega$, then $P_A(x, \cdot)$ is a sigma-finite measure for each $x \in \Omega$.*

Notice that the assumption that A can be hit with positive probability from every point x does not imply that $P_A(x, A) \equiv 1$ for all x.

Lemma 2. *Let $A \in \mathscr{A}$ be a set that can be reached with positive probability from every point $x \in \Omega$. Suppose the Markov process restricted to A has a nontrivial subinvariant (invariant) finite measure Q_A satisfying*

$$
Q_A(B) \geqslant \int_A Q_A(dx) P_A(x, B) \left(Q_A(B) = \int_A Q_A(dx) P_A(x, B) \right) \tag{4}
$$

for every set $B \subseteq A$ that is an element of $\mathscr{A}$. Then

$$
Q(B) = \int_A Q_A(dx) P_A(x, B) \tag{5}
$$

is a sigma-finite subinvariant (invariant) measure for $P(\cdot, \cdot)$, that is,

$$
Q(B) \geqslant \int Q(dx) P(x, B) \; (Q(B) = \int Q(dx) P(x, B))
$$

for all $B \in \mathscr{A}$.

The argument used in proving Lemma 1 can be simply modified to show that Q is sigma-finite. Notice that $Q(B) \leqslant Q_A(B)$ if $B \subset A$. By using (5), (4) and (2) one can show that

$$
\begin{aligned}
\int Q(dy) P(y, B) &\leqslant \int Q_A(dy) P(y, B) + \int_{A^c} \left[\int_A Q_A(dx) P_A(x, dy) \right] P(y, B) \\
&= \int_A Q_A(dx) \left[P(x, B) + \int_{A^c} P_A(x, dy) P(y, B) \right] \\
&= \int_A Q_A(dx) P_A(x, B) = Q(B).
\end{aligned}
$$

The modification of the argument required for invariant measures is obvious.

Of course, $P_A(x,B)$ for elements $B \subseteq A$ of $\mathscr{A}$ is not a Markov transition probability function unless $P_A(x,A) \equiv 1$. The following Corollary is almost immediate.

Corollary 1. *Let Q_A be a nontrivial subinvariant finite measure for $P_A(\cdot,\cdot)$, so that (4) is satisfied for all elements $B(\subset A)$ of $\mathscr{A}$. Then Q_A is an invariant measure for $P_A(\cdot,\cdot)$ if and only if $P_A(x,A) \equiv 1$ for almost all $x \in A$ (with respect to Q_A).*

If $Q_A(A) = \int\limits_A Q_A(dx) P_A(x,A)$ it is clear that $P_A(x,A) \equiv 1$ for almost all $x \in A$ with respect to Q_A. Conversely, if $P_A(x,A) \equiv 1$ for almost all $x \in A$, then $Q_A(A) = \int\limits_A Q_A(dx) P_A(x,A)$. But then (5) holds for all subsets B of A that are elements of $\mathscr{A}$.

Lemma 3. *Let Q be a subinvariant sigma-finite measure for $P(\cdot,\cdot)$. Let $A \in \mathscr{A}$ be a set with $0 < Q(A) < \infty$. Then Q restricted to subsets of A that are elements of $\mathscr{A}$ is subinvariant for $P_A(\cdot,\cdot)$.*

If $B \in \mathscr{A}$ and $B \subset A$ we have

$$Q(B) \geqslant \int Q(dy) P(y,B) = \int\limits_A + \int\limits_{A^c} .$$

However,

$$\begin{aligned}\int\limits_{A^c} Q(dy) P(y,B) &\geqslant \int\limits_{A^c} \int Q(dx) P(x,dy) P(y,B) = \int Q(dx) \int\limits_{A^c} P(x,dy) P(y,B) \\ &= \int\limits_A \int\limits_{A^c} + \int\limits_{A^c} \int\limits_{A^c} .\end{aligned}$$

Therefore

$$\begin{aligned}Q(B) \geqslant &\int\limits_A Q(dx) \left[P(x,B) + \int\limits_{A^c} P(x,dy) P(y,B) \right] \\ &+ \int\limits_{A^c} Q(dx) \int\limits_{A^c} P(x,dy) P(y,B).\end{aligned} \tag{6}$$

Continuing this argument by replacing $Q(dx)$ in the last term on the right of (6) by $\int Q(dz) P(z,dx)$, we find that

$$Q(B) \geqslant \int\limits_A Q(dx) \left[P(x,B) + \int\limits_{A^c} P(x,dy) P(y,B) + \cdots \right] = \int\limits_A Q(dx) P_A(x,B).$$

Of course, Corollary 1 implies that if $P_A(x,A) \equiv 1$ for almost all $x \in A$, then Q restricted to the subsets of A is a finite invariant measure.

The results of section 2 of Chapter 1 indicate that a Markov chain with at last one recurrent state always has at least one nontrivial invariant measure. If all the states are transient, a nontrivial invariant measure may not exist. But there will still be a nontrivial subinvariant measure. This implies that a Markov chain always has at least one nontrivial subinvariant measure. There are no results of this global character on the existence of invariant (or subinvariant) measures when we are dealing with a general transition probability function. However, by making special assumptions of a measure-theoretic or topological character about the transition function, existence of an invariant or subinvariant measure can be obtained. A result of this type making use of topological conditions will be obtained in section 3.

A few remarks should be made about $P_A(x,B)$ for $x \in A$, $B \subseteq A$ with $A \in \mathscr{A}$. Let $\{X_n(\omega); n=0,1,\ldots\}$ be a Markov process with transition probability function $P(\cdot,\cdot)$ and initial distribution μ. If one hits A with probability one from almost every initial point, then $P_A(x,B)$ corresponds to the process

$$Y_k(\omega) = X_{n_k}(\omega), \qquad k=1,2,\ldots,$$

where $n_k = n_k(\omega)$ is the kth time one hits A. We have already referred implicitly to $\{Y_k(\omega)\}$ as the process restricted to the set A in Lemma 2 with $P_A(x,B)$, $x \in A$, $B \subseteq A$, as its one step transition function. The process $\{Y_k(\omega)\}$ is in fact a Markov process with transition function $P_A(\cdot,\cdot)$, as one can verify by computing the joint probability distributions at any finite set of times for a Markov process with $P_A(\cdot,\cdot)$ as its transition function (and initial distribution μ restricted to A) and noting that these joint distributions agree with those for $\{Y_k(\omega)\}$.

2. An L^1 Ergodic Theorem

We give a proof of E. Hopf of an L^1 ergodic theorem due to Chacon and Ornstein [7]. Let Ω be a space of points x with a σ-finite measure ν on a σ-field $\mathscr{A}$ of subsets of Ω. $L^1 = L^1(\Omega)$ is the linear space of all ν-integrable functions f on Ω. The operator T is assumed to be linear and positive, that is, if $f \geqslant 0$ on Ω then $Tf \geqslant 0$ on Ω. Furthermore, T is assumed to be an operator taking L^1 into L^1 with norm less than or equal to 1 (a contraction)

$$|T| = \sup_{\substack{f \in L_1 \\ f \neq 0}} \frac{\|Tf\|_1}{\|f\|_1} \leqslant 1 .$$

It will be enough to see what the condition that T be a positive contraction on L^1 amounts to when Ω is the set of positive integers. What happens in this simple example is typical of the general situation.

The σ-finite measure $v=(v_i)$ is just a vector with finite nonnegative entries and $T=(t_{i,j})$ can be represented as a matrix with entries $t_{i,j}$. The function or vector $f=(f_i)$ is integrable if

$$\|f\|_1=\sum|f_i|\,v_i<\infty.$$

If $v_i>0$, then $T\delta^{(i)}$ is well-defined for the functions

$$\delta^{(i)}=(\delta_{i-j};j=1,2,\ldots)\qquad i=1,2,\ldots,$$

and

$$(T\delta^{(i)})_j=t_{j,i}.$$

The positivity of T simply amounts to the condition

$$t_{i,j}\geqslant 0$$

for all i,j. If T is a contraction, we must have

$$\|T\delta^{(i)}\|_1=\sum_j v_j t_{j,i}\leqslant\|\delta^{(i)}\|_1=v_i,$$

$i=1,2,\ldots$. Thus v is a subinvariant measure for the matrix T (which needn't be stochastic). Actually, the preceding inequality implies that

$$\|Tf\|_1\leqslant\|f\|_1$$

for all $f\in L^1$ with respect to v. Introduce the matrix

$$T^*=(t^*_{i,j})$$

with

$$t^*_{i,j}=\frac{v_j t_{j,i}}{v_i}.$$

T^* is the adjoint of T with respect to L^1. The subinvariant character of v implies that

$$\sum_j t^*_{i,j}\leqslant 1,\qquad i=1,2,\ldots,$$

that is, T^* is a substochastic matrix.

Although we shall obtain an ergodic theorem for positive operators in this generality, the case we are primarily interested in is that of an operator T induced by a Markov transition function

$$(Tf)(x)=\int P(x,dy)\,f(y). \tag{1}$$

Positive operators given by (1) are well-defined when acting on bounded functions f. Furthermore, $T1\equiv 1$. However, the question of existence of a σ-finite measure v relative to which T is an L^1 contraction is non-trivial. For an operator T given by (1) the existence implies that for any set $A\in\mathscr{A}$ of finite v measure

$$\int v(dx)\,P(x,A)\leqslant v(A).$$

Such a measure ν is called *subinvariant* relative to $P(\cdot,\cdot)$ (or T). Actually, T as given by (1) is an L^1-contraction relative to ν if and only if ν is subinvariant relative to T. Let us take the existence of such a measure ν for granted. Later on in section 3 some conditions under which such a measure exists for a Markov transition function will be investigated.

Let f, p be in L^1 with $p \geqslant 0$. We wish to consider the asymptotic behavior as $n \to \infty$ of the quotients

$$Q_n(f,p) = \frac{S_n(f)}{S_n(p)}, \qquad S_n(g) = \sum_0^{n-1} T^i g.$$

The L^1 ergodic theorem stated below will follow from a sequence of Lemmas.

Notice that questions on the asymptotic behavior of the quotients $Q_n(f,p)$ are naturally suggested by results obtained for ratios of sums of transition probabilities of Markov chains in Lemma 1 and 3 of section 1.2.

Theorem 1. *Let T be a positive linear operator from L^1 to L^1 with L^1-norm $|T| \leqslant 1$. If $f,p \in L^1$ with $p \geqslant 0$, the quotients $Q_n(f,p)$ have a finite limit as $n \to \infty$ almost everywhere (with respect to ν) in the set* $\left\{x : \sum_0^\infty T^i p > 0\right\}$.

Let

$$f^+ = \max(f,0), \qquad f^- = -\min(f,0).$$

Then

$$f = f^+ - f^-, \qquad |f| = f^+ + f^-.$$

Notice that one cannot have both $f^+(x)$ and $f^-(x) > 0$ at a point x. If $f \in L^1$, then both $f^+, f^- \in L^1$. Furthermore if

$$f = f_2 - f_1, \qquad f_i \geqslant 0$$

then

$$f^+ \leqslant f_2, \qquad f^- \leqslant f_1. \tag{2}$$

This implies that

$$(Tf)^+ \leqslant Tf^+, \qquad (Tf)^- \leqslant Tf^-, \qquad |Tf| \leqslant T|f|.$$

The first Lemma is now given.

Lemma 1. *If $f \in L^1$ and*

$$\sup_{n>0} \sum_0^{n-1} T^i f > 0 \tag{3}$$

holds everywhere in a set $A\in\mathscr{A}$, *then there exist, for each* $\varepsilon>0$, *functions* h, $\varphi\in L^1$ *such that*

a) $h^-\leqslant f^-$

b) $h=f+T\varphi-\varphi,\ \varphi\geqslant 0$

c) $\int h\leqslant\int f$

d) $\int_A h^-<\varepsilon.$

Let

$$h_{i+1}=Th_i^+-h_i^-,\qquad h_0=f, \tag{4}$$

$i=0,1,\ldots$. The sequence of functions h_i all belong to L^1. We will show that all the h_i's satisfy conditions a)—c) and that d) is satisfied for sufficiently large i. Relations (4), (2) imply that

$$h_{i+1}^-\leqslant h_i^- \tag{5}$$

and so a) follows. Rewrite (4) as

$$h_{i+1}=h_i+Th_i^+-h_i^+$$

and sum to obtain

$$h_i=f+T\varphi_{i-1}-\varphi_{i-1},\qquad \varphi_i=\sum_0^i h_j^+. \tag{6}$$

We have b). Condition b) implies c) since $\int T\varphi\leqslant\int\varphi$ if $\varphi\geqslant 0$. Since $\varphi_i-\varphi_{i-1}=h_i^+\geqslant h_i$,

$$\varphi_i\geqslant f+T\varphi_{i-1}$$

by equation (6). T is positive and therefore by induction

$$\varphi_i\geqslant\sum_0^{i-1}T^jf+T^i\varphi_0,\qquad i>0.$$

The inequality

$$\varphi_i=\sum_0^i h_j^+\geqslant\sum_0^i T^jf \tag{7}$$

follows because $\varphi_0=h_0^+=f^+\geqslant f$. This also holds for $i=0$. Inequality (7) and the hypothesis (3) imply that at each point of A at least one of the $h_j^+, j\geqslant 0$, is positive. Therefore, at each point at least one of the h_j^- is zero. Since $h_n^-\geqslant 0$, it follows from (5) that $h_n^-\downarrow 0$ in A and therefore $\int_A h_n^-\to 0$. The proof of the Lemma is completed.

Lemma 2. *If* $f \in L^1$ *and*

$$\sup_{n>0} \sum_{0}^{n-1} T^j f \geqslant 0 \tag{8}$$

everywhere in $A \in \mathscr{A}$, *then*

$$\int_A f + \int_{A^c} f^+ \geqslant 0, \qquad A^c = X - A. \tag{9}$$

The proof will first be given under the assumption that the sup in (8) is strictly positive in A. Fix $\varepsilon > 0$ and apply Lemma 1. From d), a) and c) of Lemma 1 it follows that

$$-\varepsilon - \int_{A^c} f^- < -\int_A h^- - \int_{A^c} h^- = -\int h^- \leqslant \int h \leqslant \int f. \tag{10}$$

The inequality (9) follows directly from (10) since (10) is valid for any $\varepsilon > 0$. To prove the Lemma under the hypothesis $\sup \geqslant 0$ in A choose a function $p \in L^1$ strictly positive on A and apply the restricted result just derived to $f + \eta p$, $\eta > 0$, in place of f. Lemma 2 is obtained by letting $\eta \downarrow 0$.

Lemma 3. *Let* $f, p \in L^1$, $p \geqslant 0$. *Then*

$$\sup_n |Q_n(f,p)| < \infty$$

almost everywhere in $\{x : p(x) > 0\}$.

Since $|Q_n(f,p)| \leqslant Q_n(|f|,p)$, it is enough to prove the Lemma for $f \geqslant 0$. Let $A = \{x : p(x) > 0, \sup Q_n(f,p) = \infty\}$. Apply Lemma 2 to $f - wp$ and A with w any fixed positive number. Then

$$\int_A (f - wp) + \int_{A^c} (f - wp)^+ \geqslant 0.$$

We also have $(f - wp)^+ \leqslant f$ since $f \geqslant 0$. Thus

$$\int_A p \leqslant \frac{1}{w} \int f.$$

Since $p > 0$ in A and w is arbitrary, A has measure zero.

Lemma 4. *If* $f, p \in L^1$, $p \geqslant 0$ *it follows that*

$$\lim_{n \to \infty} T^n f \Big/ \sum_{0}^{n-1} T^j p = 0$$

almost everywhere on $\{x : p(x) > 0\}$.

Again, it is enough to prove this for $f \geqslant 0$.
Consider the functions

$$g_n = T^n f - \varepsilon \sum_0^{n-1} T^j p, \qquad g_0 = f, \qquad \varepsilon > 0.$$

Clearly,

$$g_{n+1} + \varepsilon p = T g_n.$$

It will be enough to show that for almost every point of $\{x : p(x) > 0\}$, $g_n < 0$ for sufficiently large n, or equivalently, that $\Sigma \chi_n$ converges, where χ_n is the indicator function of $\{x : g_n(x) \geqslant 0\}$. Since $\chi_n g_n = g_n^+$, it follows that

$$\int g_{n+1}^+ + \varepsilon \int \chi_{n+1} p = \int \chi_{n+1}(g_{n+1} + \varepsilon p) = \int \chi_{n+1} T g_n \leqslant \int \chi_{n+1} T g_n^+ \\ \leqslant \int T g_n^+ \leqslant \int g_n^+ .$$

Sum the inequalities to obtain

$$\int g_n^+ + \varepsilon \int p \sum_1^n \chi_j \leqslant \int g_0^+ .$$

Then

$$\int p \sum_1^\infty \chi_j \leqslant \frac{1}{\varepsilon} \int g_0^+ .$$

Since $\limsup_{n\to\infty} T^n f \Big/ \sum_0^{n-1} T^j p \leqslant \varepsilon$ on $\{x : p(x) > 0\}$, the desired conclusion is obtained by letting $\varepsilon \downarrow 0$.

Lemma 5. *Let $f, p \in L^1, p \geqslant 0$. Assume that $p > 0, \sum_0^\infty T^j p = \infty$ and that*

$$\liminf_{n\to\infty} Q_n(f,p) < \alpha < \beta < \limsup_{n\to\infty} Q_n(f,p)$$

for some numbers α, β everywhere in a set $A \in \mathscr{A}$. Then A must have measure zero.

Let $\varepsilon > 0$. The hypothesis implies that in A

$$\sup_{n>0} Q_n(f - \beta p, p) \geqslant \limsup_{n\to\infty} Q_n(f - \beta p, p) > 0.$$

Apply Lemma 1 to $f - \beta p$. There is then a function $h \in L^1$ such that by a)

$$(h - \beta p)^- \leqslant (f - \beta p)^- \tag{11}$$

and by d)

$$\int_A (h - \beta p)^- < \varepsilon. \tag{12}$$

Inequality (11) indicates that $h \leqslant \beta p$ implies $h \geqslant f$. Since $\alpha < \beta$, $h < \alpha p$ implies $h \geqslant f$ and therefore

$$(h-\alpha p)^- \leqslant (f-\alpha p)^- . \tag{13}$$

By part b) of Lemma 1

$$Q_n(h,p) = Q_n(f,p) + \frac{T^n \varphi - \varphi}{\sum_0^{n-1} T^j p} .$$

Apply Lemma 4 to φ, p and the assumption that $\sum_0^\infty T^j p = \infty$ on A. It follows that $Q_n(h,p)$ has the same upper and lower limit as $Q_n(f,p)$ in each point of A.
Thus

$$\limsup_{n\to\infty} Q_n(\alpha p - h, p) > 0 \quad \text{in } A .$$

Apply Lemma 1 to $\alpha p - h$ in place of the function f there and let the derived function h of the Lemma be written $\alpha p - f'$. Then

$$\int (f' - \alpha p) \geqslant \int (h - \alpha p) \tag{14}$$

and by a) of the Lemma

$$(f' - \alpha p)^+ \leqslant (h - \alpha p)^+ \tag{15}$$

Inequality d) of the Lemma becomes

$$\int_A (\alpha p - f')^- = \int_A (f' - \alpha p)^+ < \varepsilon . \tag{16}$$

Inequality (14) can be written

$$\int (f' - \alpha p)^- \leqslant \int (h - \alpha p)^- + \int [(f' - \alpha p)^+ - (h - \alpha p)^+] . \tag{17}$$

Inequality (14) is still valid if the last integral is taken over A since the integrand is less than or equal to zero by (15). Now

$$(h-\alpha p)^+ \geqslant h - \alpha p = h - \beta p + (\beta - \alpha) p \geqslant -(h - \beta p)^- + (\beta - \alpha) p .$$

The right hand side of (17) is less than or equal to

$$\int (h - \alpha p)^- + \int_A (f' - \alpha p)^+ + \int_A (h - \beta p)^- - (\beta - \alpha) \int_A p .$$

Apply (13) to the first term, (16) to the second term and (12) to the third term to obtain the inequality

$$\int (f' - \alpha p)^- \leqslant \int (f - \alpha p)^- + 2\varepsilon - (\beta - \alpha) \int_A p \tag{18}$$

in which h no longer appears. Now it follows as before that $Q_n(f',p)$ has the same upper and lower limit as $Q_n(h,p)$ at each point of A. Thus for any given fixed $\varepsilon>0$ there is a function f' satisfying the hypothesis of this Lemma and also inequality (18). Apply this result to f' to obtain a function f'' satisfying the hypothesis of this Lemma and inequality

$$\int (f''-\alpha p)^- \leqslant \int (f'-\alpha p)^- + 2\varepsilon - (\beta-\alpha)\int_A p.$$

Apply this argument recursively. Adding the first n of the resulting inequalities and noting that the integral on the left is positive one finds that

$$n\left[(\beta-\alpha)\int_A p - 2\varepsilon\right] \leqslant \int (f-\alpha p)^-$$

for any integer n. Therefore, $(\beta-\alpha)\int_A p - 2\varepsilon \leqslant 0$. Since $\varepsilon>0$ is arbitrary it follows that $\int_A p=0$. But $p>0$ on A and therefore A has measure zero.

We now prove the ergodic theorem. First it will be obtained on the set $\{x:p(x)>0\}$. It is enough to prove it for $f\geqslant 0$. Lemma 3 implies that the limit must be finite almost everywhere if it exists. We prove existence of the limit. On the set $\left\{x:p(x)>0, \sum_1^\infty T^j p<\infty\right\}$ this is immediate since $\limsup_{n\to\infty} Q_n<\infty$ implies that $\sum_0^\infty T^j f<\infty$ (the terms are non-negative). On the set $\left\{x:p(x)>0, \sum_0^\infty T^j p=\infty\right\}$ positivity of the measure of the subset on which there is no limit would imply that there are numbers α, β $(\alpha<\beta)$ such that $\{x:\liminf Q_n<\alpha<\beta<\limsup Q_n\}$ has positive measure in contradiction to Lemma 5. The proof on $\{x:p(x)>0\}$ is complete. To prove the theorem on $\left\{x:\sum_0^\infty T^j p>0\right\}$ notice that this set is the union of the sets $\{x:p(x)>0\}$ and

$$\{x:S_{k+1}(p)>0,\ S_k(p)=0\}, \qquad k>0. \tag{19}$$

On the set (19), $T^k p=S_{k+1}(p)-S_k(p)>0$. Let

$$f'=T^k f, \qquad p'=T^k p.$$

Then

$$Q_{n-k}(f',p') = \frac{S_n(f)-S_k(f)}{S_n(p)} = Q_n(f,p) - \frac{S_k(f)}{S_n(p)}$$

on this set for $n>k$. By this formula and what has already been proved it is clear that $\lim_{n\to\infty} Q_n(f,p)$ exists and is finite almost everywhere in (19).

Let p be an integrable function that is positive almost everywhere. Let

$$C = \left\{x : \sum_0^\infty (T^j p)(x) = \infty\right\} \tag{20}$$

and

$$D = \left\{x : \sum_0^\infty (T^j p)(x) < \infty\right\} = C^c. \tag{21}$$

It then follows from Theorem 1 that for any integrable function $f \geqslant 0$, $\sum_0^\infty (T^j f)(x) = \infty$ almost everywhere on $C \cap \left\{x : \sum_0^\infty (T^j f)(x) > 0\right\}$ and $\sum_0^\infty (T^j f)(x) < \infty$ almost everywhere on D.

Corollary 1. *Let T be a positive linear operator from L^1 to L^1 with L^1 norm $|T| \leqslant 1$. There is then a decomposition of the space Ω into two disjoint measurable sets C (the conservative part) and D (the dissipative part) such that for any integrable function $f \geqslant 0$*

$$\sum_{j=0}^\infty (T^j f)(x) = \infty \quad \text{on } C \cap \left\{x : \sum_{j=0}^\infty (T^j f)(x) > 0\right\}$$

and

$$\sum_{j=0}^\infty (T^j f)(x) < \infty \quad \text{on } D.$$

The uniqueness up to a set of measure zero of this decomposition is immediate. Notice that in the case of a Markov chain (not necessarily irreducible), this decomposition amounts to the partition of the set of all states into recurrent states (the conservative part) and transient states (the dissipative part).

Let us now *consider the case of a transition probability function $P(\cdot,\cdot)$ with a finite invariant measure μ* which is normalized so that it is a probability measure. Notice that the argument leading to Corollary 1.1 implies that if μ is a finite nontrivial subinvariant measure, it is automatically an invariant measure. The positive linear operator T given by

$$(Tf)(x) = \int P(x, dy)\, f(y) \tag{22}$$

is an L^1 contraction, where L^1 is understood to be the space of all μ integrable functions, since the invariant character of μ implies that

$$\int |(Tf)(x)| \leqslant \int (T|f|)(x)\, \mu(dx) = \int |f(x)|\, \mu(dx).$$

The L^1 ergodic theorem of this section can be immediately applied. Since the function $1 \in L^1$ and $T1 = 1$, it follows that the whole space Ω (up to a set of μ measure zero) is conservative. Given any function $f \in L^1$, by taking $p \equiv 1$ in Theorem 1, it is clear that

$$\hat{f}(x) = \lim_{n \to \infty} \frac{1}{n} \sum_{j=1}^{n} T^j f \tag{23}$$

is well-defined almost everywhere. An interesting problem is that of identifying this limit. It is quite clear from (22) that

$$\hat{f}(x) = (T\hat{f})(x) \tag{24}$$

and we shall *call any function* $g \in L^1$ *satisfying* (24) *an invariant function.* We shall also *call a set* $A \in \mathscr{A}$ *an invariant set if its indicator function* I_A *satisfies* (24), that is, if

$$I_A(x) = (T I_A)(x). \tag{25}$$

From (25) it follows immediately that the class $\mathscr{I}$ of invariant sets is a Borel subfield of $\mathscr{A}$. Let us show that any invariant function is actually measurable with respect to the Borel field $\mathscr{I}$. The set of integrable invariant functions is a closed (under bounded pointwise convergence almost everywhere) linear space of functions. In order to show that an invariant function $f \in L^1$ is $\mathscr{I}$-measurable, it is enough to show that $\{x : f(x) > a\}$ is in $\mathscr{I}$ for each real number a. Suppose that $f \in L^1$ is invariant. It is then clear that $|f|$ is invariant since

$$|f| = |Tf| \leqslant T|f|$$

and since T is an L^1 contraction, it follows that

$$|f| = T|f|.$$

But then

$$\max(0, f) = \tfrac{1}{2}\{f + |f|\}$$

is invariant. Further, if $f, g \in L^1$ are invariant, then $\max(f,g)$ is invariant because

$$\max(f,g) = \max(f - g, 0) + g$$

and $\min(f,g)$ is invariant because $\min(f,g) = -\max(-f, -g)$. Of course, 1 is invariant. Assume $f \in L^1$ is invariant and consider any real number a. Then

$$\min(n \max(0, f - a), 1), \qquad n = 1, 2, \ldots,$$

is a sequence of invariant functions. The limit as $n \to \infty$ of this sequence is the indicator function of the set $\{x: f(x) > a\}$ which is hence invariant.

Let us now consider a bounded function f. From (22), it follows that

$$\operatorname*{ess\,sup}_x |Tf(x)| \leqslant \operatorname*{ess\,sup}_x |f(x)|.$$

Given any $A \in \mathscr{I}$, (23) implies that

$$\int_A \hat{f}(x)\mu(dx) = \int_A f(x)\mu(dx),$$

since μ restricted to the set A is an invariant measure for T.

Therefore

$$\hat{f}(x) = E(f|\mathscr{I})(x), \tag{26}$$

where $E(f|\mathscr{I})$ is the conditional expectation of f given the sigma-field of invariant sets, since $\hat{f}$ is measurable with respect to $\mathscr{I}$. However, it is now clear that (26) must be true for any $f \in L^1$. For given any such f, if it is not bounded we can approximate it to within $1/n$ by a bounded function g_n in L^1 norm, $n = 1, 2, \ldots$,

$$\int |f(x) - g_n(x)| \mu(dx) < \frac{1}{n}.$$

But then because T is an L^1 contraction

$$\int |\hat{f}(x) - \hat{g}_n(x)| \mu(dx) < \frac{1}{n}. \tag{27}$$

However, $\hat{g}_n(x) = E(g_n|\mathscr{I})(x)$ and

$$\int |E(f - g_n|\mathscr{I})| \mu(dx) \leqslant \int E(|f - g_n| \,|\mathscr{I}) \mu(dx) = \int |f - g_n| \mu(dx) < \frac{1}{n}. \tag{28}$$

From (27) and (28) on letting $n \to \infty$, it is readily seen that (26) is valid.

Corollary 2. *Let $P(\cdot, \cdot)$ be a transition probability function with the finite invariant measure μ. Then for any $f \in L^1$*

$$\hat{f}(x) = \lim_{n \to \infty} \frac{1}{n} \sum_{j=1}^{n} (T^j f)(x) = E(f|\mathscr{I})(x)$$

where $\mathscr{I}$ is the sigma-field of invariant sets, that is, the family of sets A such that

$$P(x, A) = \begin{cases} 1 & \text{if } x \in A \\ 0 & \text{otherwise} \end{cases}$$

up to a set of μ measure zero.

It seems rather natural in the context of Corollary 2 to *call the transition probability function $P(\cdot,\cdot)$ ergodic with respect to μ if*

$$\frac{1}{n}\sum_{j=1}^{n}(T^j f)(x)\to Ef=\int f(x)\mu(dx)$$

almost everywhere as $n\to\infty$ for each integrable function f. This is especially plausible in terms of the remarks made about the ergodic problem in the discussion of statistical mechanics in section 2.1. There, of course, one has a continuous group of deterministic transformations rather than a discrete (in time) semigroup of transition probability operators. We shall later make a few remarks on the case of deterministic transformations. From (26) it is clear that the transition function will be ergodic with respect to μ if and only if the sigmafield $\mathscr{I}$ of invariant sets is trivial, that is, if it consists only of the null set and the whole space Ω up to sets of μ measure zero.

The classical Birkhoff ergodic theorem is an almost immediate consequence of Corollary 2. Let φ be a measurable mapping of Ω onto itself, that is, for each $A\in\mathscr{A}$

$$\varphi^{-1}(A)=\{x:\varphi(x)\in A\}\in\mathscr{A}.$$

Further, φ is assumed to be measure-preserving with respect to a probability measure μ on $\mathscr{A}$ so that $\mu(\varphi^{-1}(A))=\mu(A)$ for all $A\in\mathscr{A}$. The transition function $P(\cdot,\cdot)$ induced by the mapping φ is

$$P(x,A)=\begin{cases}1 & \text{if } \varphi(x)\in A\\ 0 & \text{otherwise}\end{cases} \tag{29}$$

and clearly μ is an invariant measure for (29). The Birkhoff ergodic theorem is concerned with the asymptotic behavior of

$$\frac{1}{n}\sum_{j=1}^{n}(T^j f)(x)=\frac{1}{n}\sum_{j=1}^{n}f(\varphi^j(x))$$

as $n\to\infty$ where T is the L^1 contraction relative to μ induced by (29).

Corollary 3 *(Birkhoff ergodic theorem). Let μ be a probability measure on the Borel field $\mathscr{A}$ on Ω. Consider a measurable mapping φ (relative to $\mathscr{A}$) of Ω onto itself that preserves the measure μ. Then for any function $f\in L^1$ it follows that*

$$\frac{1}{n}\sum_{j=1}^{n}f(\varphi^j(x))\to E(f\,|\,\mathscr{I})(x)$$

almost everywhere as $n\to\infty$ where $\mathscr{I}$ is the family of φ invariant sets, that is, sets A such that $\varphi(x)\in A$ if and only if $x\in A$ up to an exceptional set of measure zero. The mapping φ is ergodic if and only if the sigma-field $\mathscr{I}$ of invariant sets is trivial.

The following corollary provides a convenient reformulation of the Birkhoff ergodic theorem.

Corollary 4. *Let μ be a probability measure on the Borel field $\mathscr{A}$ on Ω. If φ is a measure-preserving mapping of $\mathscr{A}$ onto itself, then φ is ergodic if and only if*

$$\frac{1}{n}\sum_{j=1}^{n}\mu(A\cap\varphi^{-j}B)\to\mu(A)\mu(B)$$

as $n\to\infty$ for each pair of sets $A,B\in\mathscr{A}$.

Set $f=I_B$, the indicator function of the set B in Corollary 3. Then

$$\frac{1}{n}\sum_{j=1}^{n}I_B(\varphi^j(x))\to\mu(B)$$

as $n\to\infty$. However, by bounded convergence

$$\frac{1}{n}\sum_{j=1}^{n}\mu(A\cap\varphi^{-j}B)=\int_A\frac{1}{n}\sum_{j=1}^{n}I_B(\varphi^j(x))\mu(dx)\to\mu(A)\mu(B)$$

as $n\to\infty$. Conversely, assume that the limit relation just obtained is satisfied. If A is invariant then

$$\mu(A)=\frac{1}{n}\sum_{j=1}^{n}\mu(A\cap\varphi^{-j}A)\to\mu^2(A)$$

so that $\mu(A)=0$ or $\mu(A)=1$. Thus, all invariant sets are trivial and φ is ergodic.

If the mapping φ is one-to-one, the probability space (the space Ω together with the Borel field $\mathscr{A}$ and the probability measure μ on $\mathscr{A}$) together with φ is sometimes referred to as an *abstract dynamical system* (see Arnold and Avez [2]).

Consider the following interesting example of an abstract dynamical system that we have already implicitly constructed. Let $P(\cdot,\cdot)$ be a transition probability function with invariant probability measure μ. As in section 1.1 let $\omega=(x_j;j=\ldots,-1,0,1,\ldots)$ be a doubly infinite vector whose entries are elements of Ω. The space of points ω is the doubly infinite product space of Ω. Given any finite set of integers

j_α; $\alpha=1,\ldots,n$, with $-\infty<j_1<\cdots<j_n<\infty$ and corresponding sets $A_{j_1},\ldots,A_{j_n}\in\mathscr{A}$ let

$$P_\mu(\{\omega: x_{j_\alpha}\in A_{j_\alpha};\alpha=1,\ldots,n\})$$
$$=\int_{A_{j_1}}\mu(dx_{j_1})\int_{A_{j_2}}P_{j_2-j_1}(x_{j_1},dx_{j_2})\ldots\int_{A_{j_{n-1}}}P_{j_{n-1}-j_{n-2}}(x_{j_{n-2}},dx_{j_{n-1}})P(x_{j_{n-1}},A_{j_n}). \tag{30}$$

This is consistent on finite product sets of the form considered in (30) because of the invariance of μ relative to $P(\cdot,\cdot)$. Extend P_μ to the Borel field $\mathscr{A}_{-\infty}^{\infty}$ of sets on the infinite product space generated by such finite product sets. Let $(\omega)_n=x_n$ be the nth component of the point ω. Then $\mathscr{A}_{-\infty}^{\infty}$ is the Borel field generated by the functions $X_n(\omega)=(\omega)_n=x_n$, $n=0,\pm1,\ldots$, that is, the smallest Borel field with respect to which all these functions are measurable. The extension of the set function P_μ to a probability measure on $\mathscr{A}_{-\infty}^{\infty}$ can be carried out by the Kolmogorov theorem under appropriate conditions. Let τ be the shift transformation, that is,

$$(\tau\omega)_n=(\omega)_{n+1}=x_{n+1}.$$

Then τ is a one-to-one mapping of the product space onto itself and is such that τ,τ^{-1} are measurable (with respect to $\mathscr{A}_{-\infty}^{\infty}$) and preserve the measure P_μ. The sequence $\{X_n(\omega); n=\ldots,-1,0,1,\ldots\}$ is called the stationary Markov process generated by transition function $P(\cdot,\cdot)$ and invariant probability measure μ. The shift transformation τ and the probability space of doubly infinite sequences on which it acts is an example of an abstract dynamical system. Now we shall show that there is essentially a one-to-one correspondence between the invariant sets of the transition function $P(\cdot,\cdot)$ relative to the invariant probability measure μ and the invariant sets of the shift operator τ on the probability space of the stationary Markov process generated by the transition function $P(\cdot,\cdot)$ and the invariant measure μ. Briefly, let $A\in\mathscr{A}$ be an invariant set of $P(\cdot,\cdot)$ so that

$$P(x,A)=\begin{cases}1 & \text{if } x\in A\\ 0 & \text{otherwise}\end{cases}$$

up to an exceptional set of μ measure zero. Then it is quite clear that

$$\{\omega: X_0(\omega)\in A\}=\{\omega: X_n(\omega)\in A\},\qquad n=0,\pm1,\ldots$$

is an invariant set of the shift transformation τ. Conversely, one will be able to show that if *$B\in\mathscr{A}_{-\infty}^{\infty}$ is an invariant set of τ, then*

$$A=\{x_n: x_n=(\omega)_n \text{ such that } \omega\in B\} \tag{31}$$

is an invariant set of $P(\cdot,\cdot)$.

Suppose $B \in \mathscr{A}_{-\infty}^{\infty}$ is an invariant set with respect to the shift transformation τ on the probability space of the stationary Markov process generated by $P(\cdot,\cdot)$ and the invariant probability measure μ. Then for each integer $k=1,2,\ldots$ there is a set $B_k \in \mathscr{A}_m^n$ for appropriately chosen integral $m=m(k)<n=n(k)$ (m sufficiently small and n sufficiently large) such that

$$P_\mu(B_k \wedge B) < \frac{1}{k}$$

where $B_k \wedge B$ is the symmetric difference of B_k and B. Since the Markov process is stationary and B is τ-invariant it follows that

$$P_\mu(\tau^{-m} B_k \wedge B) \leqslant \int |I_{\tau^{-m}B_k}(\omega) - I_B(\omega)|^2 \, dP_\mu < \frac{1}{k}$$

where I_A as usual denotes the indicator function of the set A. The best predictor of $I_{\tau^{-m}B_k}(\omega)$ given $X_j(\omega)$, $j \leqslant 0$, minimizing the mean square error of prediction is the conditional expectation

$$q_k(\omega) = E(I_{\tau^{-m}B_k}(\omega) | X_j, j \leqslant 0) = E(I_{\tau^{-m}B_k}(\omega) | X_0). \tag{32}$$

Now $q_k(\omega)$ is $\mathscr{A}_0^0$-measurable (loosely speaking, a function of X_0) because $\{X_k\}$ is a Markov process. However, $I_{\tau^{-n}B_k}$ is a possible predictor of $I_{\tau^{-m}B_k}$ and

$$E|I_{\tau^{-n}B_k} - I_B|^2 < \frac{1}{k}. \tag{33}$$

From (32) and (33) it follows that

$$E|q_k - I_B|^2 < \frac{16}{k}. \tag{34}$$

Notice that (34) implies that q_k converges to I_B in mean square as $k \to \infty$. Since all the q_k's are $\mathscr{A}_0^0$-measurable, it follows that I_B is $\mathscr{A}_0^0$-measurable. We have therefore shown that (31) is an invariant set of $P(\cdot,\cdot)$.

Corollary 5. *Let $P(\cdot,\cdot)$ be a transition probability function with invariant probability measure μ. Then $P(\cdot,\cdot)$ is ergodic with respect to μ if and only if the shift transformation τ is ergodic on the infinite product space of the Markov process generated by $P(\cdot,\cdot)$ and μ.*

The following corollary is mentioned without proof but it can be easily derived by using the ideas employed to get Corollary 4.

Corollary 6. *Let $P(\cdot,\cdot)$ be a transition probability function with invariant probability measure μ. A Markov process generated by $P(\cdot,\cdot)$ and μ is ergodic if and only if*

$$\frac{1}{n} \sum_{k=1}^{n} \int_A \mu(dx) P_k(x,B) \to \mu(A)\mu(B)$$

as $n \to \infty$ for each pair of sets $A, B \in \mathscr{A}$.

Let $P(\cdot,\cdot)$ be a transition probability function with invariant probability measure μ. Consider any two sets $A, A' \in \mathscr{A}$. The transition function $P(\cdot,\cdot)$ is said to be *mixing* with respect to μ if

$$\int_A \mu(dx)\, P_n(x, A') = \int \mu(dx)(T^n I_{A'})(x) \to \mu(A)\mu(A') \tag{35}$$

as $n\to\infty$ for every such pair of $\mathscr{A}$ measurable sets. Let $f, g \in L^2(d\mu)$. If $P(\cdot,\cdot)$ (or equivalently T) is mixing with respect to μ one can show that

$$\int \mu(dx) f(x)(T^n g)(x) \to \int \mu(dx) f(x) \int \mu(dy) g(y) = Ef\,Eg \tag{36}$$

as $n\to\infty$ by a simple approximation argument. Notice that (35) implies that (36) is valid for any two $\mathscr{A}$-measurable functions taking on at most a finite number of distinct values. Given any $\varepsilon>0$ we can find two such functions f_ε, g_ε with

$$E|f-f_\varepsilon|^2, \quad E|g-g_\varepsilon|^2 < \varepsilon^2.$$

Now

$$|Ef-Ef_\varepsilon|, \quad |Eg-Eg_\varepsilon| < \varepsilon$$

and

$$\begin{aligned} &\left|\int \mu(dx)\left[f(x)(T^n g)(x) - f_\varepsilon(x)(T^n g_\varepsilon)(x)\right]\right| \\ &\leq \{E|f-f_\varepsilon|^2 E|(T^n g)(x)|^2\}^{\frac{1}{2}} + \{E|T^n(g-g_\varepsilon)(x)|^2 E|f_\varepsilon(x)|^2\}^{\frac{1}{2}} \\ &\leq \varepsilon(\{E|g|^2\}^{\frac{1}{2}} + \{E|f_\varepsilon|^2\}^{\frac{1}{2}}). \end{aligned}$$

Since (36) holds with f_ε, g_ε in place of f, g, respectively, the desired conclusion is obtained by letting $\varepsilon\to 0$. It is also clear that an argument like that given in the case of ergodicity indicates that $P(\cdot,\cdot)$ is mixing with respect to μ if and only if the shift transformation τ is mixing on the infinite product space of a Markov process generated by $P(\cdot,\cdot)$ and μ. The shift transformation is mixing if

$$m(B \cap \tau^n B') \to m(B)m(B')$$

as $n\to\infty$ for any two sets $B, B' \in \mathscr{A}^{\infty}_{-\infty}$ with m the probability measure of a Markov process generated by $P(\cdot,\cdot)$ and μ

3. Transition Operators and Invariant Measures on a Topological Space

In this section we shall consider some conditions which will be enough to insure the existence of an invariant measure for a transition probability operator. The setting will be that of a topological space. Let us first *assume that Ω is a compact Hausdorff space with the Borel field $\mathscr{A}$ the σ-field generated by the topology.* Furthermore, *let the transition* *probability function* $P(\cdot,\cdot)$ *take continuous functions into continuous functions*, that is,

$$(Tf)(x) = \int P(x,dy)\, f(y)$$

is continuous whenever f is continuous. Under these circumstances one is always assured of the existence of invariant probability measures μ.

If T takes continuous functions into continuous functions one can assume that $P(x,\cdot)$ is for each $x \in \Omega$ a regular measure on $\mathscr{A}$ and this in turn implies that $P_n(x,\cdot)$, $n=1,2,\ldots$, will be a regular measure on $\mathscr{A}$ for each x (see Appendix 3).

It will be useful to consider the regular probability measures ν on Ω in the "weak star topology". We shall show that the set of regular probability measures $\mathscr{M}$ on Ω is a compact set by imbedding it in an appropriate product space. Let I be the closed unit interval $[0,1]$ with the usual topology of the real numbers. Consider the product $\prod_f I_f$ with the coordinate spaces I_f indexed by the continuous functions f, $0 \leqslant f \leqslant 1$, on Ω. Then ΠI_f with the usual product topology (see appendix 2) is a compact space by the Tychonov theorem since I_f is compact. Each regular probability measure ν on Ω can be regarded as a point in ΠI_f whose component in the fth coordinate space I_f is $c_f = \int f(s)\, \nu(ds)$. Let $c \in \Pi I_f$ be a limit point of the family of measures $\mathscr{M}$. Now $c_f \geqslant 0$ for all $f \geqslant 0$ and $c_1 = 1$ (1 is the function $f \equiv 1$). Given any two functions f_1, f_2 with $0 \leqslant f_1, f_2, f_1+f_2 \leqslant 1$ and any $\varepsilon > 0$ there is a measure $\mu \in \mathscr{M}$ in the neighborhood

$$\{c' : |c'_{f_1} - c_{f_1}| < \varepsilon,\ |c'_{f_2} - c_{f_2}| < \varepsilon,\ |c'_{f_1+f_2} - c_{f_1+f_2}| < \varepsilon\}$$

of c. Since this is true for all $\varepsilon > 0$

$$c_{f_1+f_2} = c_{f_1} + c_{f_2}.$$

One can show that $c_{\alpha f} = \alpha c_f$ for $\alpha \geqslant 0$ if $0 \leqslant \alpha f, f \leqslant 1$ in a similar manner. Also $|c_f| \leqslant \|f\|$ if $0 \leqslant f \leqslant 1$ where $\|\cdot\|$ is the supremum norm. Let $c_f = L(f)$. Then $L(f)$ is a functional on the continuous functions f, $0 \leqslant f \leqslant 1$. The functional L can be extended to all continuous f as follows. First assume $f \geqslant 0$. Then there is a positive constant α such that $0 \leqslant \alpha f \leqslant 1$. Set $L(f) = L(\alpha f)/\alpha$. $L(f)$ is now defined for all continuous nonnegative f. If f is not nonnegative set $f^+ = \max(0, f)$ and $f^- = f - f^+$. The functions f^+, $-f^-$ are nonnegative and continuous. Set $L(f) = L(f^+) - L(-f^-)$. The functional L is linear since $L(\alpha f) = \alpha L(f)$, $L(f_1+f_2) = L(f_1) + L(f_2)$. Because $L(f) \geqslant 0$ for $f \geqslant 0$ it follows that

$$\begin{aligned} |L(f)| &= |L(f^+) - L(-f^-)| \leqslant \max(L(f^+),\, L(-f^-)) \\ &\leqslant \max(\|f^+\|,\, \|-f^-\|) = \|f\| \end{aligned}$$

for all continuous f and therefore $\|L\| \leqslant 1$. The Riesz representation theorem implies that there is a regular probability measure μ on Ω (since $L(1)=1$) such that

$$L(f) = \int f(s)\,\mu(ds).$$

Thus $c \in \mathcal{M}$ and $\mathcal{M}$ as a point set in $\prod_f I_f$ is closed. $\mathcal{M}$ is compact because it is a closed subset of the compact set ΠI_f. We have derived the following Lemma.

Lemma 1. *Let $\mathcal{M}$ be the set of regular probability measures on the compact Hausdorff space Ω. Let $\prod_f I_f$ be the product space (with the usual product topology) whose coordinate spaces $I_f = [0,1]$ are indexed by the continuous functions f, $0 \leqslant f \leqslant 1$, on Ω. Identify the measure $v \in \mathcal{M}$ with the point $c = (c_f) \in \Pi I_f$ where $c_f = \int f(s)\,v(ds)$. Then $\mathcal{M}$ is a compact subset of $\prod_f I_f$.*

Given a point $x \in \Omega$, consider the nonnegative linear functionals

$$L_n(f;x) = \frac{1}{n} \sum_{k=1}^{n} (T^k f)(x), \qquad n = 1, 2, \ldots, \tag{1}$$

on the continuous functions on Ω. Since $L_n(1;x) = 1$, these functionals correspond to regular probability measures v_n on Ω, that is, $L_n(f;x) = \int v_n(dy)\,f(y)$. Consider any limit point μ of the set of measures $\{v_n\}$ in the weak star topology. By Lemma 1 it is clear that the set of such limit points μ is not empty and that they are all regular probability measures on Ω. We wish to show that any such limit point μ is an invariant probability measure for $P(\cdot,\cdot)$. Given any $\varepsilon > 0$ and any continuous $f, 0 \leqslant f \leqslant 1$, there is an n sufficiently large so that

$$|L_n(f;x) - \int f(y)\,\mu(dy)| < \varepsilon$$
$$|L_n(Tf;x) - \int (Tf)(y)\,\mu(dy)| < \varepsilon$$
$$|L_n(Tf;x) - L_n(f;x)| < \varepsilon.$$

Since this is true for every $\varepsilon > 0$ it follows that

$$\int f(y)\,\mu(dy) = \int (Tf)(y)\,\mu(dy)$$

and this is valid for every continuous f. We then immediately have

$$\mu(A) = \int \mu(dx)\,P(x,A)$$

for all $A \in \mathcal{A}$.

Theorem 1. *Let Ω be a compact Hausdorff space and $P(\cdot,\cdot)$ a transition probability operator taking continuous functions on Ω into continuous functions. Then for each $x \in \Omega$, the sequence of linear functionals $L_n(f;x)$, $n=1,2,\ldots$, (1) has a nonvacuous set of limit points in the weak star topology. Each of these limit points is a regular probability measure μ that is invariant with respect to $P(\cdot,\cdot)$.*

This theorem has an interesting simple application to a question concerning the existence of stationary processes with two states (say 0 and 1) and a given transition mechanism. Consider a stationary process $\{X_n; n=0,\pm 1,\ldots\}$ whose state space consists only of the two symbols 0 and 1. By stationarity we mean that

$$P(X_{n_1}=i_1,\ldots,X_{n_k}=i_k)=P(X_{n_1+m}=i_1,\ldots,X_{n_k+m}=i_k)$$

for all k-tuples $n_1,\ldots,n_k$ of integers and k-tuples $i_1,\ldots,i_k$ of 0's and 1's where m is any integral shift. This is just invariance of the probability structure under time shifts. Such processes are rarely Markovian. However, by redefining the state space one can treat this question in a Markovian framework. Suppose we consider a new process $\{Y_n; n=0,\pm 1,\ldots\}$ with $Y_n=X_n X_{n-1}\ldots$. The information carried by the new process is equivalent to the information carried by the old process $\{X_n\}$. The new state space consists of one-sided sequences of zeros and ones which are the present and the past of the original process $\{X_n\}$. The new process $\{Y_n\}$ is obviously Markovian and if $Y_n=x_n x_{n-1}\ldots$, then Y_{n+1} must be either $0x_n x_{n-1}\ldots$ or $1x_n x_{n-1}\ldots$. Given the stationary process $\{X_n\}$, the conditional probability

$$P\{X_{n+1}=0 \mid X_n X_{n-1}\ldots=y=x_n x_{n-1}\ldots\}=p(y)$$

is well-defined and measurable with respect to the Borel field $\mathscr{B}_n$ generated by X_j, $j\leqslant n$. Clearly,

$$0\leqslant p(y)\leqslant 1$$

and

$$P\{X_{n+1}=1 \mid X_n X_{n-1}\ldots=y\}=1-p(y).$$

Now $p(y)$ can be regarded as yielding a transition operator for the Markov process $\{Y_n\}$,

$$\begin{aligned} &P\{Y_{n+1}=0x_n x_{n-1}\ldots \mid Y_n=x_n x_{n-1}\ldots=y\}\\ &\qquad=p(y)=1-P\{Y_{n+1}=1x_n x_{n-1}\ldots \mid Y_n=y\}. \end{aligned}$$

The existence of a stationary process $\{Y_n\}$ (or $\{X_n\}$) with this transition function is equivalent to the existence of an invariant measure for this transition function. Sufficient conditions can be given for the existence of an invariant measure by using Theorem 1. Consider the set $I=\{0,1\}$ consisting of 0 and 1. Construct the product space $\prod_{j\leqslant n} I_j$ where $I_j=I$ with the product topology. The product space is a compact topological space with y a typical point of this space. If $p(y)$ is a continuous function of y on this compact space, the transition function generated by $p(y)$ takes continuous functions into continuous functions and Theorem 1 implies that invariant regular probability measures exist.

Under appropriate conditions one can show that a sigma-finite invariant measure exists for a transition probability function taking continuous functions into continuous functions on a locally compact Hausdorff space. The result we give is due to Foguel [28] and is obtained by reducing the problem to one covered by Theorem 1 on a compact Hausdorff space. Of course, the invariant measure will generally not be a probability measure. Basically, the argument assumes the existence of a compact set B that will be hit with probability one from any initial point in the space. However, one does not immediately look at the process on B because the property of taking continuous functions into continuous functions might not be preserved. The process of reducing the problem to the case treated in Theorem 1 is carried out somewhat more circuitously so as to retain the property that an appropriately defined transition operator takes continuous functions on a compact space into continuous functions. Let Ω now be a locally compact Hausdorff space with the Borel field $\mathscr{A}$ generated by the topology. *Assume that the transition probability function $P(\cdot,\cdot)$ takes bounded continuous functions f into bounded continuous functions Tf*:

$$(Tf)(x)=\int P(x,dy)f(y).$$

Let I_A be the indicator function of the set $A\in\mathscr{A}$:

$$I_A(x)=\begin{cases}1 & \text{if } x\in A\\ 0 & \text{otherwise}\end{cases}$$

and 1 be the function identically equal to the number 1. Now *assume that there is a compact set B such that*

$$\begin{aligned}&(I_{B^c}TI_{B^c})^n 1(x)\\ &=I_{B^c}(x)\int_{B^c}P(x,dx_1)\int_{B^c}\cdots\int_{B^c}P(x_{n-2},dx_{n-1})P(x_{n-1},B^c)\to 0 \quad \text{as } n\to\infty\end{aligned} \tag{2}$$

for all x. This is just the statement that the probability of always remaining outside of B given that one starts outside of B is zero. This immediately implies

$$(TI_{B^c})^n 1(x)\to 0 \tag{3}$$

as $n\to\infty$ for all x since $(TI_{B^c})^n 1(x)=(I_{B^c}TI_{B^c})^n 1(x)+(I_B T)(I_{B^c}TI_{B^c})^{n-1}1(x)$ and both terms tend to zero by (2). Furthermore, under assumption (2) it is clear that

$$\sum_{n=1}^{\infty}(T^n I_B)(x)>0 \tag{4}$$

for every $x\in\Omega$. For if there were a point x_0 with $(T^n I_B)(x_0)=0$ for $n=1,2,\ldots$ it would immediately follow that $(TI_{B^c})^n 1(x_0)=1$ for $n=1,2,\ldots$ contrary to (3).

Let $\beta, 0\leqslant\beta\leqslant 1$, be a continuous function equal to one on B and equal to zero outside $\bar{B}$ where $\bar{B}$ is a compact set containing B. Set $\alpha=1-\beta$. We now introduce the operator

$$T_N=\sum_{n=0}^{N}(T\alpha)^n T\beta.$$

This operator takes bounded measurable functions on $\bar{B}$ into bounded measurable functions on $\bar{B}$. Also, $(T_N 1)(x)$ is an upper bound for the probability of hitting B up to time N given that one starts at x. There are a number of obvious properties that T_N has that will be of some importance. If f is a bounded continuous function, it is clear $T_N f$ is also bounded and continuous. Furthermore, T_N is a positive operator, that is, $T_N f\geqslant 0$ if $f\geqslant 0$. The sequence T_N is nondecreasing in the sense that

$$T_{N+1}f\geqslant T_N f$$

if $f\geqslant 0$. Notice that

$$T_N I_{\bar{B}}=\sum_{n=0}^{N}(T\alpha)^n T\beta=\sum_{n=0}^{N}(T\alpha)^n(1-T\alpha)=T1-(T\alpha)^{N+1}1\leqslant 1.$$

Lemma 2. *Let $P(\cdot,\cdot)$ be a transition probability operator on a locally compact Hausdorff space that takes bounded continuous functions into bounded continuous functions and satisfies* (2). *Then the sequence of derived operators T_N acting on the continuous functions on $\bar{B}$ converges uniformly to a limiting operator T_∞. T_∞ is a positive operator taking continuous functions on $\bar{B}$ into continuous functions on $\bar{B}$ and $T_\infty 1=1$.*

First of all, $T_\infty 1=\lim_{N\to\infty}T_N 1=1-\lim_{N\to\infty}(T\alpha)^{N+1}1=1$ since

$$(T\alpha)^N 1\leqslant(TI_{B^c})^N 1\to 0$$

by (3). Now $(T\alpha)1\leqslant 1$, so that $(T\alpha)^N 1$ converges monotonically to zero on $\bar{B}$. The convergence must therefore be uniform since $\bar{B}$ is compact.

If $f \geqslant 0$ it follows that

$$\|T_{N+m}f - T_N f\| = \left\| \sum_{k=N+1}^{N+m} (T\alpha)^k T\beta f \right\| \leqslant \|f\| \cdot \|T_{N+m}1 - T_N 1\| \to 0$$

as $N \to \infty$. The positivity of T_∞ follows immediately from the positivity of the operators T_N.

It is now clear that the limit operator T_∞ corresponds to a transition probability function (say P_∞) with state space $\bar{B}$ and takes continuous functions on $\bar{B}$ into continuous functions on $\bar{B}$. By Theorem 1 there is an invariant regular probability measure μ for P_∞. Let

$$\lambda = \sum_{n=0}^{\infty} \mu(T\alpha)^n.$$

Theorem 2. *Under the assumptions of Lemma 2 λ is a sigma-finite nontrivial invariant measure for the transition probability function $P(\cdot,\cdot)$ on Ω and agrees with μ on B.*

First λ will be shown to be sigma-finite. Let

$$H = \{f : f \geqslant 0,\ \textstyle\int f(x)\lambda(dx) < \infty\}.$$

We show that $T^k\beta \in H$, $k = 1, 2, \ldots$. If $k = 1$ then

$$\int \sum_{n=0}^{\infty} \mu(T\alpha)^n(dx)(T\beta)(x) = \int \mu(dx) = 1 \tag{5}$$

since $\beta = 1 - \alpha$. The argument now continues by induction. Assume $T^k\beta \in H$. Then

$$\begin{aligned}\int \sum_{n=0}^{\infty} \mu(T\alpha)^n(dx)(T^{k+1}\beta)(x) \leqslant & \int \sum \mu(T\alpha)^{n+1}(T^k\beta)(x) \\ & + M \int \sum \mu(T\alpha)^n(dx)(T\beta 1)(x)\end{aligned} \tag{6}$$

where $M = \sup(T^k\beta)(x)$ since $\alpha + \beta = 1$. The first term on the right of (6) is finite by induction while the second on the right equals M by (5). Therefore λ is sigma-finite on

$$\bigcup_{k=1}^{\infty} \{x : (T^k\beta)(x) > 0\} \supset \bigcup_{k=1}^{\infty} \{x : (T^k I_B)(x) > 0\}$$

and by (4) sigma-finite on all of Ω.

Notice that if f is zero outside B then $(T\alpha)^n f = 0$, $n = 1, 2, \ldots$, and $\int \lambda(dx) f(x) = \int \mu(dx) f(x)$ so that λ agrees with μ on B. Now

$$\lambda T = \sum \mu(T\alpha)^n T = \sum \mu(T\alpha)^n T\alpha + \sum \mu(T\alpha)^n T\beta = \lambda - \mu + \sum \mu(T\alpha)^n T\beta.$$

Since $\mu = \sum \mu(T\alpha)^n T\beta$ it follows that

$$\lambda T = \lambda .$$

Any compact set is covered by the collection of open sets $\{x:(T^m \beta)(x) > 1/n\}$, $m, n = 1, 2, \ldots$, since the whole space Ω is covered by these sets. The compactness implies that the set is covered by a finite subcollection and this in turn by (6) indicates that every compact set has finite λ-measure.

4. Asymptotic Behavior of Powers of a Transition Probability Operator

Let Q be a σ-finite invariant measure for the transition function $P(\cdot,\cdot)$. It has already been noted (see section 1.1) that the induced operator T

$$(Tf)(x) = \int P(x, dy) f(y)$$

acting on $L^2(dQ)$ is a positive contraction, that is, $\|Tf\|_2 \leqslant \|f\|_2$. If Q is a finite measure the function one is in $L^2(dQ)$ and $T1 = 1$. The action of powers T^n, $n = 0, 1, \ldots$, of T on functions of $L^2(dQ)$ can be given a simple and interesting interpretation as a prediction or approximation problem. Let m be the stationary measure on sequences $(\ldots, x_{-1}, x_0, x_1, \ldots)$ generated by invariant measure Q and the transition operator $P(\cdot,\cdot)$. Consider a function $f(x_n)$ at time $n > 0$ in $L^2(dQ)$ that we wish to approximate by a predictor $g(x_k, k \leqslant 0)$ measurable with respect to $\mathscr{A}^0_{-\infty}$. A natural measure of approximation is given by the mean square error

$$\int |g(x_k, k \leqslant 0) - f(x_n)|^2 \, dm .$$

The Markovian character of the process (a generalized process if Q is infinite) implies that g can be taken to be a function of x_0 alone. If g is in $L^2(dQ)$

$$\begin{aligned} \int |g(x_0) - f(x_n)|^2 \, dm &= \int |g(x_0) - f(x_n)|^2 \, Q(dx_0) P_n(x_0, dx_n) \\ &= \int |g(x_0)|^2 \, Q(dx_0) - 2 \int g(x_0) \int P_n(x_0, dx_n) f(x_n) Q(dx_0) \\ &\quad + \int |f(x_n)|^2 \, Q(dx_n) = \|g - T^n f\|^2 + \|f\|^2 - \|T^n f\|^2 . \end{aligned}$$

The best predictor is given by $g(x_0) = (T^n f)(x_0)$. This is the conditional expected value of $f(x_n)$ given $\mathscr{A}^0_0$ (or $\mathscr{A}^0_{-\infty}$ using the Markov property again). A brief discussion of the conditional expectation as the best approximator in mean square is given in Appendix 1. Notice that the error of this n step prediction is

$$\|f\|^2 - \|T^n f\|^2 .$$

Let us consider the set of functions $\mathscr{U}_j$ of $L^2(dQ)$ for which T^j is norm-preserving, that is

$$\int |g(x)|^2 Q(dx) = \int |(T^j g)(x)|^2 Q(dx)$$

for $g \in \mathscr{U}_j$. If $f \in L^2(dQ)$

$$|(T^j f)(x)|^2 = \left|\int P_j(x,dy) f(y)\right|^2 \leqslant \int P_j(x,dy) |f(y)|^2 = (T^j |f|^2)(x).$$

Thus for $g \in \mathscr{U}_j$

$$(T^j |g|^2)(x) = |(T^j g)(x)|^2 \tag{1}$$

for almost all x with respect to Q. Given any x for which (1) holds, it follows that $g(y)$ is constant for almost all y with respect to $P_j(x,\cdot)$. Thus, $\mathscr{U}_j$ consists precisely of those functions with the following property: *for almost all $x(dQ)$, $g(\cdot)$ is constant for almost all y with respect to $P_j(x,\cdot)$.* $\mathscr{U}_j$ is a closed linear space of functions. Furthermore, the essentially bounded functions in $\mathscr{U}_j$ form an algebra since the product of any two essentially bounded functions in $\mathscr{U}_j$ is an essentially bounded function in $\mathscr{U}_j$.

Lemma 1. *The set $\mathscr{U}_j$ of functions in $L^2(dQ)$ whose norm is preserved by T^j is a closed linear space. The essentially bounded functions of $\mathscr{U}_j$ are an algebra.*

It immediately follows that there is a Borel field $\mathscr{C}_j$ of sets induced by $\mathscr{U}_j$. $\mathscr{U}_j$ is the set of square integrable functions (with respect to Q) that are measurable (modulo a set of Q measure zero) with respect to $\mathscr{C}_j$. Clearly $\mathscr{U}_1 \supset \mathscr{U}_2 \supset \cdots$ and $\mathscr{C} = \mathscr{C}_0 \supset \mathscr{C}_1 \supset \cdots$. Let

$$\mathscr{U}_\infty = \bigcap_{j=1}^{\infty} \mathscr{U}_j, \qquad \mathscr{C}_\infty = \bigcap_{j=1}^{\infty} \mathscr{C}_j.$$

Notice that $\mathscr{U}_j$ can also be described as the set of square integrable (with respect to Q) $\mathscr{A}_0^0$-measurable functions that are predictable j steps ahead in time without any error in mean square.

The transition operator T induces a measure-preserving set transformation on the sets of $\mathscr{C}_{j+1}$ taking the sets of $\mathscr{C}_{j+1}$ into $\mathscr{C}_j$, $j=0,1,\ldots$. For if $B \in \mathscr{C}_{j+1}$ with finite Q-measure

$$(T|I_B|^2) = (T I_B)(x) = |T I_B(x)|^2$$

so that $(T I_B)(x)$ is the indicator function of a set we shall call τB. Furthermore,

$$Q(\tau B) = \int (T I_B)(x) Q(dx) = \int I_B(x) Q(dx) = Q(B).$$

Clearly, $\tau B \in \mathscr{C}_j$ since T takes $\mathscr{U}_{j+1}$ into $\mathscr{U}_j$. If B is in $\mathscr{C}_j$ but not in $\mathscr{C}_{j+1}$ we have

$$T^k I_B = I_{\tau^k B}, \qquad k=0,1,\ldots,j$$

but the $(j+1)^{\mathrm{st}}$ application of T no longer acts as a set transformation. Consider the action of T on the characteristic functions of sets in $\mathscr{C}_\infty$. Then

$$T^k I_B = I_{\tau^k B}, \qquad k=0,1,2,\ldots$$

if $B \in \mathscr{C}_\infty$ with $\tau^k B \in \mathscr{C}_\infty$ for $k=0,1,2,\ldots$. Notice that the sets $B \in \mathscr{C}_\infty$ correspond to a class of events in the backward tail field of the stationary Markov process determined by the transition function $P(\cdot,\cdot)$ and the invariant measure $Q(\cdot)$.

Since $Q(\cdot)$ is an invariant measure we can consider the operator T^* adjoint to T

$$\int (Tf)(x)g(x)Q(dx) = \int f(x)(T^* g)(x)Q(dx)$$

for $f, g \in L^2(dQ)$. The adjoint operator T^* corresponds to the backward transition function $P^*(\cdot,\cdot)$ of the stationary Markov process determined by T and Q. The set $\mathscr{U}_j^*$ of functions in $L^2(dQ)$ whose norm is preserved by T^{*j} is a closed linear space consisting of the square integrable functions measurable with respect to some Borel field $\mathscr{C}_j^*$

$$\mathscr{U}_1^* \supset \mathscr{U}_2^* \supset \cdots$$

$$\mathscr{C} = \mathscr{C}_0^* \supset \mathscr{C}_1^* \supset \cdots .$$

Set

$$\mathscr{U}_\infty^* = \bigcap_{j=1}^{\infty} \mathscr{U}_j^*$$

$$\mathscr{C}_\infty^* = \bigcap_{j=1}^{\infty} \mathscr{C}_j^* .$$

T^* induces a measure-preserving set transformation τ^* taking the sets of $\mathscr{C}_{j+1}^*$ into $\mathscr{C}_j^*$, $j=0,1,\ldots$, and the sets of $\mathscr{C}_\infty^*$ into $\mathscr{C}_\infty^*$. The sets $B \in \mathscr{C}_\infty^*$ correspond to a class of events in the forward tail field of the stationary Markov process determined by T and $Q(\cdot)$. The Borel field

$$\overline{\mathscr{C}} = \mathscr{C}_\infty \cap \mathscr{C}_\infty^*$$

is of special interest because the set of functions $\overline{\mathscr{U}}$ square integrable and measurable with respect to $\overline{\mathscr{C}}$ consists precisely of those functions whose norm is preserved by both T^k and T^{*k}, $k=1,2,\ldots$. τ acts as a measure-preserving invertible transformation on the sets of $\overline{\mathscr{C}}$, that is

$$\tau\tau^* B = \tau^*\tau B = B$$

for $B \in \overline{\mathscr{C}}$ with

$$Q(\tau B) = Q(\tau^* B) = Q(B).$$

Lemma 2. *$\tau(\tau^*)$ acts as a measure-preserving transformation taking the sets B of $\mathscr{C}_{j+1}(\mathscr{C}^*_{j+1})$ into sets $\tau B(\tau^* B)$ of $\mathscr{C}_j(\mathscr{C}^*_j)$, $j=0,1,2,\ldots$. τ acts as a measure preserving invertible transformation on the sets of $\overline{\mathscr{C}}$ with inverse $\tau^{-1}=\tau^*$.*

The set of functions $f\in L^2(dQ)$ such that

$$\int |T^j f|^2\, Q(dx) \to 0$$

as $j\to\infty$ are of some interest. Call this set of functions $\mathscr{D}$. $\mathscr{D}$ is a closed linear space by the Schwarz inequality. Let $g\in\mathscr{D}$ and $h\in\mathscr{U}_\infty$. By the characterization of functions g in $\mathscr{U}_\infty$ it follows that

$$\begin{aligned} T^j(g\bar{h})(x) &= \int P_j(x,dy)g(y)\bar{h}(y) = \int P_j(x,dy)g(y)\int P_j(x,dy)\bar{h}(y) \\ &= (T^j g)(x)(T^j\bar{h})(x) \end{aligned}$$

for almost all x with respect to Q. Therefore

$$\begin{aligned} \left|\int g(x)\bar{h}(x)Q(dx)\right| &= \left|\int (T^j g\bar{h})(x)Q(dx)\right| = \left|\int (T^j g)(x)(T^j\bar{h})(x)Q(dx)\right| \\ &\leqslant \|T^j g\|\cdot\|T^j\bar{h}\| \to 0 \end{aligned}$$

as $j\to\infty$ so that $\mathscr{U}_\infty$ is orthogonal to $\mathscr{D}$. $\mathscr{D}^*$ is the set of functions $f\in L^2(dQ)$ such that

$$\int |(T^{*j}f)(x)|^2\, Q(dx)\to 0$$

as $j\to\infty$ and in the same manner one can show that $\mathscr{D}^*$ is orthogonal to $\mathscr{U}^*_\infty$.

For many Markov transition operators $L^2(dQ)$ is precisely the Hilbert space generated by the two orthogonal spaces $\mathscr{U}_\infty$ and $\mathscr{D}$. We shall try to characterize this situation. The operators $T^{*k}T^k$, $k=1,2,\ldots$, on $L^2(dQ)$ are a nonincreasing sequence of positive semidefinite operators and therefore converge as $k\to\infty$ to a positive semidefinite operator $M=\lim_{k\to\infty} T^{*k}T^k$.

Lemma 3. *Let Q be an invariant measure of the transition probability operator T. $L^2(dQ)$ is generated by $\mathscr{U}_\infty$ and $\mathscr{D}$ if and only if $M=\lim_{k\to\infty} T^{*k}T^k$ is a projection.*

Suppose that $\mathscr{U}_\infty$ and $\mathscr{D}$ generate $L^2(dQ)$. It is then clear that M is the projection operator leaving the functions of $\mathscr{U}_\infty$ invariant and annihilating the functions of $\mathscr{D}$. Conversely if M is a projection the functions left invariant by M are elements of $\mathscr{U}_\infty$ and the functions annihilated by M are elements of $\mathscr{D}$. The space $L^2(dQ)$ is then the direct sum of $\mathscr{U}_\infty$ and $\mathscr{D}$.

It is interesting to look at a transition probability operator irreducible with respect to Q, that is, an operator $P(\cdot,\cdot)$ for which there is no set A with $Q(A)$, $Q(A^c)>0$ and such that $P(x,A)=1$ for almost all $x\in A$. Given any A with $Q(A)>0$ it follows that for almost every x there is an integer $n=n(x)>0$ such that $P_n(x,A)>0$. For if the set $C=\{x: P_j(x,A)=0, j=1,2,\ldots\}$ has positive measure we are led to a contradiction since $P(x,C)=1$ for almost all $x\in C$ and C^c has positive measure. Thus, if $Q(A)>0$, then $\sum_j P_j(x,A)>0$ almost everywhere.

Given any set B with $Q(B)>0$, the corresponding measures

$$\int_B Q(dx) P_j(x,\cdot) \qquad j=1,2,\ldots$$

are absolutely continuous with respect to $Q(\cdot)$. Suppose there is a set A with $Q(A)<\infty$ such that

$$\sum_j P_j(x,A)=\infty \tag{2}$$

on a set C of positive Q measure. Then (2) must diverge almost everywhere since for any event B with $Q(B)>0$

$$\infty = \int_B Q(dx) \int_C P_n(x,dy) \sum_j P_j(y,A) \leqslant \int_B Q(dx) \sum_j P_j(x,A)$$

for some positive integer n.

Call an irreducible transition operator $P(\cdot,\cdot)$ with respect to Q *transient* if

$$\sum_1^\infty P_j(x,A)<\infty$$

almost everywhere for each set A with $Q(A)<\infty$ and call it *recurrent* if

$$\sum_1^\infty P_j(x,A)=\infty$$

almost everywhere for each set A with $0<Q(A)<\infty$. If $P(\cdot,\cdot)$ is irreducible transient and $Q\not\equiv 0$, we must have $Q(\Omega)=\infty$. Otherwise, by bounded convergence one would have

$$Q(A)=\int Q(dx)\frac{1}{n}\sum_1^n P_j(x,A)\to 0$$

as $n\to\infty$ for every measurable set A.

Let us now consider whether there can be any $f\not\equiv 0$ in $\mathscr{U}_\infty$ when $P(\cdot,\cdot)$ is irreducible transient. If there is such an f, we can construct a set A, $0<Q(A)<\infty$, such that $(T I_{\tau^k A})(x)=I_{\tau^{k+1}A}(x)$, $k=1,2,\ldots$, where I_B is the indicator function of the set B and τ is a measure-preserving transformation on the sequence $\tau^k A$. We shall show that the irreducibility of $P(\cdot,\cdot)$ implies that $\sum_1^\infty (T^k I_A)(x) = \sum_1^\infty I_{\tau^k A}(x) = \infty$ almost everywhere which contradicts the assumption of transience. Irreducibility implies that

$$\sum_{k=j+1}^{\infty} I_{\tau^k A}(x) \geq 1$$

and

$$\lim_{j\to\infty} \sum_{k=l}^{j} I_{\tau^k A}(x) \geq 1$$

almost everywhere. Proceeding by induction we assume that

$$\sum_{k=j+1}^{\infty} I_{\tau^k A}(x) \geq r$$

almost everywhere for some positive integer r. It then follows that

$$\sum_{k=l}^{\infty} I_{\tau^k A}(x) \geq r+1$$

almost everywhere. Therefore, $\sum I_{\tau^k A}(x) = \infty$ almost everywhere and we have the following Lemma.

Lemma 4. *If $P(\cdot,\cdot)$ is irreducible transient with respect to the invariant measure Q, $\mathcal{U}_\infty$ is trivial.*

Thus, the situation mentioned in Lemma 3 arises for an irreducible transient transition operator if and only if $\mathcal{D}$ is all of $L^2(dQ)$, or equivalently, $M=0$. Otherwise there are elements $f \in L^2(dQ)$ such that $\|T^n f\| \downarrow c > 0$. We shall give simple examples of irreducible transient chains to indicate that each of these two alternatives can arise.

Consider an irreducible Markov chain with transition matrix P and invariant measure $q=(q_i)$, $\sum q_i = \infty$. The chain is either null recurrent or transient. Consider a function $f \in l^2(q)$. Then

$$\|T^n f\|^2 = \sum_i \Big|\sum_j p_{i,j}^{(n)} f_i\Big|^2 q_i \leq \sum_i \sum_j p_{i,j}^{(n)} |f_j|^2 q_i = \sum q_i |f_i|^2 = \|f\|^2 .$$

Furthermore, $\Big|\sum_j p_{i,j}^{(n)} f_j\Big|^2 < \sum_j p_{i,j}^{(n)} |f_j|^2$ unless $f \equiv c$ on the set of points j for which $p_{i,j}^{(n)} > 0$. Since we can't have equality for all such i and n without having $f \equiv 0$ because of the irreducibility of the chain and the fact that $\sum q_i = \infty$, it follows that for some sufficiently large $n = n(f)$, $\|T^n f\| < \|f\|$ if $f \not\equiv 0$. However, for small enough n one may have $\|T^n f\| = \|f\|$.

We wish to investigate conditions under which $\|T^n f\| \downarrow 0$ as $n\to\infty$ for each $f\in l^2(q)$. This is equivalent to having

$$\sum_i q_i |p_{i,j}^{(n)}|^2 \downarrow 0$$

as $n\to\infty$ for each j since T is a contraction operator. *A sufficient condition for this is that*

$$\sup_i p_{i,j}^{(n)} \downarrow 0 \tag{3}$$

for some j. Then

$$\sum_i q_i |p_{i,j}^{(n)}|^2 \leqslant \sup_i p_{i,j}^{(n)} \sum_i q_i p_{i,j}^{(n)} = q_j \sup_i p_{i,j}^{(n)} \downarrow 0.$$

Furthermore, given any $j' \neq j$, since the chain is irreducible there is an m such that $p_{j',j}^{(m)}>0$. Hence, $p_{i,j'}^{(n)} p_{j',j}^{(m)} \leqslant p_{i,j}^{(n+m)}$. From this it follows that $\sup_i p_{i,j'}^{(n)} \downarrow 0$ as $n\to\infty$ for all j'.

Consider a random walk on the integers with spatially homogeneous transition mechanism so that $p_{i,j}=p_{j-i}$. If the random walk is irreducible there are at least two integers j for which $p_j>0$. Let $\varphi(\theta)=\sum_j p_j e^{ij\theta}$ be the generating function of the probability distribution $\{p_j\}$. The irreducibility of the random walk implies that $|\varphi(\theta)|<1$ for all $\theta\neq 0$ in $(-\pi,\pi]$. Since the $p_j^{(n)}$'s are the Fourier coefficients of $[\varphi(\theta)]^n$, we have

$$\sup_j p_j^{(n)} \leqslant \frac{1}{2\pi}\int_{-\pi}^{\pi} |\varphi(\theta)|^n \, d\theta \to 0$$

as $n\to\infty$. Condition (3) is satisfied. Thus, for an irreducible spatially homogeneous random walk on the integers, $\mathscr{D}=l^2(q)$ for arbitrary invariant measure q. Notice that the uniform measure $q_i\equiv 1$ is always an invariant measure for such a random walk.

We now consider a simple spatially inhomogeneous random walk on the integers which provides an example of an irreducible transient Markov chain for which there are $f\in l^2(q)$ (q is the invariant measure) such that $\|T^n f\| \downarrow c>0$ as $n\to\infty$. For such chains $\mathscr{D}$ is not all of $l^2(q)$. Let

$$p_{i,j}=\begin{cases} 1-\alpha_i & \text{if } j=i+1 \\ \alpha_i & \text{if } j=i-1 \\ 0 & \text{otherwise} \end{cases}$$

with $i, j=0, \pm 1, \ldots$ and $0<\alpha_i<1$, $2\sum \alpha_i<1$. The desired invariant measure is given by

$$q_j=\sum_{s=0}^{\infty} \beta_j^{(s)}, \quad \beta_j^{(0)} \equiv 1, \quad \beta_j^{(s+1)}=\alpha_j \beta_j^{(s)}+\alpha_{j+1} \beta_{j+1}^{(s)}$$

where $s=0,1,\ldots$. If $c=\sup \alpha_i$ we have $\beta_j^{(k)} \leqslant (2c)^k$, $k=1,2,\ldots$ and hence $1 \leqslant q_j \leqslant 1/(1-2c)$. Notice that $q_j=1+\alpha_j q_j+\alpha_{j+1} q_{j+1}$ so that

$$q_{j-1}(1-\alpha_{j-1})+q_{j+1}\alpha_{j+1}=$$

$$1+\alpha_{j-1}q_{j-1}+\alpha_j q_j-\alpha_{j-1}q_{j-1}+\alpha_{j+1}q_{j+1}=q_j.$$

Moreover, since $q_j(1-\alpha_j)=1+\alpha_{j+1}q_{j+1}$ it follows that

$$\lim_{|j|\to\infty} q_j=1.$$

We now construct a function $f \in l^2(q)$ such that $\|T^n f\| \downarrow a>0$. Since $\sum \alpha_i<\infty$,

$$p_{i-k,i}^{(k)}=\prod_{j=1}^{k}(1-\alpha_{i-j}) \qquad k=1,2,\ldots \tag{4}$$

and $p_{i-k,i}^{(k)} \to c>0$ as $k\to\infty$. From (4) it is clear that an example of such an f is provided by taking $f_i=1$, $f_j=0$ for $j \neq i$.

We shall now show that this phenomenon cannot occur for a null recurrent irreducible Markov chain.

Theorem 1. *In the case of a null recurrent Markov chain with invariant measure q, $\mathscr{D}=l^2(q)$.*

The theorem is proved by assuming the existence of an $f \in l^2(q)$ such that $\|T^n f\| \downarrow c>0$ as $n\to\infty$ and showing that this leads to a contradiction. The existence of such an f implies that

$$\sum_j q_j |p_{j,0}^{(n)}|^2 \downarrow c'>0$$

as $n\to\infty$. Then for some $\alpha>0$ the sets $S_n=\{j \mid p_{j,0}^{(n)} \geqslant \alpha\}$ are such that $q(S_n) \geqslant \beta>0$. For $j \in S_n$ consider $p_{j,0}^{(n)}=f_{j,0}^{(n)}+\sum_{k=1}^{n-1} f_{j,0}^{(k)} p_{0,0}^{(n-k)}$ where $f_{j,0}^{(k)}$ is the probability of going from $j \neq 0$ to 0 for the first time in precisely k steps. The null recurrence of the Markov chain implies that $p_{0,0}^{(n)} \to 0$ as $n\to\infty$ and hence $\sum_{k=n-m}^{n} f_{j,0}^{(k)} \geqslant \gamma>0$ for all $j \in S_n$ where γ is appropriately chosen and m is a fixed nonnegative integer independent of n. Notice that $\sum_{k=1}^{\infty} f_{j,0}^{(k)}=1$ because of the recurrence and irreducibility of the chain. These last inequalities are understood to hold for all n sufficiently large. Let $S_n^{(r)}$, $r=0,1,\ldots,m$ be the subsets of S_n consisting of those points $j\in S_n$ such that $f_{j,0}^{(n-r)}\geqslant\delta(=\gamma/4m)>0$ respectively. Since the union of these sets is S_n, it follows there is a fixed r such that for an infinite sequence of subscripts n, $q(S_n^{(r)})\geqslant\varepsilon>0$, where ε is a fixed small positive number. One can normalize the invariant distribution so that

$$q_j={}_0p_{0,j}^{*}=\sum_{k=1}^{\infty}{}_0p_{0,j}^{(k)}.$$

Since $\sum_{j\in S_n^{(r)}}{}_0p_{0,j}^{(k)}f_{j,0}^{(n-r)}\leqslant f_{0,0}^{(n-r+k)}$, we have

$$0<\delta\varepsilon\leqslant\sum_{j\in S_n^{(r)}}q_jf_{j,0}^{(n-r)}=\sum_k\sum_{j\in S_n^{(r)}}{}_0p_{0,j}^{(k)}f_{j,0}^{(n-r)}\leqslant\sum_k f_{0,0}^{(n-r+k)}$$

for an infinite number of n's. This states that the probability of returning to zero for the first time in more than n steps is bounded away from zero as $n\to\infty$, which is clearly a contradiction.

It is now natural to ask whether there are examples of irreducible transition functions with a finite invariant measure Q such that $L^2(dQ)$ is not generated by $\mathscr{U}_\infty$ and $\mathscr{D}$. The Markov process generated by such a transition function and measure is ergodic. We construct an example in which the constants are the only functions f for which $\|T^nf\|=\|f\|$, $n=0,1,2,\ldots$ and there are f's orthogonal to one such that $\|T^nf\|\downarrow c>0$ as $n\to\infty$. The desired transition function $P(\cdot,\cdot)$ is the composition of two transition functions $S(\cdot,\cdot)$ and $R(\cdot,\cdot)$ acting on the group of binary sequences $x=\cdot x_0x_1x_2\ldots$ with addition coordinatewise modulo 2. We set up the usual product topology so that this is a compact group. The transition function $R(\cdot,\cdot)$ corresponds to the point transformation

$$\cdot x_0x_1x_2\ldots\to\cdot x_1x_2x_3\ldots.$$

The transition function $S(\cdot,\cdot)$ is constructed as follows. Let $(x_0\ldots x_k)$ denote the set of elements x of the group whose first $(k+1)$ entries are given by $x_0,x_1,\ldots,x_k$. Take ν to be the additive set function on sets of the type $(x_0\ldots x_k)$ given by

$$\nu((x_0\ldots x_k))=\prod_{j=0}^{k}(1-\varepsilon_j)^{1-x_j}\varepsilon_j^{x_j}$$

where $0<\varepsilon_j<1$, $\sum\varepsilon_j<\infty$, and extended in the natural way to the field generated by sets of this form. This additive set function can clearly be extended as a completely additive measure to the Borel field generated by this field and we will still use the symbol ν to denote this measure. Set

$$S(x,A)=\nu(A-x)$$

for sets A in this Borel field $\mathscr{A}$. S is the translation transition function generated by ν. The transition function $P(\cdot,\cdot)$ is given by

$$P(x,A)=\int S(x,dy)\,R(y,A)$$

for $A\in\mathscr{A}$. The uniform or Haar measure Q of the group is clearly an invariant measure for both $S(\cdot,\cdot)$ and $R(\cdot,\cdot)$ and therefore is an invariant measure for $P(\cdot,\cdot)$. We first show that $S(\cdot,\cdot)$ reduces the norm of any nonconstant function $f\in L^2(dQ)$. Let

$$a(x_j)=\begin{cases}1 & \text{if } x_j=1\\ -1 & \text{if } x_j=0.\end{cases}$$

The functions 1, $a(x_0)$, $a(x_1)$, $a(x_1)\,a(x_0)$, ... are a complete orthonormal family on the probability space with probability measure Q. Any square integrable function f can be expanded as a Fourier series in terms of these functions and one gets a faithful representation of the function. Consider what the transition operator does to

$$a(x_k)\,a(x_{j_\alpha})\dots a(x_{j_1}),\qquad k>j_\alpha>\cdots>j_1.$$

It takes this function into

$$E(a(x_k+z_k)\dots a(x_{j_1}+z_{j_1})|x_k,\dots,x_{j_1})=E(a(x_k+z_k)|x_k)\dots E(a(x_{j_1}+z_{j_1})|x_{j_1})$$

where the z_k's are independent random variables (independent of the x_j sequence) and

$$z_k=\begin{cases}0 & \text{with probability } 1-\varepsilon_k\\ 1 & \text{with probability } \varepsilon_k.\end{cases}\tag{5}$$

But $E(a(x_k+z_k)|x_k)=(1-2\varepsilon_k)\,a(x_k)$ so that

$$E(a(x_k+z_k)\dots a(x_{j_1}+z_{j_1})|x_k,\dots,x_{j_1})=(1-2\varepsilon_k)\dots(1-2\varepsilon_{j_1})\,a(x_k)\dots a(x_{j_1}).$$

It is now clear that if f is not constant, then $\|Sf\|<\|f\|$. We now show that there is no nonconstant function f such that $\|Tf\|=\|f\|$ (T is the operator induced by $P(\cdot,\cdot)$). For, if there were such a function, there would have to be a set A with $0<Q(A)<1$ such that $\|TI_A\|=\|I_A\|$. I_A is the set characteristic function of A. Now

$$\begin{aligned}\|TI_A\|^2&=\int Q(dx)\left|\int P(x,dy)\,I_A(y)\right|^2\\&=\int Q(dx)\left|\int S(x,dy)\int R(y,dz)\,I_A(z)\right|^2\\&\leqslant\int Q(dx)\int S(x,dy)\left|\int R(y,dz)\,I_A(z)\right|^2\\&=\int Q(dx)\left|\int R(x,dy)\,I_A(y)\right|^2\leqslant\int Q(dx)\,|I_A(x)|^2=\|I_A\|^2.\end{aligned}$$

Since $\|TI_A\|=\|I_A\|$, we must have equality everywhere in the inequalities above. This implies that $(RI_A)(x)=\int R(x,dy)\,I_A(y)$ must be the characteristic function of some set B with $Q(B)=Q(A)$. However, since S reduces the norm of every nonconstant function, it is clear that $\|TI_A\|<\|I_A\|$ and we have a contradiction. Notice that if $f\perp 1$, then $Tf\perp 1$. An example of a function f orthogonal to one such that $\|T^n f\|\downarrow c>0$ as $n\to\infty$ is easily at hand. Let

$$g(x)=\begin{cases}0 & \text{if } x_0=0\\ 1 & \text{if } x_0=1.\end{cases}$$

Then

$$(T^k g)(x)=\begin{cases}P(z_0+\cdots+z_{k-1}=0 \bmod 2) & \text{if } x_{k+1}=1\\ P(z_0+\cdots+z_{k-1}=1 \bmod 2) & \text{if } x_{k+1}=0\end{cases}$$

where the z_j's are independent random variables with probability distribution given by (5). Moreover,

$$P(z_0+\cdots+z_{k-1}=0 \bmod 2)\to\prod_0^\infty(1-\varepsilon_k)+\sum_{j\neq k}\varepsilon_j\varepsilon_k\prod_{l\neq j,k}(1-\varepsilon_l)+\cdots$$

as $k\to\infty$ so that $(T^k g)(x)$ will not converge as $k\to\infty$. The desired function $f\perp 1$ is obtained by letting $f=g-\frac{1}{2}$. Thus, $L^2(dQ)$ is not generated by $\mathscr{D}$ and $\mathscr{U}_\infty$. Notice that $\mathscr{U}_\infty$ contains just the constant functions. However, if we look at the adjoint operator T^*, it is clear that $\mathscr{U}_\infty^*$ contains only the constants but that $L^2(dQ)$ is generated by $\mathscr{U}_\infty^*$ and $\mathscr{D}^*$. The Markov process generated by $P(\cdot,\cdot)$ and invariant measure Q has trivial forward tail field (purely nondeterministic with time reversed) but has nontrivial backward tail field. In fact, the existence here of functions $f\in L^2(dQ)$ orthogonal to one such that $\|T^n f\|\downarrow c>0$ indicates that there are events in the backward tail field which do not correspond to any subset of $\mathscr{A}$. A simple modification of the construction given above taking the basic space as the group of two-sided binary sequences $x=(x_i;\, i=\ldots,-1,0,1,\ldots)$ with addition coordinatewise modulo 2 enables one to construct an irreducible transition function $P(\cdot,\cdot)$ with finite invariant measure Q (the Haar measure) such that $L^2(dQ)$ is generated by neither $\mathscr{U}_\infty$ and $\mathscr{D}$ nor $\mathscr{U}_\infty^*$ and $\mathscr{D}^*$.

The same questions arise just as naturally in the case of a continuous time parameter semigroup $R(t)$, $t\geqslant 0$, of transition probability operators with invariant measure Q on a space W continuous in mean square, that is, for any $f\in L^2(dQ)$

$$\int|(R(t)f)(x)-f(x)|^2\,Q(dx)\to 0 \tag{6}$$

as $t\downarrow 0$. Strictly speaking, relation (6) means continuity at $t=0$ but this implies continuity for all $t>0$. $\mathscr{U}_\infty$ is still defined as the set of $f\in L^2(dQ)$ such that

$$\|R(t)f\|=\|f\| \quad \text{for all } t>0$$

and $\mathscr{D}$ as the set of $f\in L^2(dQ)$ for which

$$\|R(t)f\|\downarrow 0$$

as $t\to\infty$. The various Lemmas derived in this section are still valid for a continuous parameter semigroup of transition probability operators of this type. It is interesting to see whether one can construct a continuous parameter semigroup of irreducible transition probability operators with finite invariant measure Q with a function $f\in L^2(dQ)$ orthogonal to $\mathscr{U}_\infty$ such that $\|R(t)f\|\downarrow c>0$ as $t\to\infty$. Such an example is not a full analogue of the example constructed in the preceding paragraph since there $\mathscr{U}_\infty$ consisted only of the constant functions. The continuous time parameter example we construct now will be such that $\mathscr{U}_\infty$ contains many nonconstant functions. Let T be the transition operator of the preceding paragraph with H the corresponding invariant measure on the space X. Consider a unit interval $Y=[0,1)$. The state space of the continuous time parameter semigroup we construct consists of points $(x,y)\in W=X\times Y$. If the initial state is (x,y), the state t units thereafter is $(x,y+t)$ if $y+t<1$. For general $t>0$, the second component becomes $y+t \bmod 1$ t time units thereafter with a jump of the first component from x to x' according to transition operator T whenever the second component is congruent to zero modulo one. Therefore, if $f\in L^2(dQ)$ is a product $\alpha(x)\beta(y)$ then

$$(R(t)f)(x,y)=(T^{[y+t]}\alpha)(x)\beta(y+t \bmod 1),$$

where $[z]$ is the greatest integer less than or equal to z. The invariant measure for this semigroup is $H\times\mu$ where μ is the uniform measure on Y. $\mathscr{U}_\infty$ consists of all square integrable functions $\beta(y)$ of the one variable y. The function $g(x)-\frac{1}{2}$ of the preceding paragraph then serves as a function $f\perp\mathscr{U}_\infty$ such that $\|R(t)f\|\downarrow c>0$ as $t\to\infty$.

Notes

4.0 and 4.1 The discussion of a Markov process restricted to a set A is partially based on the exposition in T.E. Harris' paper [36]. Such a reduction is often useful because it allows one to reduce questions about the existence of infinite invariant measures to corresponding questions about the existence of finite invariant measures. Necessary and sufficient conditions for the existence of a finite invariant measure absolutely continuous with respect to an a priori given probability measure can be found in the paper of Neveu [78].

4.2 Much of the recent work in ergodic theory for positive operators was stimulated by the basic paper of E. Hopf [40]. Chacon and Ornstein [7] were the first to obtain an L^1 ergodic theorem for positive operators under the types of conditions suggested in [40]. The proof we give of the L^1 ergodic theorem proved by Chacon and Ornstein is due to E. Hopf [41].

4.3 The result on existence of an invariant measure for a transition probability function given in this section requires that the state space be a topological space and that the transition operator take bounded continuous functions into bounded continuous functions. Most of the results in the probability literature on existence of invariant measures require different types of conditions. Doeblin [17] suggested an approach which has been studied and generalized. The version of Doeblin's condition that Doob [18] has given (referred to as condition D) runs as follows: There is a finite measure φ (which could be normalized so as to be a probability measure) on the sets of $\mathscr{A}$ with $\varphi(\Omega)>0$, a positive integer m and an $\varepsilon>0$ such that

$$P_m(x,A)\leqslant 1-\varepsilon$$

for all $x\in\Omega$ if $\varphi(A)\leqslant\varepsilon$. One of the standard results of the theory is that there is then an invariant probability measure ψ for the transition function $P(x,A)$

$$\int \psi(dx)P(x,A)=\psi(A)$$

with the property that φ is absolutely continuous with respect to ψ. Actually one can show a good deal more. First of all, there are at most a finite number of minimal ergodic sets, that is, sets S such that

$$P(x,S)=\begin{cases}1 & \text{if } x\in S\\ 0 & \text{if } x\notin S.\end{cases}$$

Within each minimal ergodic set S there are at most a finite number of cyclically moving sets $A_1,\ldots,A_r$. These are sets that form a disjoint partition of S with the property that

$$P(x,A_j)=\begin{cases}1 & \text{if } x\in A_{j-1},\ j=2,\ldots,r\\ 1 & \text{if } j=1 \text{ and } x\in A_r.\end{cases}$$

If there is one ergodic set and $r=1$ then for all x

$$\|P_m(x,\cdot)-\psi(\cdot)\|\to 0$$

as $m\to\infty$ where $\|\ \|$ denotes the total variation of the set function. If there is one ergodic set and $r>1$, then for all x

$$\left\|\frac{1}{n}\sum_{m=1}^{n} P_m(x,\cdot)-\psi(\cdot)\right\|\to 0$$

as $n\to\infty$.

T.E. Harris extended the result given above on the existence of an invariant measure in an interesting paper [36]. He assumed that there was a σ-finite measure φ on $\mathscr{A}$ such that for all $A\in\mathscr{A}$ with $\varphi(A)>0$

$$P(X_n\in A \text{ infinitely often } |X_0=x)=1$$

for all $x\in\Omega$ where $\{X_n,\ n=0,1,\ldots\}$ is the Markov process with transition probability function $P(x,A)$. Harris called this "Condition C" and proved that it implied the existence of a σ-finite invariant measure ψ such that φ is absolutely continuous with respect to ψ. Basically the proof is carried out by constructing a set $B\in\mathscr{A}$ such that the Markov process restricted to B satisfies the Doeblin condition. The transition operator $P_B(x,A)$ corresponding to the process restricted to B then has an invariant probability measure. Lemma 2.2 is then applied to show that there is an invariant measure for the original transition operator $P(\cdot,\cdot)$. Other consequences of Harris' condition C have been obtained in papers by Orey [82] and Jain and Jamison [50].

Notice that initially in Section 4.3 it is assumed that $P(\cdot,\cdot)$ is a transition probability function on the Borel sets of a compact Hausdorff space that takes continuous functions into continuous functions. It is, however, worthwhile remarking that if $T1=1$ *and T is an operator taking continuous functions into continuous functions, there is then a transition probability function $P(\cdot,\cdot)$ on the Borel sets such that*

$$(Tf)(x)=\int P(x,dy)f(y)$$

for continuous f. This can be seen in the following way (essentially the same argument is given in the proof of Lemma 5.2.3). For any open set O let

$$P(x,O)=\sup_{\substack{0\leq f\leq 1\\ f=0 \text{ on } O^c}}(Tf)(x) \tag{1}$$

where f is continuous. $P(x,O)$ is first shown to be a Borel function of x when O is open. Set

$$A_\alpha=\{x:P(x,O)>\alpha\}.$$

Given any $x\in A_\alpha$, there is an $\varepsilon>0$ such that $P(x,O)>\alpha+\varepsilon$. Then there is a continuous function $f_x(\cdot)$ $(0\leq f_x\leq 1)$ with $f_x(\cdot)=0$ on O^c and

$$(Tf_x)(x)>\alpha+\frac{\varepsilon}{2}.$$

The set of points

$$\left\{z:(Tf_x)(z)>\alpha+\frac{\varepsilon}{2}\right\}$$

is an open subset of A_α containing x. Thus A_α is open. It then follows that $P(x,O)$ is a Borel function of x. $P(x,C)$ is also Borel for any closed set C. With $P(x,O)$ given by (1), the Riesz representation theorem implies that one can determine $P(x,\cdot)$ as a probability measure for each x that is regular on the Borel sets. The collection of all Borel sets A for which $P(x,A)$ is Borel measurable is closed under disjoint union, proper differences, and limit operations. The collection of such sets is therefore a monotone class. By Theorem F page 223 of Halmos' *Measure Theory*, the set of all finite disjoint unions of proper differences of closed sets is a ring R, that is, E and $F\in R$ imply that $E\cup F$ and $E-F$ are in R. Then by Theorem B page 27 of Halmos' *Measure Theorem* the smallest monotone class containing R is the minimal σ-ring S containing R, that is, S is a ring such that if $E_i\in S$ then $\bigcup_{i=1}^{\infty}E_i\in S$. Thus the collection of sets A for which $P(x,A)$ is Borel measurable contains all the Borel sets.

4.4 The development in this section is based on M. Rosenblatt: Transition probability operators. Proc. 5th Berkeley Symp. Math. Statist. Prob. 2, 473–483 (1967).

Chapter V

Random Walks and Convolution on Groups and Semigroups

0. Summary

A classical result of P. Lévy on random rotations on the circle is derived in the first section. The distribution of the product of two random factors is given by the convolution of their distributions. Convolution of measures on a compact semigroup is introduced in section 2 and some of its properties are derived. The right and left random walks induced by a measure on a semigroup are introduced. In the next section, it is shown that averages of the sequence of convolutions of a regular measure with itself on a compact semigroup converge to an idempotent measure. This provides a constructive way of deriving the Haar measure on a compact group. This method of obtaining the Haar measure on a compact group appears to be due to Ito and Kawada. The algebraic structure of compact semigroups is dealt with in section 4. The minimal right (left) and two-sided ideals in a compact semigroup are characterized. This is extremely useful in determining the form of idempotent probability measures on a compact semigroup. The ideas and results of this section are illustrated in the case of the semigroup of $n \times n$ (n finite) stochastic matrices. The last section of the chapter presents conditions under which the unaveraged sequence of convolutions of a regular probability measure with itself converge. Conditions under which stationary right and left random walks on a compact group or semigroup are ergodic or mixing are determined.

1. A Problem of P. Lévy

In a 1939 [67] paper P. Lévy discussed the addition of random variables taking values on the circumference of a circle. Think of a source of light at the center of a circular room. The light is projected on the circular boundary of the room. At time j the source (and the light projected) is rotated through a random angle X_j with probability distribution ν. The random angles X_j, $j=1,2,\ldots$, are assumed to be independent and identically distributed with common distribution ν. Of course, the randomness of the rotations can be thought of as due to small errors in alignment of a fixed rotation. The object is to find a good approximation to the probability distribution of the location of the light projected on the wall after a long time has passed. More formally, we are interested in the limiting distribution (if it exists) of $\sum_{j=1}^{n} X_j$ as $n\to\infty$.

For convenience think of the angle x measured in multiples of 2π radians so that $0\leqslant x\leqslant 1$ where 1 is identified with zero. The group addition $x+y$ is the addition of elements of the circle group (or addition of angles). Alternatively, the group addition $x+y$ can be thought of as given by ordinary addition of real numbers modulo one, that is, two real numbers are identified if their difference is an integer. The addition symbol $+$ will usually be used for group addition, at times for ordinary addition of real numbers. However, it will always be clear from the context which is intended. The open intervals on the circle group are obtained from the open intervals $\{a<x<b\}$ on the real line with identification of the real numbers modulo one. The open sets on the circle group are then obtained from the open intervals on the circle group by union of collections of open intervals. The probability distribution ν is assumed to be given on the Borel sets of the circle group.

Let $X_1, X_2, \ldots$ be a sequence of independent, identically distributed random variables taking values in the circle group with common probability distribution (or measure) ν. Let F be the distribution function corresponding to ν, that is,

$$F(b)-F(a)=\nu(\{a<x\leqslant b\}), \qquad 0<a\leqslant b\leqslant 1.$$

F can be taken to be a monotone nondecreasing right continuous function with $F(0)=0$, $F(1)=1$. Our investigation of the asymptotic behavior of $\sum_{j=1}^{n} X_j$ (under group addition on the circle) will use elements of Fourier analysis. Let $\hat{\nu}$ be the Fourier transform

$$\hat{\nu}(m)=\int_0^1 e^{2\pi imx}\,\nu(dx)=E\{e^{2\pi imX}\}=\int_0^1 e^{2\pi imx}\,dF(x)$$

of ν. The Fourier transform of the probability distribution of $\sum_{j=1}^{n} X_j$ is then

$$[\hat{\nu}(m)]^n=E\left\{e^{2\pi im\sum_1^n X_j}\right\}=[E\{e^{2\pi imX}\}]^n$$

because the random variables X_j are independent and identically distributed. The probability distribution of the sum $\sum_{j=1}^{n} X_j$ is the nth convolution of the measure ν with itself, $\nu^{(n)}$. Therefore,

$$[\hat{\nu}(m)]^n = \int_0^1 e^{2\pi i m x} \nu^{(n)}(dx).$$

Let

$$F^{(n+1)}(x) = \int_{-\infty}^{\infty} F(x-y) dF^{(n)}(y), \qquad n=1,2,\ldots,$$

with addition (and subtraction) understood to be addition for the real numbers rather than addition for the circle group. We can then still write

$$[\hat{\nu}(m)]^n = \int_{-\infty}^{\infty} e^{2\pi i m x} dF^{(n)}(x)$$

because the exponentials $e^{2\pi i m x}$ take the same value for real numbers x differing by an integer. Notice that

$$\nu^{(n)}(\{a < x \leqslant b\}) = \sum_{m=0}^{\infty} \{F^{(n)}(m+b) - F^{(n)}(m+a)\}$$

for $0 < a \leqslant b \leqslant 1$.

The Fourier transform $\hat{\nu}$ of a probability measure on the circle group satisfies

$$|\hat{\nu}(m)| \leqslant \hat{\nu}(0) = 1.$$

There are certain basic properties of Fourier transforms $\hat{\nu}$ (or of sequences of them) that will be required.

Lemma 1. *There is a one-one correspondence between probability measures ν on the Borel sets and their Fourier transforms $\hat{\nu}$.*

Suppose that there were two distinct measures ν_1, ν_2 with the same Fourier transform $\hat{\nu}$. Let F_1, F_2 be the corresponding distribution functions. Then

$$0 = \int_0^1 e^{2\pi i m x} d(F_1(x) - F_2(x)) = -2\pi i m \int_0^1 e^{2\pi i m x}(F_1(x) - F_2(x)) dx$$

for $m \neq 0$. Since $F_1 - F_2$ is square integrable, it follows that $F_1(x) - F_2(x)$ is constant almost everywhere. However this constant must be zero since F_1 and F_2 agree at zero and one and are right continuous. Since $F_1 = F_2$, it follows that $\nu_1 = \nu_2$ because measures are determined by their distribution functions.

Lemma 2. *Suppose* $|\hat{v}(m)|=1$ *for some* $m\neq 0$. *Then all the probability mass of the corresponding measure* v *must be concentrated on the points*

$$x=y+\frac{k}{m} \text{ modulo } 1, \qquad k=0,1,\ldots,m-1, \tag{1}$$

for some y, *that is, the mass is concentrated on a coset of a finite subgroup of the circle group. If* $\hat{v}(m)=1$ *then* $y=0$ *in* (1).

Let F be the cumulative distribution function corresponding to v. If $|\hat{v}(m)|=1$ there is a real number y such that

$$1=\int_0^1 e^{2\pi i m(x-y)}\,dF(x).$$

This implies that F can increase only at the points (1) since $e^{2\pi i m x}$ takes the value one only at the points k/m with k an integer.

The uniform measure on the circle group is of great interest to us because it will be the most important measure in our limit theorem. This is the probability measure v whose corresponding distribution function F is given by

$$F(x)=\int_0^x dx. \tag{2}$$

Let the set $x+A$ be the set of all points $x+y$ with $y\in A$. The uniform measure has the obvious but important property that

$$v(A)=v(x+A)$$

for all Borel sets A and all x. It is invariant under translation. Further, its Fourier transform

$$\hat{v}(m)=\delta_m,$$

where δ is the Kronecker delta function.

The notion of convergence of probability measures considered in our limit theorem is as follows. *If* v_n, $n=1,2,\ldots$, *and* v *are probability measures on the circle group, we say that* v_n *converges to* v *as* $n\to\infty$ *if*

$$\int g(x)\,v_n(dx)\to\int g(x)\,v(dx) \tag{3}$$

as $n\to\infty$ *for every continuous function* g *on the circle group.* In the literature in functional analysis this is sometimes called "weak star convergence" of v_n to v. One can show that then $v_n(O)\to v(O)$ for open intervals O. In discussions in probability theory it is usual to restate (3) in terms of distribution functions. Let F_n, $n=1,2,\ldots$, and F be the distribution functions on the unit interval $0 \leqslant x \leqslant 1$ corresponding to ν_n and ν. Then (3) can be rewritten as

$$\int_0^1 g(x)\,dF_n(x) \to \int_0^1 g(x)\,dF(x)$$

for every continuous function g on $0 \leqslant x \leqslant 1$ with $g(0)=g(1)$. The convergence is then called weak convergence of F_n to F (or of ν_n to ν). Let $\hat{\nu}_n$ and $\hat{\nu}$ be the Fourier transforms of ν_n and ν. It is clear that if ν_n converges to ν weakly (F_n converges to F weakly) that

$$\hat{\nu}_n(m) \to \hat{\nu}(m)$$

as $n \to \infty$ for all integral m. A natural converse of this statement holds because of the compactness of the circle group. A proof of this converse (usually called the continuity theorem for characteristic functions or Fourier transforms) can be found in many books on probability theory (see [10]). We shall give a brief but nonelementary sketch of a proof.

Lemma 3. *Let $\hat{\nu}_n$ be a sequence of Fourier transforms of probability measures ν_n on the circle group. If*

$$\hat{\nu}_n(m) \to \eta(m) \tag{4}$$

for all integral m as $n \to \infty$, it follows that $\eta = \hat{\nu}$ is the Fourier transform of a measure ν and ν_n converges weakly to ν.

Let $p(x) = \sum_{k=-m}^{m} c_k e^{2\pi i k x}$ be any finite trigonometric polynomial. Because of (4) it follows that

$$\int p(x)\,\nu_n(dx) = \sum_{k=-m}^{m} c_k\,\hat{\nu}_n(k)$$

converges as $n \to \infty$ to a limit

$$L(p) = \sum_{k=-m}^{m} c_k\,\eta(k)\,.$$

Since we can use the Weierstrass approximation theorem to uniformly approximate functions g continuous on the circle group, it follows that for any such g

$$\int g(x)\,\nu_n(dx) \to L(g) \tag{5}$$

as $n \to \infty$ where $L(g)$ is finite. Furthermore, the functional L is linear, that is, for any real numbers α, β and continuous functions g_1, g_2

$$L(\alpha g_1 + \beta g_2) = \alpha L(g_1) + \beta L(g_2)\,.$$

It is also clear that L is positive in the sense that

$$L(g) \geqslant 0$$

if g is a nonnegative continuous function. Finally $L(g)=1$ for the function $g(x)\equiv 1$ and

$$|L(g)| \leqslant \sup_x |g(x)|$$

for any continuous function g. However, the Riesz representation theorem (see appendix 4) implies that such a bounded positive functional L on the continuous functions on the circle group is given by

$$L(g) = \int g(x)\,\nu(dx) \tag{6}$$

where ν is a probability measure on the circle group. It follows from (5) and (6) that ν_n converges weakly to the probability measure ν.

The point x is said to belong to the *support* $\sigma(\nu)$ *of the regular measure* ν if $\nu(O)>0$ for every open interval O containing x. One can show that the support of the measure ν is a closed set.

Theorem 1. *Let* ν *be a probability measure on the circle group. If the support* $\sigma(\nu)$ *is not contained in a finite subgroup of the circle group, the averaged convolutions*

$$\frac{1}{n}\sum_{j=1}^{n} \nu^{(j)} \tag{7}$$

converge weakly to the uniform measure on the circle group as $n\to\infty$. *If the support* $\sigma(\nu)$ *is contained in a finite subgroup and*

$$\left\{\frac{k}{r},\ k=0,1,\ldots,\ r-1\right\} \tag{8}$$

is the smallest such subgroup, the averaged convolutions (7) *will converge to the uniform measure with mass* $1/r$ *at each element of the subgroup* (8).

The averaged convolutions are a sequence of measures and the Fourier transform of (7) is

$$\frac{1}{n}\sum_{k=1}^{n} [\hat{\nu}(m)]^k. \tag{9}$$

If the support $\sigma(\nu)$ of ν is not contained in a finite subgroup of the circle group, it follows from Lemma 2 that $\hat{\nu}(m)\neq 1$ for $m\neq 0$. But then

$$\frac{1}{n}\sum_{k=1}^{n} [\hat{\nu}(m)]^k = \frac{1}{n}\hat{\nu}(m)\frac{1-[\hat{\nu}(m)]^n}{1-\hat{\nu}(m)} \to 0 \tag{10}$$

for $m\neq 0$ as $n\to\infty$ with (9) equal to one for $m=0$. The limiting Fourier transform is the Kronecker delta function δ_m, the Fourier transform of the uniform distribution (2) on the circle group. Lemma 3 implies that the limiting distribution is the uniform distribution on the circle group. Assume now that the support $\sigma(v)$ is contained in the smallest finite subgroup (8). The result is obvious if $m=1$. Assume that m is greater than one. Then Lemma 2 implies that $\hat{v}(m)\neq 1$ for all $m\neq kr$. For $m\neq kr$ it is still clear that (10) holds while for $m=kr$ (9) is equal to one. However, this limiting sequence is the Fourier transform of the uniform distribution with mass $1/r$ at each of the points of (8). Lemma 3 as before implies that this distribution is the limiting distribution.

Theorem 1 dealt with the limiting behaviour of averaged convolution sequences (7). However, our main interest is in the asymptotic behavior of the sequence of probability measures of the partial sums $\sum_{j=1}^{n} X_j$, that is, of the convolution $v^{(n)}$ themselves.

Theorem 2. *Let v be a probability measure on the circle group. The sequence of convolutions $v^{(n)}$ will converge weakly to the uniform measure on the circle group if and only if the support $\sigma(v)$ of v is not contained in a coset of a finite subgroup of the circle group.*

By Lemma 2 it is clear that $|\hat{v}(m)|<1$ for all $m\neq 0$ if and only if $\sigma(v)$ is not contained in a coset of a finite subgroup of the circle group. However, $|\hat{v}(m)|<1$ for all $m\neq 0$ is a necessary and sufficient condition for the Fourier transform $[\hat{v}(m)]^n$ of $v^{(n)}$ to converge to δ_m as $n\to\infty$. By Lemma 3 this is the necessary and sufficient condition for $v^{(n)}$ to converge weakly to the uniform measure on the circle group.

Corollary 1. *Let v be a probability measure whose support $\sigma(v)$ is contained in a finite subgroup*

$$\left\{\frac{k}{r},\ k=0,1,\ldots,\ r-1\right\} \tag{11}$$

of the circle group. Then $v^{(n)}$ converges to the uniform measure on the subgroup (11) *if and only if $\sigma(v)$ is not contained in a coset of a proper subgroup of* (11).

The argument for this simple corollary is similar to that of Theorem and is left to the reader.

2. Limit Theorems and the Convolution Operation

In the previous section the asymptotic behavior of the distribution of the sum of independent and identically distributed random rotations of the circle group was studied. This suggests that one might also hope to be able to study the asymptotic behavior of the distribution of the result of n independent and identically distributed successive rotations in Euclidean 3-space. The rotations in 3-space are a group and it is now appropriate to represent the binary group operation by multiplication because it is not commutative. The rotations in 3-space can be represented by the orthogonal matrices (whose determinants is $+1$)

$$M=(m_{j,k};\ j,k=1,2,3),$$

with real entries $m_{i,j}$ satisfying

$$\sum_{j=1}^{3} m_{i,j} m_{k,j}=\delta_{i-k}.$$

The result of two successive rotations g_1, g_2 is given by $g_1 g_2$ or by the matrix product $M_1 M_2$ of the orthogonal matrices M_1, M_2 representing g_1 and g_2. A natural distance d can be set up for the orthogonal matrices

$$d(M,M')=\Big\{\sum_{j,k}(m_{j,k}-m'_{j,k})^2\Big\}^{\frac{1}{2}} \tag{1}$$

where $M=(m_{j,k})$ and $M'=(m'_{j,k})$. With the topology induced by this metric, the orthogonal matrices (or the rotation group) can be seen to be a compact Hausdorff space. Let ν be a probability measure on the group, the distribution of one rotation. If the convolution operation is properly defined for measures, we can hope that the distribution of the product of n independent identically distributed rotations with probability measure ν will be given by $\nu^{(n)}$, the nth convolution of ν with itself. The examples of rotations of the circle group and rotations in 3-space motivate in part the study of the following problem. Certain undefined terms used will be defined shortly.

Let G be a compact topological group (Hausdorff) with a regular measure ν on G. We wish to study the asymptotic behaviour of the convolution sequence $\nu^{(n)}$ as $n\to\infty$.

A *group* G is a set of elements g with a binary operation $g_1 g_2$ (denoted by multiplication) defined for each pair of elements $g_1, g_2\in G$ with the following properties:

(i) closure under multiplication:

$$g_1 g_2\in G \quad \text{if } g_1, g_2\in G;$$

(ii) *associativity:*

$$g_1(g_2 g_3)=(g_1 g_2)g_3 \quad \text{for } g_1, g_2, g_3\in G;$$

(iii) there exists an *identity* $e \in G$ such that

$$eg = ge = e;$$

(iv) for each element $g \in G$ there exists an *inverse* g^{-1}:

$$g^{-1} g = g g^{-1} = e.$$

The group G is *topological* if there is a topology (a family of open sets) given on G such that the binary operation $g_1 g_2$ is jointly continuous in g_1, g_2 and such that the inverse operation g^{-1} is continuous. A topological group G is Hausdorff or compact if it is Hausdorff or compact respectively as a topological space. Whenever one consider probability measures ν on a compact topological (Hausdorff) group it is natural to consider them on the Borel field $\mathscr{B}$ generated by the topology (the open sets). The measure ν is then said to be *regular* if for each $\varepsilon > 0$ and $B \in \mathscr{B}$ there is an open set $O \supset B$ and a closed (or compact) set $C \subset B$ (O and C depend on ε and B) such that

$$\nu(O - B) < \varepsilon, \tag{2}$$

$$\nu(B - C) < \varepsilon. \tag{3}$$

If a probability measure ν is given on the Borel field of a metric space (as in the circle group or the group of orthogonal matrices), it can be shown to be regular.

Let us now consider another example which suggests that it would be appropriate to even generalize the context in which one investigates the asymptotic behaviour of convolution sequences of measures further. Consider the set of $n \times n$ (n a positive integer) stochastic matrices $M = (m_{j,k}; j, k = 1, \ldots, n)$, that is

$$m_{j,k} \geq 0,$$

$$\sum_{k=1}^{n} m_{j,k} = 1, \qquad j = 1, \ldots, n.$$

The study of the asymptotic behavior of powers M^r as $r \to \infty$ of a fixed stochastic matrix M was examined in some detail in section 2 of Chapter 1. Suppose we now consider the situation in which a stochastic matrix is chosen at random (with given probability measure ν) from the set of $n \times n$ stochastic matrices with the choices at different times independent and identically distributed. The transition matrix corresponding to an r step transition is then the random matrix

$$M_1 M_2 \ldots M_r$$

where the matrices M_j, $j=1,2,\ldots$, are independent and identically distributed with probability measure ν. The distance d given by (1) can still be used for the set of stochastic matrices and with the topology induced by this metric we have a compact Hausdorff space. However, the $n\times n$ stochastic matrices are no longer a group but rather a semigroup since an inverse needn't exist. Again we can hope that the probability distributions of the products will be given by the convolutions of ν with itself. The problem of studying *the asymptotic behaviour of the convolution sequence* $\nu^{(n)}$ *as* $n\to\infty$ *where* ν *is a regular probability measure on a compact topological (Hausdorff) semigroup* S is suggested.

A *semigroup* S is a set of elements s with a binary operation $s_1 s_2$ defined for each pair of elements $s_1, s_2 \in S$ with the following properties:

(i) closure under multiplication:

$$s_1 s_2 \in S \quad \text{if } s_1, s_2 \in S;$$

(ii) *associativity:*

$$(s_1 s_2)s_3 = s_1(s_2 s_3) \quad \text{if } s_1, s_2, s_3 \in S.$$

The *semigroup* S is topological if there is a topology given on S such that the binary operation $s_1 s_2$ is jointly continuous in s_1, s_2. A topological semigroup is Hausdorff or compact if it is Hausdorff or compact as a topological space. As before, a probability measure ν on the Borel field $\mathscr{B}$ of subsets of S generated by the topology is regular if for each $\varepsilon>0$ and $B\in\mathscr{B}$ there are compact and open sets C and O respectively with $C\subset B\subset O$ satisfying (2) and (3).

In our discussion of the limit theorems on compact topological groups and semigroups, we will often give the derivations and results in the more general context of semigroups and specialize when necessary to groups. Let S be a compact topological (Hausdorff) semigroup with ν a regular probability measure on S. S^N is the space of sequences $\omega=(s_1, s_2, \ldots)$, $s_i\in S$. Consider the product measure on S^N generated by measure ν where ν is the probability measure of a single coordinate. The coordinate functions $X_j(\omega)=s_j$, $j=1,2,\ldots$ are the independent identically distributed random variables with values in S and common distribution ν. If S is a separable metric space, the product $X_1(\omega)X_2(\omega)=s_1 s_2$ is a random variable (that is, measurable) and the distribution of the product is given by

$$P\{X_1(\omega)X_2(\omega)\in B\} = \int I_B(s_1 s_2)\,\nu(ds_1)\,\nu(ds_2) = \nu*\nu(B) = \nu^{(2)}(B) \tag{4}$$

for any set B in the Borel field $\mathscr{B}$ on S, with I_B the set indicator function of B. Formula (4) is the definition of the convolution of ν with itself, $\nu*\nu$. However, we would like to give a discussion that will be meaningful even if S is not separable metric. Before passing on to this discussion we note that in the explicit examples noted earlier the groups and semigroups are separable metric spaces.

One of our first aims will be the definition of the convolution $\nu * \mu$ of two regular probability measures ν, μ on a general compact topological semigroup S. The qualification Hausdorff will occasionally be deleted in the sequel but it will be implicitly assumed throughout the rest of the chapter. Let f be a continuous function on S. Consider the operator T

$$(Tf)(s) = \int f(ss')\mu(ds'). \tag{5}$$

Lemma 1. *If μ is a regular probability measure on a compact topological semigroup S, the operator T given by* (5) *takes continuous functions f on S into continuous functions on S.*

The function $f(ss')$ is a continuous function on the compact space $S \times S$ since the multiplication ss' is jointly continuous in s, s'. Given a fixed $\varepsilon > 0$ and any point $(s_1, s_2) \in S \times S$ there is an open set $O_{1,2}$ in the product topology of $S \times S$ containing (s_1, s_2) such that

$$|f(ss') - f(s_1 s_2)| < \varepsilon$$

if $(s, s') \in O_{1,2}$. The sets $O_{1,2}$ can be taken to be product sets $O_1 \times O_2$ with O_1, O_2 open sets in the topology of S. Since the sets $O_{1,2} = O_1 \times O_2$ corresponding to $\varepsilon > 0$ form an open covering of the compact space $S \times S$, a finite subcovering $O_1^{(j)} \times O_2^{(j)}, j = 1, \ldots, n$, can be chosen. It then follows that if s' belongs to the intersection of the sets $O_1^{(j)}$ containing s then

$$|(Tf)(s) - (Tf)(s')| < 2\varepsilon.$$

The proof of the continuity of $(Tf)(s)$ is complete.

Let ν, μ now be two regular measures on S. Consider the iterated integral

$$\int (Tf)(s)\nu(ds) = \int \{\int f(ss')\mu(ds')\}\nu(ds) = L(f) \tag{6}$$

for a continuous function f on S. The linear functional $L(f)$ on the continuous functions (with the norm $\|f\| = \sup_x |f(x)|$) given by (6) is bounded, positive and linear. The Riesz representation theorem (see appendix 4) implies that there is a regular probability measure η on S such that

$$L(f) = \int f(s)\eta(ds) \tag{7}$$

since $L(1) = 1$. We call the regular measure η given by (7) the convolution $\nu * \mu$ of ν, μ. Notice that the convolution could equally well have been constructed by using the iterated integral

$$\int \{\int f(ss')\nu(ds)\}\mu(ds').$$

However, a simple application of the Stone-Weierstrass theorem shows that one is led to the same measure η as the convolution of ν and μ. The convolution operation will generally not be commutative ($\nu * \mu \neq \mu * \nu$) unless S is a commutative semigroup. The convolution sequence generated by a regular measure ν is given by

$$\nu^{(1)} = \nu, \qquad \nu^{(n+1)} = \nu^{(n)} * \nu = \nu * \nu^{(n)}, \qquad n = 1, 2, \ldots .$$

As already noted, $\mathscr{B} = \mathscr{B}(S)$ is the Borel field of subsets of S generated by the topology on S. Given $A \in \mathscr{B}$, let

$$s^{-1}A = \{s' : s s' \in A\}$$
$$A s^{-1} = \{s' : s' s \in A\}.$$

Lemma 2. *Let μ be a regular probability measure on the compact Hausdorff semigroup S. Then for each $s \in S$, the measures $\mu(A s^{-1})$, $\mu(s^{-1}A)$, $A \in \mathscr{B}$ are regular on S.*

The proof will be carried out for $\mu(A s^{-1})$. For each $A \in \mathscr{B}$ and $s \in S$, $A s^{-1} \in \mathscr{B}$. Given any $\varepsilon > 0$, by the regularity of μ on S, there is a closed (compact) set $C \subset A s^{-1}$ such that $\mu(A s^{-1} - C) < \varepsilon$. Let $B = C s$. By the continuity of the multiplicative operation, one sees that B is closed. Also $B \subset A$. Thus $\mu((A - B) s^{-1}) < \varepsilon$ since $C \subset B s^{-1} \subset A s^{-1}$. The corresponding outer approximation by open sets is obtained by complementation.

Lemma 3. *If $A \in \mathscr{B}$ and μ is a regular measure on the compact Hausdorff semigroup S, $\mu(A s^{-1})$ and $\mu(s^{-1}A)$ are Borel measurable functions of s.*

Let $\mathscr{C}$ denote the class of continuous functions f on S with $0 \leqslant f \leqslant 1$. For any given set A let A^c denote the complement of A. Let O be an open set and set

$$A_\alpha = \{s : \mu(O s^{-1}) > \alpha\}.$$

Since the measure $\mu(\cdot\, s^{-1})$ is regular (see appendix 2) it follows that

$$\mu(O s^{-1}) = \sup_{\substack{f \in \mathscr{C} \\ f = 0 \text{ on } O^c}} \int f(u)\, \mu(du\, s^{-1}) = \sup_{\substack{f \in \mathscr{C} \\ f = 0 \text{ on } O^c}} \int f(u s)\, \mu(du).$$

Given any $s \in A_\alpha$ there is an $\varepsilon > 0$ such that $\mu(O s^{-1}) > \alpha + \varepsilon$. But there is then a function $f_s \in \mathscr{C}$ with $f_s = 0$ on O^c such that

$$\int f_s(u s)\, \mu(du) > \alpha + \frac{\varepsilon}{2}.$$

The set of points $\{z : \int f_s(u z)\, \mu(du) > \alpha + \varepsilon/2\}$ is an open set containing s and is a subset of A_α. Therefore A_α is open. This implies that $\mu(O s^{-1})$ is Borel measurable in s. The class of Borel sets A such that $\mu(A s^{-1})$

is Borel measurable is closed under disjoint union and proper differences. It is closed under limit operations and hence is a monotone class. By Theorem B page 27 and Theorem F page 223 of Halmos' *Measure Theory* (1950), the class must contain all of $\mathscr{B}$ since it contains the open sets. Of course, the argument is the same for $\mu(s^{-1}A)$.

Lemma 4. *Given any set $A\in\mathscr{B}$ and ν,μ regular probability measures on the compact Hausdorff semigroup S,*

$$\nu*\mu(A)=\int\mu(s^{-1}A)\,\nu(ds)=\int\nu(A s^{-1})\,\mu(ds). \tag{8}$$

It will be enough to prove the first equality in (8) since the second follows by a similar argument. If we set

$$P(s,A)=\mu(s^{-1}A),$$

$P(s,A)$ is a transition probability function which is a regular probability measure on $\mathscr{B}$ for each $s\in S$ and a Borel function of s for each $A\in\mathscr{B}$ by Lemmas 2 and 3. Most important of all

$$(Tf)(s)=\int P(s,ds')\,f(s')=\int f(s s')\,\mu(ds')$$

is continuous for each continuous f by Lemma 1. However, by a remark in appendix 3, any such transition function on a compact Hausdorff space takes regular measures into regular measures. The proof of the first equality is complete.

Let us consider again the space S^N of sequences $\omega=(s_1,s_2,\ldots)$ where S is a compact Hausdorff semigroup. Assume that ν is a regular measure on S. Consider a Markov process on S^N (state space S) with transition probability function

$$P(s,A)=\nu(s^{-1}A)(=\nu(A s^{-1})),$$

$A\in\mathscr{B}$. The random variables are again the coordinate functions $X_j(\omega)=s_j$, $j=1,2,\ldots$. A Markov process with transition function $\nu(s^{-1}A)(\nu(A s^{-1}))$ is called a left (right) random walk generated by the measure ν. If the initial distribution (of $X_1(\omega)$) is itself given by ν, the probability measure of $X_n(\omega)$ in both the left and right random walks generated by ν is $\nu^{(n)}$. Thus, we are then interested in the asymptotic distribution of $X_n(\omega)$ as $n\to\infty$. Notice that in this construction one needn't assume that S is a separable metric space.

3. Idempotent Measures as Limiting Distributions

Let ν be a regular probability measure on the compact Hausdorff semigroup S. Our object in this section is to characterize in some natural way the limiting measures (if they exist) that the convolution sequence $\nu^{(n)}$, $n=1,2,\ldots$, or the averaged convolution sequence $\frac{1}{n}\sum_{j=1}^{n}\nu^{(j)}$, $n=1,2,\ldots$, converge to as $n\to\infty$. *A sequence of regular probability measures μ_n will be said to converge to the regular probability measure η as $n\to\infty$ if*

$$\int f(s)\,\mu_n(ds)\to\int f(s)\,\eta(ds)$$

as $n\to\infty$ for each continuous function on S. This is the "weak star" (or weak) convergence of measures spoken of earlier. One will show that every limit measure η that a convolution sequence $\nu^{(n)}$ (or the averaged convolution sequence) converges to is an idempotent measure under convolution, that is,

$$\eta*\eta=\eta.$$

As a byproduct of our results, we will be able to construct the Haar (or uniform) measure on a compact Hausdorff group as a limiting distribution of an averaged convolution sequence. This can be thought of as a probabilistic derivation of the Haar measure.

The support $\sigma(\nu)$ of the regular measure ν is the set of all points s with the property that every open set O containing s has positive ν measure. By Urysohn's Lemma (see appendix 2) it is clear that $s\in\sigma(\nu)$ if and only if

$$\int f(s')\,\nu(ds')>0$$

for every nonnegative continuous function f on S which is positive at s. The support $\sigma(\nu)$ is a closed set.

Lemma 1. *Let ν,μ be two regular probability measures on the compact Hausdorff semigroup S with supports $\sigma(\nu),\sigma(\mu)$. The support of the convolution $\eta=\nu*\mu$ is the product $\sigma(\nu)\,\sigma(\mu)$.*

We know that

$$\int f(s)\,\eta(ds)=\int\{\int f(s_1 s_2)\,\nu(ds_1)\}\,\mu(ds_2) \tag{1}$$

for every continuous function f on S. Suppose $c\notin\sigma(\nu)\sigma(\mu)$. Let f be a nonnegative continuous function on S which is positive at c and zero on $\sigma(\nu)\sigma(\mu)$. Then $f(s_1 s_2)$ as a function of s_1 is zero on $\sigma(\nu)$ for fixed $s_2\in\sigma(\mu)$. Thus $g(s_2)=\int f(s_1 s_2)\,\nu(ds_1)=0$ for $s_2\in\sigma(\mu)$. This implies that (1) is zero for such a nonnegative function f since g is continuous by Lemma 2.1. Therefore $c\notin\sigma(\eta)$. Conversely let $c=ab$ with $a\in\sigma(\nu)$, $b\in\sigma(\mu)$. Let f be any nonnegative continuous function that is positive at c. Then $f(s_1 b)$ is positive at $s_1=a$ and hence $g(b)=\int f(s_1 b)\,\nu(ds_1)>0$. This implies that (1) is positive for such a function f and so $c\in\sigma(\eta)$.

Lemma 1 implies that the support of $\nu^{(n)}$ is $[\sigma(\nu)]^n$. *The closed semigroup generated by the support of ν is therefore*

$$\overline{\left(\bigcup_{n=1}^{\infty}[\sigma(\nu)]^n\right)}=S(\nu).$$

This is a compact Hausdorff semigroup with respect to its topology relative to S (see appendix 2) since it is a closed subset of S. It is natural to assume that S is the closed semigroup generated by the support of ν, $S(\nu)$, since any possible limit measure to which a convolution sequence $\nu^{(n)}$ (or the averaged convolution sequence) may converge will have measure zero on the complement of $S(\nu)$. We shall therefore assume from now on that $S=S(\nu)$. Notice that if η is a limit measure to which the convolution sequence $\nu^{(n)}$, $n=1,2,\ldots$, converges, then the averaged convolution sequence $\frac{1}{n}\sum_{j=1}^{n}\nu^{(j)}$ converges automatically to the limit measure η.

We shall call a family $\mathscr{C}$ of continuous functions f on S *equicontinuous* if for each $\varepsilon>0$ there is a finite covering $\{O_j; j=1,\ldots,n\}$ (depending on ε) of S consisting of open sets such that

$$\sup_{s,s'\in O_i}|f(s)-f(s')|<\varepsilon,\qquad i=1,\ldots,n,$$

for all $f\in\mathscr{C}$.

Lemma 2. *Let f be a fixed continuous function on the compact Hausdorff semigroup S. Consider the family of continuous functions derived from f by*

$$\int f(s,s')\,\nu(ds')$$

where ν is any regular probability measure. The family is bounded and equicontinuous.

In Lemma 2.1, for each $\varepsilon>0$ a finite open covering $\{O_j; j=1,\ldots,n\}$ of S was obtained such that

$$\sup_{s_1,s_2\in O_i}|f(s_1 s)-f(s_2 s)|<\varepsilon,\qquad i=1,\ldots,n$$

for all $s\in S$. This implies that

$$\sup_{s_1,s_2\in O_i}\left|\int f(s_1 s')\,\nu(ds')-\int f(s_2 s')\,\nu(ds')\right|<\varepsilon,$$

$i=1,2,\ldots,n$, for all regular probability measures ν.

Let us call a bounded linear operator T taking continuous functions into continuous functions an *equicontinuous operator* if for each continuous f, the sequence $\{T^n f, n=1,2,\ldots\}$ is equicontinuous and bounded.

Lemma 3. *A bounded equicontinuous sequence of continuous functions on a compact Hausdorff space has a uniformly convergent subsequence.*

Let f_n be the bounded equicontinuous sequence of continuous functions. It is clear that if any subsequence $f_{n'}$ converges, it must converge uniformly to a continuous limiting function. Our object is to construct the convergent subsequence. For each integer $j=1,2,\ldots$ there is a finite open covering $\mathscr{C}_j=\{O_{j,k}; k=1,\ldots,j_m\}$ of the space such that

$$\sup_n \max_k \sup_{s,s'\in O_{j,k}} |f_n(s)-f_n(s')| < \frac{1}{2^j}. \tag{2}$$

In each open set $O_{j,k}$ choose a point $s_{j,k}$. We can then choose by the usual diagonal procedure a subsequence of functions $f_{n'}$ that converges at all the points $s_{j,k}$, $k=1,\ldots,j_m$, $j=1,2,\ldots$. But then (2) implies that the subsequence $f_{n'}$ converges everywhere.

Our object is to prove a simple ergodic theorem for a positive equicontinuous operator T acting on the continuous functions of a compact space and apply this theorem to the limiting behaviour of averaged convolution sequences. The operator T is said to be *positive* if Tf is a nonnegative function whenever f is a nonnegative function. The sequence

$$T_n = \frac{1}{n}\sum_{k=1}^{n} T^k$$

will be said to converge to a limiting operator $\bar{T}$ taking continuous functions into continuous functions if $T_n f \to \bar{T} f$ in the supremum norm as $n\to\infty$ for each continuous function f, that is,

$$\|(T_n-\bar{T})f\| = \sup_x |(T_n f)(x)-(\bar{T}f)(x)| \to 0$$

as $n\to\infty$.

Theorem 1. *Let T be a positive equicontinuous operator acting on the continuous functions of a compact Hausdorff space. Then $T_n = \frac{1}{n}\sum_{k=1}^{n} T^k$ converges to a limiting operator $\bar{T}$ such that*

$$\bar{T}T=T\bar{T}=\bar{T}^2=\bar{T}. \tag{3}$$

Notice that the operator norms

$$|T^n| = \sup_{0\leqslant f\leqslant 1} \|T^n f\| \leqslant \sup_x |(T^n 1)(x)| \leqslant c < \infty$$

are uniformly bounded because of the positivity and equicontinuity of T. Consider $(T-I)f = Tf - f$ for any continuous function f. It is clear that

$$T_n(T-I)f \to 0$$

as $n \to \infty$ since

$$\|T_n(T-I)f\| = \|(T-I)T_nf\| = \frac{1}{n}\|(T^{n+1}-T)f\| \to 0$$

as $n \to \infty$. Thus

$$(T-I)T_n = T_n(T-I) \to 0$$

as $n \to \infty$. Let $\mathcal{N}$ be the set of all functions of the form $(T-I)f$ with f continuous and of limits in supremum norm of sequences of such functions. The functions $g \in \mathcal{N}$ are continuous since they are limits of uniformly convergent sequences of continuous functions. Notice that

$$T_n g \to 0 \tag{4}$$

for all $g \in \mathcal{N}$. For if $g \in \mathcal{N}$ for each $\varepsilon > 0$ there is a continuous function f such that $\|g-(T-I)f\| < \varepsilon$. Then

$$\|T_n g\| \leqslant \|T_n(g-Tf+f)\| + \|T_n(T-I)f\| \leqslant c\varepsilon + \|T_n(T-I)f\|\,.$$

The limiting property (4) is obtained by first letting $n \to \infty$ and then $\varepsilon \to 0$. Consider any fixed continuous function f. The equicontinuity of T implies that T_nf, $n=1,2,\ldots$, is an equicontinuous sequence of functions. There is then a subsequence $T_{n'}f$ converging to a continuous function $\bar{f}$ as $n \to \infty$ by Lemma 3. But then $(T-I)T_{n'}f \to (T-I)\bar{f}$. Thus, $T\bar{f} = \bar{f}$ and $T_n\bar{f} = \bar{f}$ for all n. Set $f = (f-\bar{f}) + \bar{f}$. If we can show that $T_n(f-\bar{f}) \to 0$ then this will imply that

$$T_n f \to \bar{f} \tag{5}$$

as $n \to \infty$. To do this it will be enough to verify that $f - \bar{f} \in \mathcal{N}$. Now

$$(T_n - I)f = (T-I)\sum_{j=0}^{n-1}\left(\frac{n-j}{n}\right)T^j f$$

so that $(T_n - I)f \in \mathcal{N}$ for all n. However, $(T_{n'} - I)f \to \bar{f} - f$ for a subsequence $n' \to \infty$. Hence $f - \bar{f} \in \mathcal{N}$ and (5) is valid. Let $\bar{T}$ be the linear operator taking continuous f into $\bar{f}$

$$\bar{T}f = \bar{f}.$$

It is now obvious that $T_n \to \bar{T}$ and that $\bar{T}$ satisfies (3).

Corollary 1. *Let ν be a regular probability measure on the compact Hausdorff semigroup S. Then the averaged convolution sequence*

$$\frac{1}{n}\sum_{j=1}^{n} \nu^{(j)}$$

converges to a regular probability measure η.

If for f continuous

$$(Af)(s)=\int f(ss')\eta(ds')$$
$$(Bf)(s)=\int f(s's)\eta(ds'),$$

then

$$(A^2f)(s)=(Af)(s)$$
$$(B^2f)(s)=(Bf)(s).$$

Further

$$v*\eta=\eta*v=\eta*\eta=\eta. \tag{6}$$

Let T be the operator

$$(Tf)(s)=\int f(ss')v(ds') \tag{7}$$

acting on the continuous functions f on S. By Lemma 2 T is an equicontinuous operator. Theorem 1 implies that for each continuous f

$$T_n f=\frac{1}{n}\sum_{j=1}^{n} T^j f\to\bar{f}$$

where $\bar{f}$ is continuous $T\bar{f}=\bar{f}$. Therefore,

$$\begin{aligned}\lim_{n\to\infty}\frac{1}{n}\sum_{j=1}^{n}\int f(s)v^{(n)}(ds)&=\lim_{n\to\infty}\int(T_n f)(s)v(ds)=\int(\bar{T}f)(s)v(ds)\\&=\int\bar{f}(s)v(ds)=L(f)\end{aligned} \tag{8}$$

where L is a positive continuous functional on the continuous functions f with $L(1)=1$. The Riesz representation theorem implies there is a regular probability measure η such that

$$L(f)=\int f(s)\eta(ds).$$

Thus the averaged convolution sequence $\frac{1}{n}\sum_{j=1}^{n} v^{(j)}$ converges to η as $n\to\infty$. For fixed s, $f(ss')$ is a continuous function of s' if f is continuous. Replace $f(s')$ by $f(ss')$ in (8) so as to obtain

$$\lim_{n\to\infty}(T_n f)(s)=(\bar{T}f)(s)=\bar{f}(s)=\int f(ss')\eta(ds'). \tag{9}$$

Since $\bar{f}$ is left invariant by $\bar{T}$ and $\bar{T}$ is given in terms of η by (9) it follows that $(A^2f)(s)=(Af)(s)$. If we integrate (9) with respect to v, we obtain

$$\int f(s)\eta(ds)=\int(\bar{T}f)(s)v(ds)=\int\{\int f(ss')\eta(ds')\}v(ds)=\int f(s)(v*\eta)(ds)$$

which implies that $\nu * \eta = \eta$. But then $\nu^{(n)} * \eta = \eta$ for all n and hence $\eta * \eta = \eta$. A similar argument with the equicontinuous operator

$$(T'f)(s) = \int f(s's)\nu(ds') \tag{10}$$

indicates that $(B^2 f)(s) = (Bf)(s)$ and $\eta * \nu = \eta$.

The corollary derived above characterizes limiting measures η of convolution sequences (or averaged convolution sequences) as idempotent measures (6). A more detailed analysis of idempotent measures on compact semigroups requires information on the structure of such semigroups. In the next section we shall develop the required results on the structure of compact semigroups and apply them in the analysis of idempotent measures. However, one can already say a good deal further about idempotent measures on compact groups. It will be helpful to have the following Lemma.

Lemma 4. *Let ν be a regular probability measure on the compact Hausdorff group G. Then the closed semigroup $S(\nu)$ generated by the support $\sigma(\nu)$ of ν is a compact Hausdorff subgroup of G. If η is the limit measure of the averaged convolution sequence $\frac{1}{n}\sum_{j=1}^{n}\nu^{(j)}$ as $n \to \infty$, then $\sigma(\eta) = S(\nu)$.*

Let $\eta = \lim \frac{1}{n}\sum_{j=1}^{n}\nu^{(j)}$. Then $\eta^{(2)} = \eta$ and Lemma 1 implies that $\sigma(\eta)^2 = \sigma(\eta)$. Obviously $\sigma(\eta)$ is a compact Hausdorff subsemigroup of G since it is a closed subset. To show that it is a compact subgroup it will be enough to establish that if $g \in \sigma(\eta)$ then $g^{-1} \in \sigma(\eta)$ since compactness then follows by the continuity of the inverse. If $g \in \sigma(\eta)$ then $g^n \in \sigma(\eta)$ for $n = 1, 2, \ldots$. If e were not in the closure of $\{g^n\}$ there would be a neighborhood of e disjoint from $\overline{\{g^n\}}$. But then there would be a countably infinite collection of disjoint open sets O_n with $g^n \in O_n$ (because of the continuity of multiplication) contradicting the compactness of $\overline{\{g^n\}}$. However, if e is in $\overline{\{g^n\}}$, it follows that g^{-1} is also. Now that it has been established that $\sigma(\eta)$ is a compact subgroup let us make use of the fact that $\nu * \eta = \eta$. This implies by Lemma 1 that $\sigma(\nu)\sigma(\eta) = \sigma(\eta)$. Since $\sigma(\eta)$ is a group, $\sigma(\nu) \subset \sigma(\eta)$. Clearly, then $S(\nu) \subset \sigma(\eta)$. However, since $\sigma(\eta) \subset S(\nu)$ because η is the limit of the averaged convolution sequence generated from ν, we have

$$S(\nu) = \sigma(\eta).$$

The proof of the Lemma is complete.

Theorem 2. *Let ν be a regular probability measure on the compact Hausdorff group G with $S(\nu)=G$. Then the regular probability measure*

$$\eta = \lim_{n\to\infty} \frac{1}{n} \sum_{j=1}^{n} \nu^{(j)}$$

is the normalized "uniform" measure (or Haar measure) on G, that is, for each Borel set $B\in\mathscr{B}$ and each $g\in G$

$$\eta(g^{-1}B) = \eta(Bg^{-1}) = \eta(B).$$

Let the operator T be given by (7). Then

$$(\overline{T}f)(s) = \int f(ss')\eta(ds')$$

for continuous f and $(\overline{T}^2 f)(s)=(\overline{T}f)(s)$. We wish to show that in the case of a compact group G with $\sigma(\eta)=G$ the function $(\overline{T}f)(s)$ is a constant. Let

$$\alpha = \max_g \overline{T}f(g), \qquad \beta = \min_g \overline{T}f(g)$$

and suppose that $\alpha>\beta$. Let g_1, g_2 be elements of G such that $\alpha=\overline{T}f(g_1)$, $\beta=\overline{T}f(g_2)$. Then there is an open set O such that $(\overline{T}f)(g_1 g)<\beta+(\alpha-\beta)/2$ for $g\in g_1^{-1}O$ and $\eta(O)>0$. But then

$$(\overline{T}f)(g_1) = \int (\overline{T}f)(g_1 g)\eta(dg) < (\overline{T}f)(g_1)$$

so that we have a contradiction. Thus, for each continuous f, $\overline{T}f$ is a constant function. In fact,

$$(\overline{T}f)(g) \equiv \int (\overline{T}f)(g)\nu(dg) = L(f) = \int f(g)\eta(dg).$$

This implies that $\eta(g^{-1}B)=\eta(B)$ for each $g\in G$ and $B\in\mathscr{B}$. A similar discussion with T' as given by (10) implies that $\eta(Bg^{-1})=\eta(B)$. We note that the regular probability measure with these left and right invariance properties is uniquely determined. For if η, ζ are two such measures, then

$$\begin{aligned}\int \{\int f(gg')\eta(dg)\}\zeta(dg') = \int f(g)\eta(dg) &= \int \{\int f(gg')\zeta(dg')\}\eta(dg)\\ &= \int f(g)\zeta(dg)\end{aligned}$$

for each continuous function f.

Theorem 2 in effect provides us with a constructive definition of the uniform or Haar measure on a compact Hausdorff group.

4. The Structure of Compact Semigroups

The principal aim of this section is to obtain information about the structure of compact Hausdorff semigroups. This information will then be used to get a more detailed description of the asymptotic behaviour of convolution sequences and the form of idempotent probability measures on compact semigroups.

Let S be a compact Hausdorff semigroup. A *nonempty subset I of S is called a left (right) ideal if* $SI \subset I$ $(IS \subset I)$. If I is both a left and right ideal of S it is called a *two-sided ideal*. Algebraically speaking, an ideal of S is a subsemigroup of S. If the ideal is closed, it is a compact topological semigroup. A *minimal left (right) ideal* of S is a left (right) ideal containing no other left (right) ideal of S. We first show that minimal left and right ideals of a compact Hausdorff semigroup exist.

Lemma 1. *Let S be a compact Hausdorff semigroup. Each left (right) ideal of S contains at least one minimal left (right) ideal and each minimal left (right) ideal is closed.*

The argument need only be given for left ideals since it is quite analogous for right ideals. Let I be any left ideal of S. Consider the collection of Q of all closed left ideals of S contained in the ideal I. Q is not empty since if $s \in I$, then Ss is a closed left ideal (by the continuity of multiplication) contained in I. Given any two closed left ideals $J, J' \in Q$ we shall say that J precedes J' if $J \subset J'$. This sets up a partial ordering in Q. Let Q' be a subcollection of Q that is linearly ordered, that is, for any $J, J' \in Q'$, either J precedes J' or J' precedes J. Then the intersection of the ideals of Q' is nonempty because of the compactness of S. The intersection is a closed left ideal in Q that is contained in every ideal J of Q'. Therefore, every linearly ordered subset of Q has a smallest element. By Zorn's Lemma (see [20] p. 6) there is a minimal element J_0 in Q, that is, there is no element J in Q that precedes J_0. Let us now show that J_0 is a minimal left ideal. Assume that J_1 is a left ideal contained in J_0 and that $\sigma \in J_1$. The set $S\sigma \subset J_1 \subset J_0$ is a closed left ideal of S. Since J_0 is minimal in Q it follows that $S\sigma = J_1 = J_0$. Therefore, J_0 is a minimal left ideal. However, any minimal left ideal J is closed. For if $\sigma \in J$ then $S\sigma \subset J$ is a closed left ideal which by the minimality of J must equal J. The proof is complete.

The intersection of all two-sided ideals of a semigroup S is called the *kernel* K of S. If K is not empty, it is the smallest two-sided ideal of S. We shall show that a compact Hausdorff semigroup has a nonempty kernel and give its structure in terms of the minimal left and right ideals. The following Lemma will be required for these results.

Lemma 2. *Let G be a compact Hausdorff space that is a group algebraically with the multiplication jointly continuous in both variables. Then G is a compact topological (Hausdorff) group, that is, the inverse operation is continuous.*

The Lemma will follow if one can show that for any closed set A of G, the set $A^{-1}=\{g:g^{-1}\in A\}$ is closed. Let $g\in\overline{A^{-1}}$ and let $\{g_\alpha\}$ be a net (see appendix 2) in A^{-1} with limit g. Since $g_\alpha\in A^{-1}$ it follows that $g_\alpha^{-1}\in A$. By the compactness of G there is a cluster point $h\in A$ of the points g_α^{-1}. Suppose $gh=l\neq e$ where e is the identity of the group G. Consider an open set O such that $l\in O$, $e\notin O$. By the joint continuity in both variables of the product, there are open sets O_1 and O_2 with $g\in O_1$, $h\in O_2$ such that $O_1O_2\subset O$. However, $g_\alpha g_\alpha^{-1}=e$ for $g_\alpha\in O_1$, $g_\alpha^{-1}\in O_2$ which leads to a contradiction. Thus, $h=g^{-1}$ and A^{-1} is closed.

An element e of a semigroup for which $e^2=e$ is called an idempotent. Idempotents will be of some interest in analyzing the structure of the kernel of a semigroup.

Theorem 1. *Let S be a compact Hausdorff semigroup. Then the kernel K of S is nonempty. Further,*

(i) *If J_1 and J_2 are both minimal left (right) ideals and $J_1\cap J_2$ is nonempty, then $J_1=J_2$.*

(ii) *If J is a minimal left (right) ideal, then $J\sigma=J$ $(\sigma J=J)$ for all $\sigma\in J$.*

(iii) *K is the union of all minimal left (right) ideals.*

(iv) *If J_1 and J_2 are minimal left and right ideals respectively, then $J_1\cap J_2$ contains a unique idempotent. If the idempotent is e, then $J_1\cap J_2=J_2J_1=eSe$ and with e as identity J_2J_1 is a compact Hausdorff group.*

(v) *The minimal left (right) ideals of S are the sets Se (eS), where e is some idempotent of the kernel K. The kernel K is a compact Hausdorff subsemigroup of S and for any pair J_1, J_2 of minimal left and right ideals $K=J_1J_2$.*

We need only give the arguments for left ideals in (i), (ii), and (iii) since the arguments for right ideals are the same. If J_1 and J_2 are minimal left ideals, then $J_1\cap J_2$ is a left ideal and by the minimality $J_1=J_1\cap J_2=J_2$. If σ belongs to the minimal left ideal J_1, then $J_1\sigma\subset J_1$ is a minimal left ideal and so $J_1\sigma=J_1$. Let us now prove (iii). If I is a minimal left ideal and $\tau\in S$, then $I\tau$ is a minimal left ideal. For if there were a left ideal J properly contained in $I\tau$, then $I\cap\{\sigma:\sigma\tau\in J\}$ would be a left ideal properly contained in I. Therefore $\bigcup_{\tau\in S} I\tau$ is a union of minimal left ideals and is a two-sided ideal. If I_1 is any two-sided ideal of S and I is a minimal left ideal, since $I=I_1 I\subset I_1$ (by the minimality of I) it follows that $\bigcup_{\tau\in S} I\tau\subset I_1$. Thus, $K=\bigcup_{\tau\in S} I\tau$. Any minimal left ideal I_2 must be contained in K since K is a two-sided ideal and so I_2 must be an $I\tau$. The proof of (iii) is complete. From (i) and (iii) it follows that the kernel K is the disjoint union

$$K=\bigcup_{I,J} I\cap J=\bigcup_{I,J} JI \tag{1}$$

where the I and J are minimal left and right ideals respectively. Let I and J be minimal left and right ideals. Then since $JI\subset I\cap J$ it follows that $I\cap J$ is not empty. Further, (1) implies that $JI=I\cap J$. If $\sigma\in I$, $\tau\in J$ then

$$(JI)\sigma=JI,\qquad \tau(JI)=JI. \tag{2}$$

We now use (2) to show that $JI=I\cap J$ is a group. It is clear that $I\cap J$ is a semigroup. If $\sigma\in JI$, there is an element $e\in JI$ such that $e\sigma=\sigma$. The relation $\sigma JI=JI$ implies that e is a left identity. There is similarly a right identity e'. But $e=ee'=e'$ so that e is the identity. Left inverses σ^{-1} exist since $JI\sigma=JI$ for each σ in JI. There is similarly a right inverse σ'^{-1} of σ. But $\sigma^{-1}=\sigma^{-1}\sigma\sigma'^{-1}$ so that σ^{-1} is the inverse of σ. The continuity of multiplication jointly in both variables and Lemma 2 imply that JI is a compact Hausdorff group with e as the identity. Since $I=Se$ and $J=eS$ by (ii), it follows that $JI=eSe$. Let J_1, J_2 be given minimal left and right ideals, respectively. Consider any minimal left ideal I. If $\sigma\in I$, then $J_1\sigma=I$. However, $I\cap J_2$ is nonempty for all I. Therefore $K=J_1 J_2$ by (iii).

Theorem 1 has given us information about the structure of the kernel of a compact semigroup. These structural results on the kernel will presently be rephrased in a form more convenient for our purposes. However, before proceeding in this direction we shall derive the following Lemma which indicates how important the kernel of the semigroup is in describing the asymptotic behavior of a convolution sequence on the semigroup.

Lemma 3. *Let v be a regular probability measure on the compact Hausdorff semigroup S. Assume that the support of v, $\sigma(v)$, generates S. Then the mass of the convolution sequence $v^{(n)}$ concentrates on the kernel K of S as $n\to\infty$, that is, given any open set O with $K\subset O$ and any fixed $\varepsilon>0$, there is a sufficiently large positive integer m such that for $n>m$*

$$v^{(n)}(O)>1-\varepsilon.$$

Let $N(k)$ be any given neighborhood of an element $k\in K$. Since $\sigma(v)$ generates S, there are elements $s_1,\dots,s_m\in\sigma(v)$ (m finite) and neighborhoods $N(s_i)$ of s_i, $i=1,\ldots,m$, such that $\prod_{i=1}^{m} N(s_i) \subset N(k)$. Let O be an open set containing K. Then there is a neighborhood $N(k)$ of k such that $SN(k)S \subset O$. For if s, $s' \in S$ there are neighborhoods $N_1(s)$, $N_2(s')$ of s and s', and there is a neighborhood $N_{s,s'}(k)$ of k such that

$$N_1(s) N_{s,s'}(k) N_2(s') \subset O.$$

Because of the compactness of S there are a finite number of open sets $N_1(s_j) \times N_2(s_j')$, $j=1,\ldots,n$, that cover $S \times S$. Set

$$N(k) = \bigcap_{j=1}^{n} N_{s_j,s_j'}(k).$$

It is clear that $N(k)$ is such that $SN(k)S \subset O$. Look at the product space S^N of sequences $(t_1, t_2, \ldots)$ of elements $t_i \in S$ with product measure generated by ν. Notice that $\nu(N(s_i)) > 0$ for $i=1,\ldots,m$. By the Borel-Cantelli Lemma (see Appendix 1) there is at least one block of m successive elements from $N(s_1),\ldots,N(s_m)$ respectively in any sequence $(t_2, t_3, \ldots)$ with measure one. Choose n large enough so that the measure of the set with a block of m such successive elements in truncated sequences of length n is greater than $1-\varepsilon$. But this implies that $\nu^{(n+1)}(O) > 1-\varepsilon$. The proof is complete.

Lemma 1 tells us that any limit measure of a convolution sequence $\nu^{(n)}$ on a compact semigroup must be concentrated on the kernel of the semigroup. It is natural to try to determine the structure of idempotent probability measures η $(\eta * \eta = \eta)$ to which convolution sequences converge in terms of the kernel that contains the support of the measure η. Semigroups that have the properties determined in Theorem 1 for the kernel are commonly called *kernel semigroups* or *completely simple semigroups* (see [69] for an extended discussion). We will now construct rather special compact semigroups called *Rees products* and then show that every compact kernel semigroup can be given a representation as a Rees product. Let X, Y be compact Hausdorff spaces with G a compact Hausdorff group. Let $\varphi(x,y)$ be a continuous mapping of $X \times Y$ into G. Then $X \times G \times Y$ with the product topology is a compact Hausdorff semigroup when given the multiplication

$$(x,g,y)(x',g',y') = (x, g\,\varphi(x',y)\,g', y'). \tag{3}$$

Notice that X is the space of minimal right ideal labels and Y the space of minimal left ideal labels. The semigroup $X \times G \times Y$ with the multiplication given by (3) is called a Rees product of the group G over the spaces X, Y. It is clear by inspection that it is a kernel semigroup.

Theorem 2. *A compact Hausdorff semigroup K is a kernel semigroup if and only if it is isomorphic algebraically and topologically to a compact Rees product $X \times G \times Y$ with G a group.*

The "if" part of this result is already obvious. Suppose K is a kernel semigroup. Choose the specific minimal left ideal I_0 and minimal right ideal J_0. By (iv) of Theorem 1 $I_0 \cap J_0$ is a compact Hausdorff group G. The idempotent $e \in G$ is the identity of G. Let $X(Y)$ be the set of idempotents in $Ke(eK)$. Then $X(Y)$ is compact Hausdorff since $Ke(eK)$ is a compact semigroup and the set of idempotents in a compact Hausdorff semigroup is closed. Notice that $G = eKe$. We now construct a mapping of K onto $X \times G \times Y$ which is one-to-one. Given $k \in K$, let $x(k)$ $(y(k))$ be the idempotent of kKe (eKk) and $g(k) = eke$. The mapping takes k into $(x(k), g(k), y(k)) \in X \times G \times Y$. In the multiplication on $X \times G \times Y$

$$\begin{aligned}(x(k), g(k), y(k))\,(x(k'), g(k'), y(k'))\\ = (x(k), g(k)\,\varphi(x(k'), y(k))\,g(k'), y(k')),\end{aligned} \tag{4}$$

the mapping

$$\varphi(xy) = yx$$

since

$$g(k)y(k)x(k')g(k') = eky(k)x(k')k'e = ekk'e = g(kk'). \tag{5}$$

The inverse mapping of $X \times G \times Y$ onto K is given by

$$x(k)g(k)y(k) = x(k)key(k) = ky(k). \tag{6}$$

Since $eky(k) = ek$ and there is an element $k' \in K$ such that $k'ek = k$, it follows that

$$ky(k) = k. \tag{7}$$

A similar argument shows that

$$x(k)k = k.$$

In any case from (6) and (7) we see that

$$x(k)g(k)y(k) = k.$$

Since the mapping xgy of the compact Hausdorff space $X \times G \times Y$ (with the induced product topology) onto the compact Hausdorff semigroup K is continuous, the mapping is a homeomorphism. The homeomorphism preserves multiplication as is seen from (4) and (5). Thus, we can always assume that a compact kernel semigroup is a Rees product and it will sometimes be convenient to do this in some derivations.

Lemma 4. *Let ν be a regular probability measure on the compact Hausdorff semigroup S whose support generates S. If the convolution sequence $\nu^{(n)}$, $n=1,2,\ldots$, converges to the limit measure η, then η is an idempotent measure whose support is the kernel K of S.*

We already know that the measure η is an idempotent measure from Corollary 3.1. Lemma 3 implies that the support of η is contained in the kernel K of S. Further, Corollary 3.1 tells us that $\nu^{(n)}*\eta=\eta*\nu^{(n)}=\eta$, $n=1,2,\ldots$. Lemma 3.1 and Theorem 1 indicate that the support of η must in fact be K.

Of course, the conclusion of Lemma 4 is valid for the limit η of the averaged convolution sequence $\frac{1}{n}\sum_{j=1}^{n}\nu^{(j)}$ as $n\to\infty$. In fact, Lemma 4 directly implies that any regular idempotent probability measure η must have a kernel semigroup as its support. Our object now is to describe the structure of idempotent probability measures η in terms of the kernel semigroup that supports them.

Lemma 5. *Let η be a regular idempotent probability measure with support the kernel semigroup K. Then $\eta(k^{-1}A)$ $(\eta(A k^{-1}))$ for A a Borel subset of the minimal right (left) ideal I is independent of $k\in I$.*

The argument is given for right ideals since it is the same for left ideals. Let f be any continuous function on the minimal right ideal I. The support of $\eta(k^{-1}\cdot)$ for $k\in I$ is the ideal I. Consider

$$\bar{f}(k)=\int\eta(k^{-1}dk')\,f(k').$$

Now

$$\bar{f}(k)=\int\eta(k^{-1}dk')\,\bar{f}(k')$$

since η is an idempotent measure. Because the support of $\eta(k^{-1}\cdot)$ is I, by an argument like that given in Theorem 3.2, one can show that $\bar{f}(k)$ is constant on I. However, this implies the conclusion of the Lemma.

Let us now consider a regular idempotent probability measure η on a compact Hausdorff kernel semigroup K. The Borel field $\mathscr{B}$ on K is as usual the Borel field generated by the topology on K. It is convenient to take the representation of K as a Rees product $X\times G\times Y$ where X, Y are compact Hausdorff spaces and G is a compact Hausdorff group. Let $\mathscr{B}(X)$, $\mathscr{B}(G)$, and $\mathscr{B}(Y)$ be the Borel fields on X, G and Y respectively generated by the topologies on these spaces. The structure of η will be described on the product Borel field $\mathscr{B}(X)\times\mathscr{B}(G)\times\mathscr{B}(Y)$.

Theorem 3. *Let η be a regular idempotent probability measure on the compact Hausdorff kernel semigroup $K=X\times G\times Y$ with support K. Then η is a product measure*

$$\alpha\times\chi\times\beta$$

on the Borel field $\mathscr{B}(X)\times\mathscr{B}(G)\times\mathscr{B}(Y)$ *where* χ *is the normed Haar measure of the compact group* $G(\chi(G)=1)$ *and* α, β *are regular probability measures on* X, Y *with supports* X, Y *respectively.*

Let us look at

$$\eta(k^{-1}(U\times V\times W)) \tag{8}$$

where $U\times V\times W$ is a product set with $U\in\mathscr{B}(X)$, $V\in\mathscr{B}(G)$, $W\in\mathscr{B}(Y)$. If (8) is positive one must have $x(k)\in U$. Now

$$k^{-1}(U\times V\times W)=\{k':g(k')\in\varphi(x(k'),y(k))^{-1}g(k)^{-1}V, y(k')\in W\}. \tag{9}$$

Lemma 5 implies that (8) is constant for all k belonging to the same minimal right ideal. But then by (9)

$$\eta(k^{-1}(U\times V\times W))=\eta(k^{-1}(U\times gV\times W)) \tag{10}$$

for all $g\in G$ if $x(k)\in U$. Expression (10) is zero if $x(k)\notin U$. Therefore,

$$\eta(k^{-1}(U\times V\times W))=\chi(V)\,\eta(k^{-1}(U\times G\times W))$$

where χ is the normed Haar measure of G.

However,

$$\eta(k^{-1}(U\times G\times W))=\beta(W)$$

if $x(k)\in U$ and zero otherwise. Now

$$\begin{aligned}\eta((U\times V\times W))&=\int\eta(k^{-1}(U\times V\times W))\,\eta(dk)\\&=\int_{x(k)\in U}\chi(V)\beta(W)\eta(dk)=\alpha(U)\chi(V)\beta(W).\end{aligned} \tag{11}$$

The proof is complete.

If ν is a regular probability measure on a compact Hausdorff semigroup S whose support generates S, it is in general difficult to determine the idempotent measure η to which the averaged convolution sequence $\frac{1}{n}\sum_{j=1}^{n}\nu^{(j)}$ converges as $n\to\infty$. However, if S is already a kernel semigroup K itself, the determination of η is trivial. Using the representation of $S=K$ as a Rees product $X\times G\times Y$, it follows from (11) that

$$\eta(U\times V\times W)=\alpha(U)\chi(V)\beta(W)$$

for $U\in\mathscr{B}(X)$, $V\in\mathscr{B}(G)$, $W\in\mathscr{B}(Y)$ with

$$\alpha(U)=\nu(U\times G\times Y)$$

and

$$\beta(W)=\nu(X\times G\times W).$$

Given a compact Hausdorff semigroup S, it is clearly of interest to determine the kernel K of S. Let us consider the semigroup S_m of $m\times m$

(m finite) stochastic matrices introduced in section 2. Every stochastic $m \times m$ matrix U of the form

$$U = \begin{pmatrix} u_1 \, u_2 \dots u_m \\ \cdots\cdots\cdots \\ u_1 \, u_2 \dots u_m \end{pmatrix} \tag{12}$$

with $u_i \geqslant 0$, $\sum_{i=1}^{m} u_i = 1$, is an idempotent, that is,

$$U^2 = U .$$

These are the stochastic matrices all of whose rows are the same. For every stochastic matrix M

$$M U = U .$$

This implies that each minimal left ideal has one and only one element which is a matrix of the form (12). The set of all matrices of the form (12) is the unique minimal right ideal since every stochastic matrix M under left multiplication by a matrix U of the form (12) is taken into another matrix of the same form. The kernel of the semigroup of all $m \times m$ stochastic matrices is therefore the set of all matrices of the form (12). Notice that in the Rees product representation (3) of this kernel K the sets X and G are degenerate since they each contain exactly one point.

If v is any given regular probability measure on the semigroup S_m of $m \times m$ stochastic matrices, the closed semigroup $S(v)$ generated by v may be properly contained in S_m. In that case the kernel of $S(v)$ will be different from the kernel semigroup determined in the preceding paragraph. This suggests *the problem of determining all closed subsemigroups of the semigroup of $m \times m$ stochastic matrices that are kernel semigroups.* Theorem 2 or the Rees product representation tells us that a kernel semigroup can be regarded as a collection of isomorphic groups bound together in an appropriate manner. With this in mind let us first determine the form of a subgroup of S_m and especially that of the identity of the group, an idempotent. An idempotent $m \times m$ stochastic matrix U takes the following form. There is a partition of the integers $1, 2, \dots, m$ into disjoint classes of integers $T, C_1, \dots, C_r$ of $n_0, n_1, \dots, n_r$ integers respectively with $\sum_{\alpha=0}^{r} n_\alpha = m$. It is convenient to assume that the integers $1, 2, \dots, m$ first run through the elements of T and then through the elements of $C_1, \dots, C_r$ respectively. There are column vectors

$$\rho^{(\alpha)} = (\rho_j^{(\alpha)}; j \in T), \quad \rho_j^{(\alpha)} \geqslant 0, \quad \sum_{\alpha=1}^{r} \rho_j^{(\alpha)} = 1, \tag{13}$$

$$u^{(\alpha)} = (u_j^{(\alpha)}; j \in C_\alpha), \quad u_j^{(\alpha)} > 0, \quad \sum_j u_j^{(\alpha)} = 1, \tag{14}$$

$$1^{(\alpha)} = (1; j \in C_\alpha), \tag{15}$$

$\alpha = 1, \ldots, r$. The idempotent U can then be written in block matrix form (see [95])

$$U = \begin{pmatrix} 0 & \rho^{(1)} u^{(1)\prime} & \rho^{(2)} u^{(2)\prime} & \ldots & \rho^{(r)} u^{(r)\prime} \\ 0 & 1^{(1)} u^{(1)\prime} & 0 & \ldots & 0 \\ 0 & 0 & 1^{(2)} u^{(2)\prime} & \ldots & 0 \\ \ldots & \ldots & \ldots & \ldots & \ldots \\ 0 & 0 & 0 & \ldots & 1^{(r)} u^{(r)\prime} \end{pmatrix}. \tag{16}$$

Let $M = (m_{i,j})$ be any other element of a group of stochastic $m \times m$ matrices with the idempotent U as identity.

Then

$$MU = UM = M.$$

Clearly,

$$m_{ij} = 0 \quad \text{if } j \in T.$$

Moreover, if $i \in C_\alpha, j \in C_\beta$, then

$$\begin{aligned} m_{i,j} &= \sum_k m_{i,k} u_{k,j} = \sum_{k \in C_\beta} m_{i,k} u_j^{(\beta)} = m_{i,C_\beta} u_j^{(\beta)} \\ &= \sum_k u_{i,k} m_{k,j} = \sum_{k \in C_\alpha} u_k^{(\alpha)} m_{k,j}. \end{aligned}$$

This implies that

$$m_{i,j} = m(\alpha, \beta) u_j^{(\beta)} \quad \text{if } i \in C_\alpha, \ j \in C_\beta.$$

Further, if $i \in T$ and $j \in C_\alpha$ then

$$\begin{aligned} m_{i,j} &= \sum_k u_{i,k} m_{k,j} = \sum_\beta \rho_i^{(\beta)} \sum_{k \in C_\beta} u_k^{(\beta)} m_{k,j} = \sum_\beta \rho_i^{(\beta)} \sum_{k \in C_\beta} u_k^{(\beta)} m(\beta, \alpha) u_j^{(\alpha)} \\ &= \Big\{ \sum_\beta \rho_i^{(\beta)} m(\beta, \alpha) \ u_j^{(\alpha)} \Big\}. \end{aligned}$$

Let $\bar{M} = \{\bar{m}_{i,j}\}$ be the stochastic matrix that is the inverse of M in the group with identity U. Then by the argument just given

$$\begin{aligned} \bar{m}_{i,j} &= 0 \quad \text{if } j \in T, \\ \bar{m}_{i,j} &= \bar{m}(\alpha, \beta) u_j^{(\beta)} \quad \text{if } i \in C_\alpha, \ j \in C_\beta. \end{aligned} \tag{17}$$

Since

$$\bar{M}M = M\bar{M} = U$$

it follows from (17) that

$$\sum_{\beta} m(\alpha,\beta)\bar{m}(\beta,\gamma) = \sum_{\beta} \bar{m}(\alpha,\beta)m(\beta,\gamma) = \delta_{\alpha-\gamma}.$$

If $m(\alpha,\beta)>0$ it follows that $\bar{m}(\beta,\gamma)=0$ for all $\gamma\neq\alpha$. Hence $\bar{m}(\beta,\alpha)=m(\alpha,\beta)=1$. To each α there is a unique $\beta(\alpha)$ such that $m(\alpha,\beta(\alpha))=\bar{m}(\beta(\alpha),\alpha)=1$. The matrix $\{m(\alpha,\beta)\}$ is a permutation matrix and $\{\bar{m}(\alpha,\beta)\}$ is its inverse. The following Lemma is an immediate consequence.

Lemma 6. *Let $T, C_1, \ldots, C_r$ be any given partition of the integers $1, \ldots, m$ with the C_α nonempty. Take fixed vectors $\rho^{(\alpha)}$, $u^{(\alpha)}$, $1^{(\alpha)}$ of the form* (13), (14) *and* (15) *for $\alpha=1,\ldots,r$. Consider a subgroup H of the permutation group on r integers. Then the set of matrices*

$$\begin{pmatrix} 0 & \rho^{(h1)}u^{(1)\prime} & \ldots & \rho^{(hr)}u^{(r)\prime} \\ 0 & \delta_{1-h1}1^{(1)}u^{(1)\prime} & \ldots & \delta_{1-hr}1^{(1)}u^{(r)\prime} \\ \cdots & \cdots & \cdots & \cdots \\ 0 & \delta_{r-h1}1^{(r)}u^{(1)\prime} & \ldots & \delta_{r-hr}1^{(r)}u^{(r)\prime} \end{pmatrix} \tag{18}$$

for $h\in H$ is a group of stochastic matrices isomorphic to the permutation group H. Further, every group of stochastic $m\times m$ matrices can be described in such a manner by specifying the partition, the vectors $\rho^{(\alpha)}$ and $u^{(\alpha)}$, as well as the group of permutations H.

Let us now consider a closed subsemigroup K of S_m that is a kernel semigroup. From (16) it follows that a specific maximal subgroup of K is of the form given in (18) for an appropriate choice of the partition, the vectors $\rho^{(\alpha)}$, $u^{(\alpha)}$ and the group H. Let us call this specific group the distinguished group for convenience in the derivation. Let M be any matrix of K not in the distinguished group. Write M in the block matrix form

$$M=(M^{(\alpha,\beta)};\ \alpha,\beta=0,1,\ldots,r)$$

with $M^{(\alpha,\beta)}$ an $n_\alpha\times n_\beta$ matrix. If M is premultiplied and postmultiplied by elements of the distinguished group, the resulting product is again an element of the distinguished group by (3). This implies that

$$\sum_{\beta=1}^{r} \delta_{\alpha-h\beta}1^{(\alpha)}u^{(\beta)\prime}M^{(\beta,0)}\rho^{(h'\varepsilon)}u^{(\varepsilon)\prime} + \sum_{\beta,\gamma=1}^{r} \delta_{\alpha-h\beta}1^{(\alpha)}u^{(\beta)\prime}M^{(\beta,\gamma)}\delta_{\gamma-h'\varepsilon}1^{(\gamma)}u^{(\varepsilon)\prime}$$
$$= \delta_{\alpha-h''\varepsilon}1^{(\alpha)}u^{(\varepsilon)\prime} \tag{19}$$

$\alpha, \varepsilon = 1, \ldots, r$ with h, h', h'' appropriate elements of the group H. Since the right hand side of equation (19) is the null matrix if $\alpha \neq h''\varepsilon$, it follows that $M^{(\beta,0)} = 0$ for $\beta \neq 0$. Also

$$M^{(\beta,\gamma)} = \delta_{\beta - g\gamma} Q^{(\beta,\gamma)}, \qquad \beta, \gamma = 1, \ldots, r$$

with $g = h^{-1} h'' h'^{-1}$ and $Q^{(\beta,\gamma)}$ a matrix with nonnegative entries and row sums one. An argument like that given in the proof of Lemma 6 implies that the row vectors of $Q^{(\beta,\gamma)}$ are all the same. From (4) it is clear that if an appropriate element of the distinguished group is premultiplied and postmultiplied by M, the resulting product is M. This implies that $M^{(0,0)} = 0$. The matrix M must therefore be of the form (18) with the same partition $T, C_1, \ldots, C_r$ and the same group H but with possibly different vectors $u^{(\alpha)}$, $\rho^{(\alpha)}$. Further, some of the entries in the $u^{(\alpha)}$ vectors may be zero. The proof of the following theorem is complete.

Theorem 4. *Every closed subsemigroup of the* $m \times m$ *stochastic matrices that is a kernel semigroup K can be described in the following way. Let* $T, C_1, \ldots, C_r$ *be a partition of the integers* $1, \ldots, m$ *and H a subgroup of the permutation group on* $1, \ldots, r$. *The elements of the kernel semigroup K are matrices of the form* (18) *with*

$$u_j^{(\alpha)} \geq 0, \qquad j = 1, \ldots, n_\alpha,$$

varying over some closed subset of the direct product of

$$\sum_{j=1}^{n_\alpha} u_j^{(\alpha)} = 1, \qquad \alpha = 1, \ldots, r,$$

and

$$\rho_j^{(\alpha)} \geq 0, \qquad \alpha = 1, \ldots, r$$

varying over some closed subset of the direct product of

$$\sum_{\alpha} \rho_j^{(\alpha)} = 1, \qquad j = 1, \ldots, n_0.$$

The vectors $\{u^{(\alpha)}; \alpha = 1, \ldots, r\}$ *correspond to the left ideal structure of K and the vectors* $\{\rho^{(\alpha)}; \alpha = 1, \ldots, r\}$ *to the right ideal structure of K. The continuous map* $\varphi(x,y)$ *on* $X \times Y$ *referred to in* (3) *is trivial and takes everything into the identity element of the group* $H = G$. *The right ideal structure is trivial if* $r = 1$.

5. Convergent Convolution Sequences

Corollary 1 in section 3 tells us that the averaged convolution sequences on compact semigroups always converge. The main purpose of this section is to obtain necessary and sufficient conditions for the convergence of convolution sequences. We restate Lemma 4.4.1 as Lemma 1 of this section.

Lemma 1. *Let $\mathcal{M}$ be the set of regular probability measures on the compact Hausdorff semigroup S. Let $\prod_f I_f$ be the product space (with the usual product topology) whose coordinates spaces $I_f=[0,1]$ are indexed by the continuous functions f, $0\leqslant f\leqslant 1$, on S. Identify the measure $\nu\in\mathcal{M}$ with the point $c=(c_f)\in\prod_f I_f$ where $c_f=\int f(s)\nu(ds)$. Then $\mathcal{M}$ is a compact subset of $\prod_f I_f$.*

Notice that in the proof of Lemma 1 a topology (open sets) has been set up for the regular probability measures on S generated by sets of the form

$$\{\mu\in\mathcal{M}: \left|\int f(s)[\nu(ds)-\mu(ds)]\right|<\varepsilon\} \tag{1}$$

where f is a continuous function on S, ν a regular probability measure and ε a positive number. Lemma 1 states that $\mathcal{M}$ is a compact set in this topology. The convolution operation $\nu*\mu$ is a mapping of $\mathcal{M}\times\mathcal{M}$ into $\mathcal{M}$. In fact, convolution can be formally considered as a multiplication operation for measures. Our object is to now show that the convolution $\nu*\mu$ on regular probability measures is jointly continuous in ν, μ with respect to the topology generated by the sets (1).

Lemma 2. *The convolution operation $\nu*\mu$ for regular probability measures ν, μ on the compact Hausdorff semigroup S is jointly continuous in ν, μ with respect to the topology generated by the sets* (1).

Let ν, μ be any two measures in $\mathcal{M}$. Consider a continuous function f and a positive number ε. The neighborhood of $\nu*\mu$

$$\{\eta\in\mathcal{M}: \left|\int f(s)[\nu*\mu(ds)-\eta(ds)]\right|<\varepsilon\} \tag{2}$$

is of the form (1). To prove the joint continuity of the convolution in ν, μ jointly it will be enough to exhibit neighborhoods $N(\nu)$, $N(\mu)$ of ν, μ respectively such that for all pairs ν', μ' with $\nu'\in N(\nu)$, $\mu'\in N(\mu)$ the convolution $\nu'*\mu'$ is in (2). Given any $\varepsilon>0$, there is a positive integer $N(\varepsilon)$ and there are continuous functions $g_j(s)$, $h_j(s)$, $j=1,\dots,N(\varepsilon)$ such that

$$u_N(s,s')=\sum_{j=1}^{N} g_j(s)h_j(s')$$

uniformly approximates $f(ss')$

$$\sup_{s,s'}|u_N(s,s')-f(ss')|<\frac{\varepsilon}{4}.$$

Then

$$\left|\int\left\{\int\left[u_N(s,s')-f(s s')\right]v'(ds)\right\}\mu'(ds')\right| < \frac{\varepsilon}{4}$$

for all $v', \mu' \in \mathscr{M}$. Let $M < \infty$ be the upper bound

$$M = \sup_{i=1,\dots,N} \sup_s (|g_i(s)|, |h_i(s)|).$$

Consider the following neighborhoods of v, μ

$$N(v) = \{v' : \left|\int g_i(s)[v(ds) - v'(ds)]\right| < \delta;\ i=1,\dots,N\}$$
$$N(\mu) = \{\mu' : \left|\int h_i(s)[\mu(ds) - \mu'(ds)]\right| < \delta;\ i=1,\dots,N\}.$$

If $v' \in N(v)$, $\mu' \in N(\mu)$, then

$$\left|\int\left\{\int u_N(s,s')\,v(ds)\right\}\mu(ds') - \int\left\{\int u_N(s,s')\,v'(ds)\right\}\mu'(ds')\right|$$
$$\leqslant \sum_{i=1}^{N}\left|\int g_i(s)\,v(ds)\int h_i(s')\,\mu(ds') - \int g_i(s)\,v'(ds)\int h_i(s')\,\mu'(ds')\right| \leqslant 2N\delta M.$$

If $\delta < \varepsilon/4NM$, it follows that

$$\left|\int f(s)\left[(v*\mu)(ds) - v'*\mu'(ds)\right]\right| < \varepsilon.$$

The proof is complete.

Lemmas 1 and 2 imply that the set $\mathscr{M}$ of regular probability measures on S is a compact topological semigroup with the multiplication the convolution operation. We are especially interested in the semigroup $\Sigma(v) = \{v^{(n)}, n=1,2,\dots\}$ generated by a single measure $v \in \mathscr{M}$ and its closure $\overline{\Sigma(v)}$ with respect to the topology generated by the sets (1). The semigroup $\overline{\Sigma(v)}$ is compact since it is a closed subset of the compact set $\mathscr{M}$. Furthermore, it is a commutative semigroup. Theorem 1 of section 4 implies that *the kernel of* $\overline{\Sigma(v)}$ *is a group*. It is now clear that the convolution sequence $v^{(n)}$ will converge as $n \to \infty$ if and only if the kernel of $\overline{\Sigma(v)}$ contains precisely one element.

Lemma 3. *The closure* $\overline{\Sigma(v)}$ *of the semigroup* $\Sigma(v) = \{v^{(n)}, n=1,2,\dots\}$ *generated by a regular probability measure* v *on the compact Hausdorff semigroup* S *is a compact commutative semigroup. The kernel of* $\overline{\Sigma(v)}$ *is a group. The convolution sequence* $v^{(n)}$ *converges as* $n \to \infty$ *if and only if the kernel of* $\overline{\Sigma(v)}$ *contains exactly one element.*

We shall now derive necessary and sufficient conditions for the convergence of a convolution sequence $v^{(n)}$ as $n \to \infty$. The compact Hausdorff semigroup S is assumed to be generated by the support $\sigma(v)$ of v. The conditions will be given in terms of the Rees product representation of the kernel $K = X \times G \times Y$ of S.

Theorem 1. *Let ν be a regular probability measure on the compact Hausdorff semigroup S whose support generates S. The sequence of measures $\nu^{(n)}$, $n=1,2,\ldots$, will not converge as $n\to\infty$ if and only if there is a proper closed normal subgroup G' (proper inclusion) of G such that $YX\subset G'$ and the support of ν is contained in*

$$(X\times G'\times Y)^{-1}(X\times g\,G'\times Y) \tag{3}$$

where $g'\notin G'$. Further, $\overline{\bigcup_{j=1}^{\infty} g^j G'}=G$.

Suppose the support of ν is contained in (3) with G' a proper closed normal subgroup of G and $g'\notin G'$. We shall construct a continuous function f on S such that $\int f(s)\,\nu^{(n)}(ds)$ does not converge as $n\to\infty$. Consider the factor group $H=G/G'$. Since the support of ν generates S, it follows that H is homeomorphic to $\{g'^n;\, n=1,2,\ldots\}$ and so is compact and commutative. Let a_0 be a nonnegative continuous function on H with

$$0=\min_{h\in H} a_0(h)<\max_{h\in H} a_0(h)=1\,.$$

Consider the canonical map a_1 of G onto H and set $a_2(g)=a_0(a_1(g))$, $g\in G$, so that we now have a continuous function a_2 on G with $0=\min a_2<\max a_2=1$. Define $a_3((x,g,y))=a_3(k)=a_2(g)$ for $k=(x,g,y)$ in the kernel K of S and finally set

$$f(s)=a_3(s\,k_0) \tag{4}$$

for all $s\in S$ where k_0 is a fixed element of K. The function f is a real continuous function on S with $0=\min f<\max f=1$ such that

$$0=\liminf_{n\to\infty}\int f(s)\,\nu^{(n)}(ds)<\limsup_{n\to\infty}\int f(s)\,\nu^{(n)}(ds)=1$$

because of (4).

Let us now assume that $\nu^{(n)}$ does not converge as $n\to\infty$. Call the kernel of the closure $\overline{\Sigma(\nu)}$ of the semigroup of measures $\{\nu^{(n)};\, n=1,2,\ldots\}$, $K(\nu)$. By Lemma 3 $K(\nu)$ is a group. Let the idempotent element (the identity of $K(\nu)$ as a group) of $K(\nu)$ be denoted by η_e. Lemma 3 also tells us that there must be measures η in $K(\nu)$ other than η_e since $\nu^{(n)}$ does not converge as $n\to\infty$. Since η_e is an idempotent measure in $\overline{\Sigma(\nu)}$, it must have the form

$$\alpha\times\chi_{G'}\times\beta$$

with α, β regular probability measures on X, Y respectively and $\chi_{G'}$ the normed Haar measure of a closed subgroup G' of G. The supports of α, β must be X, Y respectively since the support of ν generates S. But then one must have $YX \subset G'$. Now G' *must be a proper subgroup of* G. Otherwise, we would have

$$\eta = \eta_e * \eta * \eta_e = \eta_e$$

for every $\eta \in K(\nu)$ since such a measure η has its support in the kernel K of S by Lemma 4.3. However, this contradicts the fact that there are measures $\eta \in K(\nu)$ other than η_e. For any η in $K(\nu)$

$$\eta * \eta_e = \eta_e * \eta = \eta .$$

This implies that the support of η must be contained in a set of the form $X \times CG' \times Y = X \times G'C \times Y$ where C is a closed subset of elements of G. In fact, the projection of the support of η on G is a union of cosets of G'. There is a measure $\eta' \in K(\nu)$ such that

$$\eta' * \eta = \eta * \eta' = \eta_e .$$

However, this implies that the projection of the support of η on G must be a simple coset $g'G' = G'g'$ with $g' \notin G'$. Let $\eta = \nu * \eta_e$. Since the semigroup S is generated by the support of ν, the set $\bigcup_{k=1} g'^k G'$ must be dense in G. This implies that G' is a proper closed normal subgroup of G. It is also clear that the support of ν is contained in $(X \times G' \times Y)^{-1}(X \times g'G' \times Y)$. We have therefore shown that *if* ν *is a regular probability measure on* S *whose support generates* S, *then* $\nu^{(n)}$ *will converge as* $n \to \infty$ *if and only if there is no proper closed normal subgroup* G' *of* G *with* $YX \subset G'$ *such that the support of* ν *is contained in*

$$(X \times G' \times Y)^{-1}(X \times g'G' \times Y)$$

with g' *an element of* G *outside* G'.

As a corollary of Theorem 1 we have the following elegant result of Kawada and Ito [53].

Corollary 1. *Let* ν *be a regular probability measure on the compact topological group* G *whose support generates* G. *Then the convolution sequence* $\nu^{(n)}$ *will converge as* $n \to \infty$ *if and only if there is no proper closed normal subgroup* G' *of* G *such that for some element* g' *of* G *outside* G' *the support of* ν *is contained in the coset* $g'G'$.

Consider now a compact Hausdorff group G with its normalized Haar measure η. Let ν be any regular probability measure on G. The left (right) random walk on G generated by ν has transition function $\nu(g^{-1}A)(\nu(Ag^{-1}))$ for $A \in \mathscr{B}$, $g \in G$. Since η is Haar measure

$$\int \eta(dg)\,\nu(g^{-1}A) = \int \eta(Ag^{-1})\,\nu(dg) = \eta(A)$$

so that η is an invariant probability measure with respect to the transition function $\nu(g^{-1}A)$. Similarly η is seen to be an invariant probability measure for the right random walk generated by ν. Notice that here we do not assume that G is the group generated by the support of ν. Because η is invariant, we can generate the stationary Markov process generated by η and transition function $\nu(g^{-1}A)(\nu(Ag^{-1}))$. Our object is to determine the conditions under which the Markov process is ergodic, mixing or purely nondeterministic.

Theorem 2. *Let G be a compact Hausdorff group with normalized Haar measure η. Consider a regular probability measure ν on G and the stationary left (right) random walk on G generated by ν with invariant measure η. The invariant sets of the left (right) random walk are the measurable sets which are unions of left (right) cosets $gH(Hg)$ where H is the closed subgroup of G generated by the support of ν. Thus the random walks are ergodic if and only if $G=H$. The random walks are mixing if and only if the support of ν is not contained in the coset of a proper closed normal subgroup. The random walks are mixing if and only if they are purely nondeterministic.*

It will be enough to give the arguments for left random walks only. First consider ergodicity. If H is a proper closed subgroup, there is obviously no ergodicity. On the other hand, if $H=G$ we can use Corollary 4.2.6 to show that the process is ergodic. First, given any two continuous functions h and f, we will show that

$$\frac{1}{n}\sum_{j=1}^{n}\int \eta(dg)h(g)\nu^{(j)}(g^{-1}dg')f(g') \to \int \eta(dg)h(g)\int \eta(dg)f(g) \tag{5}$$

as $n\to\infty$. By Theorem 3.2 and the fact that η is Haar measure, it follows that

$$\begin{aligned}\frac{1}{n}\sum_{j=1}^{n}\int \nu^{(j)}(g^{-1}dg')f(g') &= \frac{1}{n}\sum_{j=1}^{n}\int \nu^{(j)}(dg')f(gg')\\ &\to \int \eta(dg')f(gg') = \int \eta(dg)f(g)\end{aligned} \tag{6}$$

as $n\to\infty$. On multiplying (6) by $h(g)$ and integrating with respect to η, (5) is obtained. Let A, B be any two sets of $\mathscr{B}$. Given any $\varepsilon>0$ one can approximate the set indicator functions I_A and I_B in L^1 by continuous functions h and f, respectively, so that

$$\int \eta(dg)\,|h(g)-I_A(g)|, \qquad \int \eta(dg)\,|f(g)-I_B(g)| < \varepsilon \tag{7}$$

with $\max|f|$, $\max|g| \leqslant 2$. But then

$$\begin{aligned} &\left|\int \eta(dg)\, h(g)\, \nu^{(j)}(g^{-1}dg')\, f(g') - \int_A \eta(dg)\, \nu^{(j)}(g^{-1}B)\right| \\ &\leqslant 2\left\{\int \eta(dg)\, |h(g) - I_A(g)| + \int \eta(dg)\, |f(g) - I_B(g)|\right\} < 4\varepsilon . \end{aligned} \tag{8}$$

The possibility of making the approximations (7), (8) together with the limiting relation (5) imply that for each pair of sets $A, B \in \mathscr{B}$

$$\frac{1}{n} \sum_{j=1}^{n} \int_A \eta(dg)\, \nu^{(j)}(g^{-1}B) \to \eta(A)\eta(B)$$

as $n \to \infty$. However, by Corollary 4.2.6 this is equivalent to ergodicity of the random walk. If the left random walk is not ergodic, all the measurable sets which are unions of left cosets gH are clearly invariant sets of the process. A simple application of Theorem 3.2 indicates that these are the only invariant sets (up to exceptional sets of η measure zero).

The characterization of mixing follows from Corollary 1. By formula (4.2.35) the left random walk is mixing if and only if for each pair of sets $A, B \in \mathscr{B}$

$$\int_A \eta(dg)\, \nu^{(n)}(g^{-1}B) \to \eta(A)\eta(B) \tag{9}$$

as $n \to \infty$. However, the same type of approximation argument as that used in the discussion of ergodicity implies that (9) will hold for each pair of sets $A, B \in \mathscr{B}$ if and only if

$$\int \eta(dg)\, h(g)\, \nu^{(n)}(g^{-1}dg')\, f(g') \to \int \eta(dg)\, h(g) \int \eta(dg)\, f(g) \tag{10}$$

as $n \to \infty$ for each pair of continuous functions h and f. Of course, the support of ν must generate G if the random walk is mixing. By Corollary 1

$$\int \nu^{(n)}(dg)\, f(g) \to \int \eta(dg)\, f(g) \tag{11}$$

as $n \to \infty$ for each continuous function f if and only if the support of ν is not contained in a coset of a proper closed normal subgroup of G. However, (11) is equivalent to

$$\int \nu^{(n)}(g^{-1}dg')\, f(g') \to \int \eta(dg)\, f(g) \tag{12}$$

as $n \to \infty$ for each continuous f. But (12) is equivalent to (10) because of the equicontinuity of the sequence of functions

$$\int \nu^{(n)}(g^{-1}dg')\, f(g')$$

when f is continuous. Thus the random walk is mixing if and only if the support of ν is not contained in a coset of a proper closed normal subgroup of G. It is now easy to show that for the random walks mixing is equivalent to being purely nondeterministic. We have already seen that (12) is equivalent to mixing. However, given any set $A \in \mathscr{B}$ and any $\varepsilon > 0$ there is a continuous function f such that

$$\int \eta(dg)|I_A - f| < \varepsilon. \tag{13}$$

Then

$$\int \eta(dg) \left| \int v^{(n)}(g^{-1} dg') \{f(g') - I_A(g')\} \right| < \varepsilon. \tag{14}$$

From (12), (13) and (14) it follows that

$$v^{(n)}(g^{-1} A) \to \eta(A) \tag{15}$$

in the mean (with respect to η) as $n \to \infty$. But this means that the random walk is purely nondeterministic. It is easy to show that if (15) is satisfied for every $A \in \mathscr{A}$, then the random walk is mixing.

Stationary random walks on groups have been discussed in some detail. Suppose we now consider stationary random walks on compact Hausdorff semigroups. Lemma 4.3 implies that one could just as well restrict oneself to completely simple semigroups K. Let v be a regular probability measure on K. Assume that the support of v generates K. The transition function $v(k^{-1}A)(v(Ak^{-1}))$, $A \in \mathscr{A}$, generates a left (right) random walk on K. Corollary 3.1 indicates that

$$\eta = \lim_{n \to \infty} \frac{1}{n} \sum_{j=1}^{n} v^{(j)} \tag{16}$$

exists and is an idempotent probability measure that is an invariant measure for the transition function $v(k^{-1}A)$ $(v(Ak^{-1}))$. We shall consider the stationary left (right) random walk on K with transition function $v(k^{-1}A)$ $(v(Ak^{-1}))$ and invariant measure η.

Theorem 4. *Let K be a compact completely simple semigroup. Let v be a regular measure on K whose support generates K. Consider the stationary left (right) random walk on K generated by v with invariant measure η given by* (16). *The left (right) random walk is ergodic if and only if in the Rees product representation $X \times G \times Y$ of K, $X(Y)$ is trivial (that is, the right (left) ideal structure of K is trivial) and the probability distribution for $g\varphi(y)$ $(\varphi(x)g)$ induced by v has a support generating the whole group G. The left (right) random walk is mixing if and only if the probability distribution for $g\varphi(y)$ $(\varphi(x)g)$ induced by v has a support which is not contained in a coset of a proper closed normal subgroup of G.*

Again it is enough to give the proofs for left random walks. It is clear that the left random walk is not ergodic if X is not trivial, that is, if X is a one point space. If X is trivial one can write $K = G \times Y$ with

$kk'=(g,y)(g',y')=(g\varphi(y)g',y')$ when $k=(g,y)$, $k'=(g',y')$. To prove ergodicity, it is enough to show that

$$\frac{1}{n}\sum_{j=1}^{n}\int v^{(j)}(k^{-1}dk')f(k')\to\int \eta(dk')f(k') \tag{17}$$

as $n\to\infty$ for each continuous function f. Then, given any set $A\in\mathscr{B}$, by approximating the indicator I_A in L^1 with respect to η by continuous functions, one can show that

$$\frac{1}{n}\sum_{j=1}^{n} v^{(j)}(k^{-1}A)\to\eta(A)$$

in L^1 with respect to η as $n\to\infty$. But this certainly implies ergodicity. Let us now prove (17). It is enough to establish (17) for product functions

$$f(k)=\psi(g\varphi(y))h(y) \tag{18}$$

with ψ, h continuous. This follows since a general continuous function can be approximated uniformly by finite linear combinations of such products using the Stone-Weierstrass theorem. By using Corollary 3.1 it follows that

$$\frac{1}{n}\sum_{j=1}^{n}\int v^{(j)}(k^{-1}dk')f(k')=\frac{1}{n}\sum_{j=1}^{n}\int v^{(j)}(dk')f(kk')\to\int\eta(dk')f(kk')$$

as $n\to\infty$. However, since X is trivial

$$\eta(dk')=\chi(dg')\alpha(dy') \tag{19}$$

when $k'=(g',y')$ where χ is the normalized Haar measure of G and α is a regular probability measure on Y. If f is given by (18)

$$\begin{aligned}\int\eta(dk')f(kk')&=\int\chi(dg')\psi(g\varphi(y)g'\varphi(y'))h(y')\alpha(dy')\\&=\int\chi(dg')\chi(g'\varphi(y'))h(y')\alpha(dy')=\int\eta(dk')f(k')\end{aligned}$$

because χ is Haar measure. The proof of ergodicity is complete. In this argument we have implicitly used the fact that the probability distribution for $g\varphi(y)$ induced by v has a support generating the whole group G. But this follows from the assumption that the support of v generates K. The ideas already employed can be used to establish the desired results on mixing. It is clear that the process cannot be mixing if the probability distribution of $g\varphi(y)$ induced by v has its support contained in the coset of a proper closed subgroup of G. Suppose now that the distribution of $g\varphi(y)$ induced by v doesn't have its support contained in such a coset. Then by Corollary 1

$$\int v^{(n)}(dk)f(k)\to\int\eta(dk)f(k) \tag{20}$$

as $n \to \infty$ for each continuous function f. Just as in the discussion of ergodicity, to prove mixing it is enough to show that

$$\int \nu^{(n)}(k^{-1} dk') f(k') \to \int \eta(dk') f(k') \tag{21}$$

for each continuous function f as $n \to \infty$. Also it will be enough to establish (21) for continuous product functions of the form (18). But then by (20)

$$\int \nu^{(n)}(k^{-1} dk') f(k') = \int \nu^{(n)}(dk') f(kk') \to \int \eta(dk') f(kk')$$

and (19) indicates that

$$\int \eta(dk') f(kk') = \int \eta(dk') f(k')$$

using the fact that f is of product form. The characterization is essentially complete. The limit relation (21) can also be used to show that in this case mixing is again equivalent to the process being purely nondeterministic.

Notes

5.1 The techniques used in this section are Fourier analytic. They are the analogues for the circle of results on the real line that are familiar in probability theory. The first of these is the one-one correspondence between distribution functions $F(x) = P\{X \leqslant x\}$ of real-valued random variables X and the corresponding Fourier-Stieltjes transforms (or characteristic functions) $\varphi(t) = \int_{-\infty}^{\infty} e^{itx} dF(x)$. The distribution function of the sum of two independent real-valued random variables X and Y with distribution functions F, G respectively is the convolution

$$F * G(x) = \int_{-\infty}^{\infty} F(x-z) dG(z).$$

The corresponding characteristic function of the sum $X+Y$ is the product $\varphi(t)\psi(t)$ of the characteristic functions φ, ψ of X, Y respectively. The convolution in the domain of distribution functions corresponds to the simpler product for characteristic functions and this explains in part the relevance of Fourier analytic methods when dealing with sums of independent random variables. The second result is the continuity theorem for characteristic functions. If the distribution functions $F_n(x)$ converge to a limiting distribution function G in the sense that

$$\int_{-\infty}^{\infty} h(x) dF_n(x) \to \int_{-\infty}^{\infty} h(x) dG(x)$$

for each bounded continuous function h, then clearly the corresponding sequence of characteristic functions $\varphi_n(t) = \int_{-\infty}^{\infty} e^{itx} dF_n(x)$ converge to the characteristic function $\psi(t) = \int_{-\infty}^{\infty} e^{itx} dG(x)$ of G for all t as $n \to \infty$. This convergence of distribution functions is usually called weak convergence (rather than weak star convergence as it should be) and is stated in the equivalent form of convergence of $F_n(x)$ to $G(x)$ at each continuity point x of G. The continuity theorem for characteristic functions states that if the characteristic functions $\varphi_n(t)$ converge to a limit function ψ which is continuous at zero, then ψ is a characteristic function and the distribution functions F_n converge weakly to the distribution function G corresponding to ψ. A detailed derivation of these results can be found in [31] or [27].

5.2 and 5.3 The discussion of the convolution for regular probability measures on a compact semigroup given in section 5.2 is similar to that given in Heble and Rosenblatt [37]. This development can be extended so as to be valid for convolution of regular probability measures on locally compact semigroups. Pym [87] discussed convolution on locally compact semigroups satisfying an additional condition. This additional requirement he called an $R(L)$ condition and it amounted to demanding that AB^{-1} $(B^{-1}A)$ be compact for any pair of compact sets A, B. Both H.S. Collins [13] and Kloss [58] considered groups of regular probability measures (with multiplication for the measures the convolution operation) on a compact group G. We shall say that G is generated by the group of measures if the closure of the union of the supports of the measures is G and shall assume this is so. The identity of the group of measures is shown to be the Haar measure of some closed normal subgroup H of G. If the group of measures has one element, then $H=G$. The support of any given measure of the group of measures is some coset gH of H. The cosets supporting distinct measures are disjoint. If gH is the support of the measure ν of the group, then $\nu(A)=\chi_H(g^{-1}A)$, $A\in\mathscr{A}$, where χ_H is the normalized Haar measure of H. The survey paper of J.H. Williamson [112] is a useful reference to related literature.

5.4 The derivation of results on the structure of compact semigroups is similar to that given in de Leeuw and Glicksberg [64]. The names of Rees and Suschkewitsch are associated with these results. The characterization of the form of idempotent measures on compact semigroups was obtained by Pym [87] and Heble and Rosenblatt [37]. The discussion of semigroups of finite stochastic matrices is an amplification of that given in M. Rosenblatt [95]. It is worthwhile looking at the apparently simple case of 2×2 stochastic matrices

$$\begin{pmatrix} a & 1-a \\ b & 1-b \end{pmatrix}$$

with $0\leqslant a\leqslant 1$, $0\leqslant b\leqslant 1$. Suppose we have a regular measure ν on these matrices with the topology given by (5.2.1). The support of ν can be any closed set of pairs (a,b) in $[0,1]\times[0,1]$. If there is a matrix with $0<a,b<1$ in the support of ν, the limiting measure η of $\nu^{(n)}$ as $n\to\infty$ is idempotent with support on the matrices of the form

$$\begin{pmatrix} c & 1-c \\ c & 1-c \end{pmatrix}.$$

The limit measure η is characterized by $\eta*\nu=\eta$. In fact, η is determined by this equation. Let F be the distribution function on pairs (a, b) determined by ν and G the distribution function on real numbers c determined by η. If the ν measure of the set $\{(a, b): a=b\}$ is zero $\eta*\nu=\eta$ can be written as the equation

$$\int_{a>b} G\left(\frac{x-b}{a-b}\right) dF(a,b) + \int_{a<b} \left[1 - G\left(\frac{x-b}{b-a} - \right)\right] dF(a,b) = G(x).$$

It is not clear how one would solve this equation to determine G in terms of F in a neat and effective way. It would even be interesting to determine sufficient conditions (on F) for G to be discrete, continuous singular or absolutely continuous. One can see that these three possibilities arise even when the F distribution is discrete. If the v measure is such that $a-b=\frac{1}{2}$ with probability one and $b=0, \frac{1}{2}$ with probabilty $\frac{1}{2}$ each, then the F distribution is the uniform distribution on $[0, 1]$. If v is such that $a-b=\frac{1}{3}$ with probability one and $b=\frac{1}{3}, \frac{2}{3}$ with probability $\frac{1}{2}$, then F is the wellknown singular continuous Cantor distribution on $[0,1]$. However, if $a-b=0$ with probability one and b has a given distribution then F is the same distribution. A simple procedure can be given that allows one to write out the moments of the F distribution in terms of moments of the G distribution. Let (a_j, b_j), $j=1,2,\ldots$ be independent and identically distributed pairs with common distribution G. Then F is the distribution of the random variable given by the series

$$b_1+(a_1-b_1)b_2+(a_1-b_1)(a_2-b_2)b_3+\cdots.$$

Moments of this random variable can be read off in terms of the G distribution by simple but for higher moments tedious computations.

5.5 Martin-Löf [73] has considered idempotent probability measures on countable semigroups. The support of the measure is again a completely simple semigroup with the structure of the measure still given by Theorem 4.3. The group G in the representation of the completely simple semigroup is finite but the right and left ideal indicators X and Y may be infinite. Tortrat [109] has also considered convolution on locally compact semigroups but many questions still remain open. Martin-Löf also obtained conditions for convergence of the convolution sequence $v^{(n)}$ to an idempotent measure on a countable semigroup. The results parallel those given in Theorem 5.1 for the compact case. Notice that in this situation the associated random walk has positive recurrent states. Theorem 5.1 was obtained in [94].

Chapter VI

Nonlinear Representations in Terms of Independent Random Variables

0. Summary

This chapter is concerned with the question of one-sided nonlinear representations of Markov sequences in terms of independent random variables. Such representations could be phrased in the broader context of stationary (possibly non-Markovian) processes and our investigation in the Markovian context is a question of convenience. One way of motivating such problems is by considering the linear Wold representation of a weakly stationary sequence. The Wold representation is briefly derived in section 1. It arises in the context of the linear prediction problem and in the purely nondeterministic case gives a linear resolution of the process in terms of orthonormal random variables. Section 2 describes how the nonlinear problem is suggested by Wold's result if linearity is replaced by nonlinearity and orthogonality by independence. Wiener regarded such a nonlinear representation as providing a re-encoding of the process. The following section indicates that such a representation can be valid if and only if the process is purely nondeterministic in a nonlinear sense and also poses the question of invertibility of such a representation. The existence of such one-sided nonlinear representations is examined at some length for finite state Markov chains in section 4. Partial results for the more general real-valued Markov process are derived in the last section.

1. The Linear Prediction Problem for Stationary Sequences

Consider a weakly stationary sequence of real-valued random variables $\{X_n(\omega); n = 0, \pm 1, \ldots\}$ on a probability space, that is, a sequence with constant first and second moments

$$E X_n \equiv m, \qquad E X_n^2 \equiv r_0 \tag{1}$$

and second order mixed moments

$$E X_n X_m = r_{n-m} \tag{2}$$

depending only on the time difference $n-m$. The linear prediction problem is concerned with predicting (or approximating) X_n, the present, by that linear expression in terms of the past $\{X_k, k<n\}$ which is best in terms of minimizing the mean square error of prediction. It is natural to look at this problem in terms of Hilbert spaces generated by linear combinations of finite numbers of the random variables. Let $\mathcal{M}_n$ be the closed linear manifold generated by X_k, $k \leqslant n$,

$$\mathcal{M}_n = \mathcal{M}\{X_k, k \leqslant n\}.$$

The best linear predictor X_n^* of X_n in terms of the past is the projection of X_n on $\mathcal{M}_{n-1}$

$$X_n^* = \mathcal{P}\{X_n; \mathcal{M}_{n-1}\}$$

and the error in prediction is

$$\eta_n = X_n - X_n^*.$$

If $\eta_n = 0$ for some n, it is clear that $\eta_n \equiv 0$ for all n because of the stationarity. This is usually referred to as the purely deterministic case since then

$$\cdots = \mathcal{M}_n = \mathcal{M}_{n-1} = \cdots = \mathcal{M}_{-\infty} = \bigcap_n \mathcal{M}_n,$$

that is, the full history of the process is determined by the infinite past.

If $\eta_n \neq 0$ with probability one, it is convenient to renormalize η_n and obtain ξ_n with unit variance

$$\xi_n = \eta_n \{E|\eta_n|^2\}^{-\frac{1}{2}}.$$

By the construction of the η_n's, it follows that the ξ_n's are orthonormal and orthogonal to the infinite past

$$\begin{aligned} E\xi_n \xi_m &= \delta_{n-m} \\ \xi_n &\perp \mathcal{M}_{-\infty}. \end{aligned} \tag{3}$$

Let

$$a_k = E X_n \xi_{n-k} \tag{4}$$

be the Fourier coefficient of X_n with respect to ξ_{n-k} where $n = 0, \pm 1, \ldots$ and $k = 0, 1, 2, \ldots$. The weak stationarity implies that a_k is independent of n. The process X_n can then be written

$$X_n = \sum_{k=0}^{\infty} a_k \xi_{n-k} + v_n \tag{5}$$

where v_n is the projection of X_n on the infinite past

$$v_n = \mathscr{P}\{X_n; \mathscr{M}_{-\infty}\}.$$

The process is called purely nondeterministic if the closed linear manifold $\mathscr{M}_{-\infty}$ is trivial, that is, consists only of the random variable zero. In the purely nondeterministic case there is no information in the infinite past. The representation of the process X_n given in (5) is due to Wold and is a linear decomposition of the process into a purely nondeterministic part

$$Y_n = \sum_{k=0}^{\infty} a_k \xi_{n-k}, \qquad n = 0, \pm 1, \dots \tag{6}$$

and a purely deterministic part $\{v_n; n = 0, \pm 1, \dots\}$. That $\{Y_n\}$ is purely nondeterministic follows from the orthogonality of the ξ_n's. We shall show that $\mathscr{M}_{-\infty}$ is contained in the closed linear manifold $\mathscr{M}\{v_k; k \leqslant n\}$ for any choice of n. This directly implies that the weakly stationary sequence $\{v_n\}$ is purely deterministic. Consider any element $w \in \mathscr{M}_{-\infty}$. Since w is orthogonal to the sequence $\{\xi_n; n = 0, \pm 1, \dots\}$ and is an element of $\mathscr{M}_n = \mathscr{M}\{v_k, \xi_k; k \leqslant n\}$, w must belong to $\mathscr{M}\{v_k; k \leqslant n\}$.

The linear decomposition $X_n = Y_n + v_n$ of $\{X_n\}$ into a sum of two orthogonal components $Y_n \in \mathscr{M}_n$, $v_n \in \mathscr{M}_{-\infty}$, with $\{Y_n\}$ purely nondeterministic and $\{v_n\}$ purely deterministic, is unique. The random variable η_n, sometimes called the innovation at time n, represents the additional random input to the process at time n orthogonal to the past.

There are analytic characterizations of the purely nondeterministic and purely deterministic components of the process $\{X_n\}$ in terms of the spectral distribution function of the process. This spectral distribution function is obtained by a harmonic analysis of the second order moment sequence $\{r_n, n = 0, \pm 1, \dots\}$ (see [98]). We shall not pursue at this point the question of the relation between the Wold decomposition and harmonic analysis of the process.

2. A Nonlinear Prediction Problem

It is natural to look for a nonlinear analogue of the Wold representation for a strictly stationary sequence of random variables $\{X_n; n = 0, \pm 1, \dots\}$. Most of the examples and explicit results obtained are for Markovian processes. However, the process $\{Y_n; n = \dots, -1, 0, 1, \dots\}$ with $Y_n = \cdot X_n X_{n-1} \dots$ the formal one-sided sequence of X values is a Markov process obtained by a horrendous enlarging of the state space when the original X process is not Markov. Since this presents no additional difficulty, our initial discussion will be carried out for general stationary sequences. There is a natural nonlinear prediction problem (or set of nonlinear prediction problems) corresponding to the linear prediction considered earlier. Assume that all moments $E X_n^k$, $k=0,1,2,\dots$, are finite. Even if this is not the case initially, one can always apply an instantaneous nonlinear transformation f to obtain a derived process $Y_n=f(X_n)$ with all moments finite to which one can apply our analysis. Let $\mathscr{H}_n$ be the closure in mean square of the algebra generated by X_k, $k\leqslant n$. The Borel field of sets $\mathscr{B}_n$ generated by the random variables X_k, $k\leqslant n$, is the smallest Borel field containing all sublevel sets $\{\omega: X_k(\omega)\leqslant\alpha\}$, $-\infty<\alpha<\infty$, for $k\leqslant n$. A standard argument shows that $\mathscr{H}_n$ is the Hilbert space of square integrable random variables measurable with respect to $\mathscr{B}_n$. $\mathscr{B}_n$ represents the information available by observing X_k, $k\leqslant n$. Let Z be a random variable in $\mathscr{H}_n$ such as, for example, X_n. The (possibly nonlinear) predictor Z^* of Z in terms of the past $\{X_k, k<n\}$ that is best in terms of minimizing mean square error is given by the conditional expectation of Z given $\mathscr{B}_{n-1}$

$$Z^*=E(Z|\mathscr{B}_{n-1})=E(Z|X_k, k<n). \tag{1}$$

This is so because (1) is the projection of Z on $\mathscr{H}_{n-1}$. A process is said to be *purely deterministic* (in the nonlinear sense) if $Z^*=Z$ for all $Z\in\mathscr{H}_n$. For it is then apparent that

$$\cdots=\mathscr{H}_n=\mathscr{H}_{n-1}=\cdots=\bigcap_n \mathscr{H}_n=\mathscr{H}_{-\infty},$$

or equivalently,

$$\cdots=\mathscr{B}_n=\mathscr{B}_{n-1}=\cdots=\bigcap_n \mathscr{B}_n=\mathscr{B}_{-\infty}.$$

The process is completely determined by the infinite past. A process is said to be *purely nondeterministic* (in the nonlinear sense) if $\mathscr{H}_{-\infty}=\bigcap_n \mathscr{H}_n$ is the trivial Hilbert space consisting only of constant functions or equivalently if $\mathscr{B}_{-\infty}=\bigcap_n \mathscr{B}_n$ is the trivial Borel field consisting only of the empty and universal sets.

Suppose the strictly stationary process $\{X_n\}$ is purely nondeterministic in the nonlinear sense. A nonlinear analogue of the linear Wold representation of a purely nondeterministic (linear) process in terms of innovation variables ξ_n

$$X_n=\sum_{k=0}^{\infty} a_k \xi_{n-k} \tag{2}$$

was considered by Wiener in his book on nonlinear problems in random theory [111]. The object was to construct a stationary sequence ξ_n of independent random variables (say uniformly distributed on [0, 1]) with ξ_n measurable with respect to $\mathscr{B}_n$ (dependent on the present and past $X_k, k \leqslant n$) and independent of $\mathscr{B}_{n-1}$. The sequence ξ_n is to be constructed by time shifts of a time-invariant representation

$$\xi_n = g(\tau^n X). \tag{3}$$

In (3) g is a function independent of n, X the process $\{X_n\}$ and τ the shift operator. Furthermore, this transformation from the process $\{X_n\}$ to the process of random variables $\{\xi_n\}$ is to be invertible so that X_n is measurable with respect to the Borel field $\mathscr{B}\{\xi_k, k \leqslant n\}$ generated by the random variables $\xi_k, k \leqslant n$. Thus $\mathscr{B}_n = \mathscr{B}\{X_k, k \leqslant n\}$ and $\mathscr{B}\{\xi_k, k \leqslant n\}$ are the same. Wiener refers to the representation (2) as a recoding of the process $\{X_n\}$. The representation

$$X_n = f(\tau^n \xi) \tag{4}$$

implied by X_n measurable with respect to $\mathscr{B}\{\xi_k, k \leqslant n\}$ is the nonlinear analogue of the Wold representation (2). Notice that f is measurable with respect to $\mathscr{B}\{\xi_k, k \leqslant 0\}$. The existence of an invertible mapping from the process $\{X_n\}$ to the sequence of independent uniformly distributed random variables $\{\xi_n\}$ implies that the two sequences are isomorphic in the sense of ergodic theory [2].

An invertible mapping from $X = \{X_n\}$ to $\xi = \{\xi_n\}$ does not generally exist even if X is purely nondeterministic. Wiener suggested conditions insuring the existence of such a mapping and a possible construction. Consider the conditional distribution function of X_n given the past

$$\begin{aligned} F(x_n|x_k, k \leqslant n-1) &= P\{X_n(\omega) \leqslant x_n | X_k(\omega) = x_k, k \leqslant n-1\} \\ &= P\{X_n(\omega) \leqslant x_n | \mathscr{B}_{n-1}\}. \end{aligned} \tag{5}$$

Wiener's conditions were

(i) $\{X_n\}$ is purely nondeterministic.

(ii) $F(x_n|x_k, k \leqslant n-1)$ is a strictly increasing continuous function of x_n for almost every past.

For technical reasons, the condition that the version of a conditional distribution function $F(x_n|x_k, k \leqslant n-1)$ taken be a Borel (or Baire) function of its arguments $x_k, k \leqslant n$, should be imposed. This is because composition of functions will be required and Borel measurability will keep the resulting functions measurable. The construction of the uniformly distributed random variables ξ_n runs as follows. Set

$$\xi_n = F(X_n|X_k, k \leqslant n-1). \tag{6}$$

ξ_n is a random variable (because of $\mathscr{B}_n$ measurability). Let

$$F^{-1}(\alpha|x_k, k \leqslant n-1)$$

be the inverse function of $F(\cdot\,|\,\cdot)$ which is continuous and strictly increasing for almost every past. Consider $0<\alpha<1$ and any $B\in\mathscr{B}_{n-1}$. The probability

$$P(\{\xi_n\leqslant\alpha\}\cap B)=\int_B P(\xi_n\leqslant\alpha|\mathscr{B}_{n-1})\,dP \tag{7}$$

with

$$\begin{aligned} P(\xi_n\leqslant\alpha|\mathscr{B}_{n-1}) &= P(F(X_n|X_k,\,k\leqslant n-1)\leqslant\alpha|\mathscr{B}_{n-1}) \\ &= P(X_n\leqslant F^{-1}(\alpha|X_k,\,k\leqslant n-1)|\mathscr{B}_{n-1})=\alpha \end{aligned} \tag{8}$$

for almost every past. Thus (7) equals $\alpha P(B)$. The random variable ξ_n is uniformly distributed on $[0,1]$ and independent of the past $\mathscr{B}_{n-1}$. The sequence $\{\xi_n\}$ consists of independent uniformly distributed random variables on $[0,1]$. The sequence $\{\xi_n\}$ could have been constructed with the properties just derived even if $F(\cdot\,|\,\cdot)$ were not strictly increasing for almost every past. In the Markovian case $F(x_n|x_k,\,k\leqslant n-1)=F(x_n|x_{n-1})$.

The construction and the conditions imposed on $F(\cdot\,|\,\cdot)$ imply that X_n is determined by $\xi_n,\ldots,\xi_{n-k+1},X_{n-k},X_{n-k-1},\ldots$ if $k>0$. It is intuitively plausible that one should be able to formally let $k\to\infty$ and obtain a representation of X_n in terms of ξ_k, $k\leqslant n$. However, this argument as used by Wiener is not generally valid under the conditions he cites. The following example points up the difficulties that may arise. Let $\{X_n\}$ be the strictly stationary Markov process with transition probability density

$$p(x|x')=\begin{cases} p_i>0 & \text{if } i<x\leqslant i+1,\ x'\in S \\ & \text{or if } i-1<x\leqslant i,\ x'\in S^c \end{cases} \tag{9}$$
$$i=0,\,\pm1,\ldots,\ \sum_i p_i=1$$

where $S=\bigcup_{i=-\infty}^{\infty}\{x|2i<x\leqslant 2i+1\}$ with S^c the complementary set on the real line. There is a unique stationary Markov process with (9) as transition probability density and the process is purely nondeterministic. The conditional distribution function $F(x\,|\,x')$ is given by

$$\begin{aligned} F(x\,|\,x') &= \begin{cases} F(x) & \text{if } x'\in S \\ F(x-1) & \text{if } x'\in S^c, \quad \text{where} \end{cases} \\ F(x) &= \sum_{i<\{x\}} p_i+(x-\{x\})p_i, \end{aligned} \tag{10}$$

and $\{x\}$ is the greatest integer less than or equal to x. $F(x\,|\,x')$ is strictly increasing and continuous in x for every past. Knowledge of

$$\xi_n=F(X_n\,|\,X_{n-1})$$

indicates that either $X_n = F^{-1}(\xi_n)$ or $F^{-1}(\xi_n)+1$. The $\xi_k, k \leqslant n$, give no additional information about X_n. Specifically, knowledge of $\xi_k, k \leqslant n$, will not tell us whether X_n is in S or S^c. The construction of the ξ_n given by Wiener does not determine the original X_n sequence. Nonetheless, another construction of independent uniformly distributed η_n, with $\eta_n \in \mathscr{B}_n$, will enable us to reconstruct the original process given in this example. A partial answer to the question of when and how a nonlinear analogue of the Wold representation can be set up in terms of independent random variables will be taken up in the following sections.

3. Questions for Markov Processes

There are a number of related questions suggested by the notion of a nonlinear analogue of a Wold representation of a strictly stationary process as it has been formulated earlier. Let $\{X_n; n=0, \pm 1, \ldots\}$ be a strictly stationary Markov process. The first question is concerned with the possibility of splitting off a sub-Borel field of past and present that is independent of the past and yet together with the past determines the present, that is

(i) Is there a random variable ξ_n measurable with respect to $\mathscr{B}_n$ (generated by $X_k, k \leqslant n$) such that ξ_n is independent of $\mathscr{B}_{n-1}$ and yet $\mathscr{B}_n$ is generated by $\mathscr{B}_{n-1}$ and the Borel field generated by ξ_n?

It is clear that if the answer to (i) is affirmative, then the random variables ξ_n can be taken to be a strictly stationary sequence with $\xi_n = \tau^n \xi_0$ where τ is the shift transformation. Furthermore, X_n can then be represented as a function

$$X_n = f_k(\xi_n, \ldots, \xi_{n-k}, X_{n-k-1}) \tag{1}$$

for every finite integer $k \geqslant 0$ by making use of the property mentioned in (i) and the fact that $\{X_n\}$ is Markovian. Intuitively it seems plausible that one ought to be able to let k go to infinity formally and obtain a representation

$$X_n = f(\xi_n, \xi_{n-1}, \ldots) \tag{2}$$

in terms of the ξ sequence, but we've already seen that this needn't be the case.

(ii) If the answer to question (i) is affirmative, can one obtain a one-sided representation (2) for the process $\{X_n\}$ in terms of a stationary sequence $\{\xi_n\}$ of independent random variables satisfying the conditions of (i)?

Even if the answer to (ii) is negative, it may still be possible to represent a Markov process $\{Y_n\}$ with the same probability structure as $\{X_n\}$ as a one-sided function and its shifts

$$Y_n = f(\eta_n, \eta_{n-1}, \ldots) \tag{3}$$

of a sequence of independent, identically distributed random variables $\{\eta_n\}$. This may be due to the fact that the probability space of a sequence of independent random variables $\{\xi_n\}$ satisfying the conditions of (i) is not sufficiently large to support a representation of the process $\{X_n\}$.

(iii) Can a Markov process $\{Y_n\}$ with the same probability structure as $\{X_n\}$ be given as a one-sided function and its shifts $Y_n = f(\eta_n, \eta_{n-1}, \ldots)$ of a sequence of independent identically distributed (for convenience, uniformly distributed on $[0,1]$) random variables $\{\eta_n\}$?

Lemma 1. *If the answer to problem* (iii) *is affirmative, then the process* $\{X_n\}$ *must be purely nondeterministic.*

One can assume that the random variables Y_n are bounded. For if they are not, with the aid of a monotone one-to-one function $h(\cdot)$ taking $(-\infty, \infty)$ into $(0,1)$ a new bounded process $Y'_n = h(Y_n)$ is defined. A representation of the desired type is valid for $\{Y'_n\}$ if and only if it is valid for $\{Y_n\}$. We now show that the backward tail field $\mathscr{B}_{-\infty}$ of $\{Y_n\}$ must be trivial. It is enough to show that any bounded random variable Z measurable with respect to $\mathscr{B}_{-\infty}$ is a constant. Since the Y_n are bounded, given any fixed $\varepsilon > 0$ there is a polynomial $p(Y_n, \ldots, Y_{n-k})$ in a large but finite number of variables $Y_n, \ldots, Y_{n-k}$ such that

$$E|Z - p| < \varepsilon. \tag{4}$$

The Y_n's are one-sided functions of the η_n's and so there is a Borel function $q(\eta_n, \ldots, \eta_{n-j})$ with j finite such that

$$E|q - p| < \varepsilon. \tag{5}$$

Thus $E|Z - q| < 2\varepsilon$. But

$$E(q|\mathscr{B}_{-\infty}) = E(q|\mathscr{B}_m) = E(q|\eta_m, \eta_{m-1}, \ldots) = Eq \tag{6}$$

for $m < n - j$ and $E(Z|\mathscr{B}_m) \to E(Z|\mathscr{B}_{-\infty})$ as $m \to -\infty$ in the mean by a martingale convergence theorem (see Appendix 1). Since

$$E\big|E(Z|\mathscr{B}_m) - E(q|\mathscr{B}_m)\big| \leqslant E|Z - q| < 2\varepsilon$$

this implies that $E\big|E(Z|\mathscr{B}_{-\infty}) - E(Z)\big| < 4\varepsilon$. This is true for every $\varepsilon > 0$ and so $E(Z|\mathscr{B}_{-\infty}) = E(Z)$ with probability one.

Actually it would have been enough to show that the tail field of a sequence of independent identically distributed random variables is trivial and we have implicitly proven this. It is natural to conjecture that a necessary and sufficient condition for an affirmative answer to (iii) is that the process $\{X_n\}$ be purely nondeterministic. We cannot establish this at present. However, this conjecture is valid for Markov chains. All three questions will be discussed in detail for the case of finite state Markov chains where a fairly complete answer can be given. Later, a result will be given for a class of general state Markov processes.

Under appropriate conditions it is easy to give a satisfactory answer to (i). Given two distribution functions F, G we shall say that they are *equivalent* if the jumps of F and G can be mapped onto each other in a one-to-one manner so that jump sizes are preserved. Clearly, any two continuous distribution functions F, G are equivalent in this sense.

Lemma 2. *Let $\{X_n\}$ be a stationary real-valued Markov process with conditional distribution function $F(x|x')=P\{X_n \leqslant x | X_{n-1}=x'\}$ a Borel function of the two variables x, x'. The answer to question* (i) *is affirmative if and only if the distribution functions $F(x|x')$ (as functions of x) are equivalent for almost all x' with respect to the probability measure of the process $\{X_n\}$.*

The random variable ξ_n referred to in (i) is to be statistically independent of X_k, $k<n$, and so the conditional distribution of ξ_n given X_k, $k<n$, as a function is independent of X_k, $k<n$. Let $\mathscr{C}_n$ be the Borel field generated by ξ_n. Since $\mathscr{B}_n=\mathscr{B}_{n-1}\times\mathscr{C}_n$, the distribution function of ξ_n is equivalent, with probability one, to the conditional distribution of X_n given X_k, $k<n$, that is, $F(x_n|x_{n-1})$.

Conversely, given the equivalence of the distribution functions $F(x|x')$ for almost all x', a random variable ξ_n must be constructed. If the distribution function $F(x|x')$ is continuous in x for almost all x', set

$$\xi_n = F(X_n | X_{n-1}). \tag{7}$$

An argument due to P. Lévy [66] shows that ξ_n is uniformly distributed on $[0,1]$ and independent of $X_k, k<n$. Further, if

$$\psi(\xi,x')=\inf\{b: F(b|x')>\xi\}$$

then $X_n=\psi(\xi_n, X_{n-1})$ with probability one. Now consider what happens if $F(x|x')$ has a discrete part in x with probability one. Take the jumps in the discrete part to be $p_1 \geqslant p_2 \geqslant \cdots$ ordered according to magnitude. We can assume strict inequality for the p_i since a simple elaboration of the discussion given for the case of strict inequality allows the argument to be carried out in general. With probability one the size of the jumps $p_1, p_2, \ldots$ do not depend on x' because of the equivalence of the distribution functions $F(x|x')$. Set

$$A_j=\{(x,x'): \Delta F(x|x')=p_j\} \tag{8}$$

where ΔF denotes the size of the jump provided one occurs at the argument indicated. Let

$$S(x,x')=\{j:(y,x')\in A_j, y\leqslant x\} \tag{9}$$

and

$$\xi_n=\begin{cases} j & \text{if } (X_n,X_{n-1})\in A_j \\ F(X_n|X_{n-1})-\sum\limits_{j\in S(X_n,X_{n-1})} p_j & \text{if } (X_n,X_{n-1})\notin\bigcup A_j. \end{cases}$$

Then ξ_n is independent of X_k, $k<n$. Note that ξ_n is uniformly distributed on $(0,1-\sum p_j)$ with density one and takes the integer value j with probability p_j. From the construction, it is clear that X_n is determined with probability one by knowledge of ξ_n, and X_k, $k<n$.

4. Finite State Markov Chains

Questions (i), (ii) and (iii) will now be considered in the case of finite state Markov chains. Let $\{X_n\}$ be a strictly stationary Markov chain with transition probability matrix

$$P=(p_{i,j}),\quad p_{i,j}=P[X_{n+1}=j\,|\,X_n=i],\quad i,j=1,2,\ldots,m<\infty \tag{1}$$

and stationary vector of probabilities

$$p_i=P[x_n=i]>0. \tag{2}$$

Lemma 3.2 indicates that a random variable ξ_n satisfying the conditions of question (i) exists if and only if the distributions given by the row vectors of P are equivalent. This can alternatively be stated in the following convenient way. Given any i, i' there is a permutation $M(i,i')$ of the integers $1,2,\ldots,m$ such that

$$p_{i,j}=p_{i',M(i,i')j} \tag{3}$$

for all j. We shall call such special Markov chains *uniform chains* because the row distributions are the same except for permutation. Label the positive probability masses common to the row distributions $\{p_{i,j}, j=1,2,\ldots\}$ $q_1\geqslant q_2\geqslant\cdots\geqslant q_r>0$ in order of magnitude, $\sum q_i=1$. If the q_i's are all distinct, the random variable ξ_n can be easily described.

Lemma 1. *If $\{X_n\}$ is a uniform stationary finite-state Markov chain with the q_i's distinct, then the random variable ξ_n of question* (i) *is determined up to a one-to-one transformation modulo sets of measure zero and is given by $\xi_n=\xi(X_n,X_{n-1})=k$ if (X_n,X_{n-1}) is such that $p_{X_{n-1},X_n}=q_k$.*

Let us write $\xi_n=\xi(X_n,X_{n-1},\ldots)$. ξ takes on r distinct values with probability one and we label these $1,2,\ldots,r$. Let $\eta(\xi,x_{n-1},x_{n-2},\ldots)$ be the inverse function of ξ as a function of x_n for fixed x_k, $k<n$. ξ_n is independent of X_k, $k<n$, *so that*

$$\mathrm{P}[\xi_n=j\,|\,X_{n-k}=i_k, k\geqslant 1]=P[\xi_n=j]=p_{i_1,\eta(j,i_1,i_2,\ldots)}>0. \tag{4}$$

The q_i's are distinct and so for fixed i_1, j the function $\eta(j,i_1,i_2,\ldots)$ is independent of $i_2,i_3,\ldots$. Therefore, $\eta(j,i_1,i_2,\ldots)=\eta(j,i_1)$ and

$$\xi(x_n,x_{n-1},\ldots)=\xi(x_n,x_{n-1}).$$

The function ξ is constant on the set $\{(i,j)\,|\,p_{i,j}=q_k\}$ and we set $\xi=k$ there. Clearly, any other ξ satisfying the requirements of question (i) must be a one-to-one function of the ξ just constructed.

We now look at the problem posed by question (ii) for stationary uniform finite state Markov chains. Let M_k be the matrix

$$M_k=\{e_{ij}(k)\} \tag{5}$$

with

$$e_{ij}(k) = \begin{cases} 1 & \text{if } p_{i,j}=q_k \\ 0 & \text{otherwise}. \end{cases} \tag{6}$$

There are $r(\leqslant m)$ matrices M_k determined by the Markov chain $\{X_n\}$. Every matrix M_k has exactly one element equal to one in each row and all other elements zero. No two M_k's have a one in the same entry. Furthermore, every product of a finite number of the M_k's also has these properties. The matrices $\{M_k\}$ correspond to a family of mappings of the set of states of the process into itself induced by the transition probability matrix of $\{X_n\}$. Construct the semigroup under multiplication generated by the matrices $\{M_k\}$. If the semigroup contains a matrix with a column of ones we say that the semigroup is *point collapsing* since the mapping corresponding to such a matrix maps all the states into one state.

Theorem 1. *Let $\{X_n\}$ be a finite state stationary uniform Markov chain with the q_i's distinct. If $\{M_k\}$ is the family of mapping matrices generated by $\{X_n\}$, the answer to question* (ii) *is affirmative if and only if the semigroup generated by $\{M_k\}$ is point collapsing.*

This result can be interpreted in terms of the structural results on semigroups developed in section 5.4. It states that the answer to question (ii) is positive if and only if the kernel of the semigroup generated by the matrices $\{M_k\}$ has trivial group and left ideal structure. Only the right ideal structure is to be (possibly) nontrivial.

Let the semigroup generated by the matrices $\{M_k\}$ be point collapsing. A finite product $M_{k_1}\ldots M_{k_s}$ of matrices selected from the set $\{M_k\}$ (with replacement) has all the elements in the ith column equal to one. The random variable ξ_n satisfying the requirements of question (i) is determined in Lemma 1 and given by

$$\xi_n = \xi(X_n, X_{n-1}) = k \quad \text{if } e_{X_{n-1}, X_n}(k) = 1\,. \tag{7}$$

Let A_j be the event

$$A_j = \{\xi_{j+1} = k_1, \ldots, \xi_{j+s} = k_s\}\,. \tag{8}$$

If A_j occurs then $X_{j+s} = i$ with probability one. By the Borel-Cantelli Lemma, at least one of the events $A_{n-js}, j = 1, 2, \ldots,$ must occur with probability one since they are independent. Hence, at least one of the events $A_j, j \leqslant n-s+1$, must occur with probability one. Let T_k be the mapping corresponding to M_k so that $T_k i = j$ where j is the unique integer such that $e_{ij}(k) = 1$. A function $g(\xi_n, \xi_{n-1}, \ldots) = X_n$ with probability one will now be constructed. Let S_j be the set of sequences $(\ldots, \xi_{n-1}, \xi_n)$ for which some A_i, $i \leqslant n-s+1$, occurs and A_j is the one with largest index less than or equal to $n-s+1$. The sets S_j, $j \leqslant n-s+1$, are disjoint and $P(\bigcup S_j) = 1$. Set

$$g(\xi_n, \xi_{n-1}, \ldots) = \begin{cases} T_{\xi_n} \cdots T_{\xi_{j+s+1}} i & \text{if } (\ldots, \xi_{n-1}, \xi_n) \in S_j \\ 0 \text{ otherwise}. \end{cases} \tag{9}$$

Then $X_n = g(\xi_n, \xi_{n-1}, \ldots)$ with probability one.

Assume now that the semigroup generated by the set of matrices $\{M_k\}$ is not point collapsing. Every matrix of the semigroup has a minimal number $r > 1$ of distinct columns containing a one as a column entry. But this means that a knowledge of all the ξ_i's, $i \leqslant n$, can indicate no more than that X_n takes on one of r distinct values. The answer to question (ii) must then be negative.

Theorem 1 implies that there is a large class of purely nondeterministic finite state stationary uniform Markov chains for which the answer to question (ii) is negative. A very simple example exhibiting this pathology can easily be constructed. Let $\{X_n\}$ be the stationary Markov chain with states 1, 2, instantaneous stationary distribution

$$P(X_n = 1) = P(X_n = 2) = \tfrac{1}{2} \tag{10}$$

and transition probability matrix

$$P = \begin{pmatrix} p & q \\ q & p \end{pmatrix}, \quad 0 < p \neq q = 1 - p < 1\,. \tag{11}$$

Then

$$M_1 = \begin{pmatrix} 1 & 0 \\ 0 & 1 \end{pmatrix}, \quad M_2 = \begin{pmatrix} 0 & 1 \\ 1 & 0 \end{pmatrix} \tag{12}$$

so that

$$\xi_n = \xi(X_n, X_{n-1}) = \begin{cases} 1 & \text{if } (X_{n-1}, X_n) = (1,1), (2,2) \\ 2 & \text{otherwise} \end{cases} \tag{13}$$

is measurable with respect to $\mathscr{B}_n$ but independent of $X_{n-1}, X_{n-2}, \ldots$. Lemma 1 indicates that ξ_n is essentially the only random variable with these properties. The process $\{X_n\}$ is purely nondeterministic but X_n is not a function of $\xi_n, \xi_{n-1}, \ldots$ since $E(X_n|\xi_n, \xi_{n-1}, \ldots) = EX_n$.

Theorem 1 gives a reasonable description of the range within which there is an affirmative answer to question (ii) for Markov chains. It is a limited range because one can only rarely define random variables ξ_n with the properties required in (i) (and therefore (ii)) on the initially given probability space. However, question (iii) does not require such a narrow specification for the random variables ξ_n and consequently its answer has much broader scope.

Theorem 2. *The probability distribution of a finite state stationary Markov chain has a representation of the type considered in question* (iii) *if and only if the Markov chain is purely nondeterministic.*

That the chain must be purely nondeterministic if there is such a representation is already clear from Lemma 3.1. Let us therefore assume this to be the case. The Markov chain is then irreducible with no cyclically moving sets of states. Let $P = (p_{ij})$ (say $m \times m$) be the transition probability matrix of the chain. Let M be an $m \times m$ *mapping matrix*, that is, having exactly one element equal to one in each row and all other elements zero. Call M a mapping matrix *consistent* with P if its entry $m_{ij} = 1$ only if $p_{ij} > 0$. Let $\{M_k\}$ be the set of mapping matrices consistent with P. Then P can be written as a linear combination with positive weights of these matrices

$$P = \sum_k \alpha_k M_k, \quad \alpha_k > 0, \quad \sum_k \alpha_k = 1. \tag{14}$$

Notice that the representation (14) is not unique. Let T_k be the mapping of the integers $(1, \ldots, m)$ into themselves determined by M_k. Consider $p = (p_i)$ the invariant instantaneous distribution of the chain. Then if X is a random variable with distribution $p = (p_i)$ and ξ is independent of X with distribution (α_k), it follows that

$$P(X_n = i, X_{n+1} = j) = P(X = i, T_\xi X = j).$$

This implies that one can construct jointly stationary sequences $\{Y_n, \xi_n, n = 0, \pm 1, \ldots\}$ with the following properties:

a) (Y_n) is a stationary Markov chain with the same probability structure as (X_n).

b) (ξ_n) is a sequence of independent identically distributed random variables with common distribution (α_k). Also, ξ_k, $k \geqslant n$ are independent of X_k, $k<n$ for each k.

c) $Y_n = T_{\xi_n} Y_{n-1}$ with probability one.

The proof is now very much like that of Theorem 1. Consider the semigroup generated by the mapping matrices $\{M_k\}$ consistent with P. This semigroup must be point-collapsing since the Markov chain is purely nondeterministic. Let $M_{k_1} \dots M_{k_s}$ be a finite product of matrices from the set $\{M_k\}$ having all elements in the ith row equal to one. Let A_j be the event given by (8) so that in view of the fact that $Y_n = T_{\xi_n} Y_{n-1}$, it follows that if A_j occurs, then $X_{j+s} = i$ with probability one. Define the function g as in the proof of Theorem 1. The argument given there shows that $Y_n = g(\xi_n, \xi_{n-1}, \dots)$ with probability one.

5. Real-Valued Markov Processes

In this section we wish to modify and extend the basic ideas in the proofs of section 4 so as to obtain a representation of the type considered in question (iii) for a much larger class of real-valued stationary Markov processes. The main result is given in the following theorem.

Theorem 1. *Let $\{Z_n\}$, $n=0, \pm 1, \dots$ be a real-valued strictly stationary Markov process with*

(i) *trivial tail field $\mathscr{B}_{-\infty}$*

(ii) *Borel sets A, $B \in \mathscr{A}$ on the real line and a non-negative measure φ on the real line for which $P(B)$, $\varphi(A)>0$ and such that for all $x \in B$ and $A' \subset A$ one has $P(x, A') \geqslant \varphi(A')$. Then, if $\{\xi_n\}$, $n=0, \pm 1, \dots$, is a sequence of independent random variables uniformly distributed on $[0, 1]$, there is a one-sided function $g(\dots, \xi_{-1}, \xi_0)$ such that $\{Z_n\}$ and $\{g(\tau^n \xi)\}$ have the same probability structure.* Here $\xi = (\dots, \xi_{-1}, \xi_0, \xi_1, \dots)$ and τ is the shift operator.

Lemma 1. *Let A, $B \in \mathscr{A}$ be events such that $P(A)$, $P(B)>0$ and $P(x, A)>0$ for every $x \in B$. Then, if $\varepsilon>0$ there is a $\delta>0$ such that for $A' \in \mathscr{A}$ with $A' \subset A$ and $P(A-A')<\delta$, it follows that*

$$P\{x : x \in B,\ P(x, A')>0\} > P(B) - \varepsilon\,. \tag{1}$$

Let η be such that $P\{x : 0<P(x, A) \leqslant \eta\} < \varepsilon/2$ and $0<\delta<\varepsilon\eta/2$. If $P(A-A')<\delta$, then

$$\frac{\varepsilon}{2} > \frac{\delta}{\eta} > \frac{P(A-A')}{\eta} = \frac{1}{\eta} \int_{\{x: P(x, A-A')>0\}} P(x, A-A')\,dP$$

$$\geqslant \frac{1}{\eta} \int_{\{x: P(x,A')=0,\, P(x,A)>\eta\}} P(x,A)\,dP \geqslant P\{x: P(x,A')=0,\, P(x,A)>\eta\} \tag{2}$$

$$\geqslant P\{x: P(x,A')=0,\, x\in B\} - \frac{\varepsilon}{2}.$$

Thus, $\varepsilon > P\{x: x\in B,\, P(x,A)>0\}$ and the proof is complete.

Lemma 2. *Let $\mathscr{B}_{-\infty}$ be trivial with A_i, $i=0,1,\ldots$, a sequence of events of $\mathscr{A}$ such that $P(A_0)>0$ and $A_{i+1}=\{x: P(x,A_i)>0\}$. Then $P(A_i)$ approaches one monotonically as $i\to\infty$.*

Monotonicity is a consequence of

$$P(A_i) = \int_{A_{i+1}} P(x, A_i)\,dP \leqslant P(A_i). \tag{3}$$

Since $\int |P_n(x,A) - P(A)|\,dP \to 0$ as $n\to\infty$, it follows that

$$P\{x: P_i(x, A_0)=0\} \to 0 \tag{4}$$

as $i\to\infty$. $A_i = \{x: P_i(x,A_0)>0\}$ except possibly for a set of measure zero and so $P(A_i)\to 1$ as $i\to\infty$.

Lemma 3. *Let $\mathscr{B}_{-\infty}$ be trivial with $A_0\in\mathscr{A}$, $P(A_0)>0$. Then there is a sequence $\{A_i\}$ of events of $\mathscr{A}$ and a sequence of numbers $\{\varepsilon_i\}$ $i=1,2,\ldots$ such that*

(a) *$P(A_i)\uparrow 1$ as $i\to\infty$*

(b) *$P(x, A_{i-1}) \geqslant \varepsilon_i > 0$ uniformly for $x\in A_i$ $i=1,2,\ldots$.*

Our object is to exhibit an integer N and sequences $\{A_i\}$ of sets of $\mathscr{A}$ and $\{\varepsilon_i\}$ of positive numbers, $i=1,\ldots,N$, such that

(a′) $0 < P(A_0) \leqslant P(A_1) \leqslant \cdots \leqslant P(A_N) \geqslant \dfrac{1+P(A_0)}{2}$

(b′) $P(x, A_{i-1}) \geqslant \varepsilon_i > 0$ uniformly for $x\in A_i$ and

$$i=1,\ldots,N.$$

This would immediately imply (a) and (b). If $P(A_0)=1$ defining

$$A_i = \{x: P(x, A_{i-1})=1\}$$

and $\varepsilon_i = 1$ gives us sequences with properties (a) and (b). Let $0<P(A_0)<1$. Lemma 2 implies that there is an integer N and sets $B_0,\ldots,B_N$ of $\mathscr{A}$ with $B_0 = A_0$ satisfying

(a″) $0 < P(B_0) \leqslant \cdots \leqslant P(B_N) > a = \frac{1}{2}[1+P(B_0)] > 0$

(b″) $P(x, B_{i-1}) > 0$ for all $x\in B_i$, $i=1,\ldots,N$.

We shall show that the existence of a sequence $B_0, \dots, B_N$ satisfying conditions a″), b″) implies that there is a sequence of sets $C_0, \dots, C_N$ of $\mathscr{A}$ with the following properties: (i) $C_0 = B_0$; (ii) $P(C_0) \leqslant \cdots \leqslant P(C_N) > a$; (iii) $P(x, C_{i-1}) > 0$ for all $x \in C_i$, $i = 1, \dots, N$; (iv) $P(x, C_0) \geqslant \varepsilon > 0$ for all $x \in C_1$. Let $P(B_N) = a + \delta_N$, $\delta_N > 0$. By Lemma 1 one can recursively obtain positive numbers δ_i, $i = 1, \dots, N-1$, such that if $B_i' \subset B_i$ (B_i' in $\mathscr{A}$) and $P(B_i - B_i') < \delta_i$ then

$$P(x \in B_{i+1}, P(x, B_i') > 0\} > P(B_{i+1}) - \delta_{i+1}, \quad i = 1, \dots, N-1 \tag{5}$$

and for some $\varepsilon > 0$,

$$P\{x \in B_1, P(x, B_0) \geqslant \varepsilon\} > P(B_1) - \delta_1 . \tag{6}$$

Set $C_0 = B_0$, $C_1 = \{x \in B_1 : P(x, B_0) \geqslant \varepsilon\}$ and $C_i = \{x : P(x, C_{i-1}) > 0\}$ for $i \geqslant 2$. Conditions (i), (ii), (iii), (iv) are satisfied by the C sequence except possibly for $P(C_N) > a$. However, $C_{n+1} \subset B_{n+1}$ since

$$C_{n+1} \subseteq \{x : P(x, C_n) > 0\} .$$

An induction argument using the definition of the δ_i's implies that $P(B_n - C_n) < \delta_n$. Thus $P(C_N) > a$. The possibility of constructing sets $C_0, \dots, C_N$ satisfying conditions (i)—(iv) implies that one can construct sequences $\{A_i\}$, $\{\varepsilon_i\}$ satisfying (a′), (b′). The proof is complete.

Lemma 4. *Consider the Borel subsets of the interval* $[0,1]$. *Let* m *be Lebesgue measure on the Borel sets. For each* $x \in [0,1]$ *let* $Q(x, \cdot)$ *be a probability measure on the Borel subsets of* $[0,1]$ *and for each Borel subset* A *of* $[0,1]$ *let* $Q(\cdot, A)$ *be a Borel function of* x. *Set*

$$\psi(\xi, x) = \begin{cases} \inf\limits_{0 \leqslant b \leqslant 1} \{b : Q(x, [0, b]) > \xi\} & \text{if } 0 \leqslant \xi < 1 \\ 1 \quad \text{if } \xi = 1 . \end{cases} \tag{7}$$

Then

1. $\psi(\xi, x)$ *is jointly Borel measurable in* ξ *and* x
2. $m(\xi : \psi(\xi, x) \in A\} = Q(x, A)$ *for each* $x \in [0,1]$ *and Borel subset* A *of* $[0,1]$.

Notice that $\psi(\xi, x) < b$ if and only if for some rational $a \in [0, b]$ one has $Q(x, [0, a]) > \xi$.
Therefore,

$$\{x : \psi(\xi, x) < b\} = \bigcup_{\substack{\text{rational} \\ a \in [0, b)}} \{x : Q(x, [0, a]) > \xi\} \tag{8}$$

is a Borel set. The function ψ is right continuous in ξ for each fixed x. This implies that

$$\{(x, \xi) : \psi(\xi, x) < b\} = \bigcup_{\text{rational } \eta} \{(x, \xi) : \psi(\eta, x) < b, \ \xi \in [0, \eta]\} \tag{9}$$

and so ψ is jointly Borel measurable in ξ and x. Consider

$$\psi_x(A)=\{\xi:\psi(\xi,x)\in A,\ \xi\in[0,1]\}.$$

Since $m[\psi_x(A)]=Q(x,A)$ for A an interval $[0,a]$, it follows that (2) is valid for A a Borel subset of $[0,1]$.

It is enough to prove Theorem 1 for Markov processes with state space the interval $[0,1]$ and we will assume this to be the case. The following Lemma will be required as a main step in the proof.

Lemma 5. *There is a Borel function $h(\xi,x)$ of $\xi,x\in[0,1]$ taking values in $[0,1]$ and there are sequences $\{C_i\}$ of positive numbers and Borel subsets $\{A_i\}$ of $[0,1]$, $i=0,1,\ldots$, such that*

1. $m\{\xi:h(\xi,x)\in A\}=P(x,A)$ *for all* $x\in[0,1]$ *and all Borel subsets* A *of* $[0,1]$,
2. $0=C_0<C_1<\cdots<1$,
3. $A_0=A$ *and* $A_1=B$ *where* A, B *are the sets of condition* (ii) *in Theorem* 1,
4. $P(A_0)>0<P(A_1)\leqslant P(A_2)\leqslant\cdots$ *with* $P(A_i)\to 1$ *as* $i\to\infty$,
5. $C_{i-1}\leqslant\xi<C_i$ *and* $x\in A_i$ *imply that* $h(\xi,x)\in A_{i-1}$ *except for a* (ξ,x) *set of measure zero,*
6. $0\leqslant\xi<C_1$ *with* $x,y\in A_1$ *imply that* $h(\xi,x)=h(\xi,y)$.

By Lemma 3 there is a sequence $\{A_i\}$ satisfying conditions (4), (5) above and a sequence of positive numbers $\{\varepsilon_i\}$ for which $P(x,A_{i-1})\geqslant\varepsilon_i$ for all $x\in A_i$, $i=2,3,\ldots$. Set

$$C_1=\tfrac{1}{2}\varphi(A_0) \tag{10}$$

(φ from condition (ii) of Theorem 1) and

$$C_n=C_{n-1}+\frac{\varepsilon_n}{2^n},\qquad n=2,3,\ldots. \tag{11}$$

Condition (2) is clearly satisfied. Set

$$T^{(1)}(x,S)=\begin{cases}\tfrac{1}{2}\varphi(S\cap A_0) & \text{if } x\in A_1\\ C_1P(x,S) & \text{otherwise}\end{cases} \tag{12}$$

and

$$T^{(n)}(x,S)=\begin{cases}\dfrac{\varepsilon_n}{2^n}\,\dfrac{P(x,S\cap A_{n-1})}{P(x,A_{n-1})} & \text{if } x\in A_n\\[2ex] \dfrac{\varepsilon_n}{2^n}\,P(x,S) & \text{otherwise}.\end{cases} \tag{13}$$

Notice that

$$T(x,\cdot)=P(x,\cdot)-\sum_{n=1}^{\infty}T^{(n)}(x,\cdot) \tag{14}$$

is a nonnegative measure on the Borel sets for each $x \in [0,1]$ and $T(x,[0,1]) = 1 - \lim_{n\to\infty} C_n$. Further, $T(\cdot, A)$ and $T^{(n)}(\cdot, A)$ are Borel functions of x for each Borel subset A of $[0,1]$. Set up the following correspondence by using Lemma 4 and norming appropriately

$$\begin{aligned} T(x,\cdot) &\leftrightarrow \psi(\xi,x) \qquad 0 \leqslant \xi \leqslant 1 - \lim_{n\to\infty} C_n \\ T^{(n)}(x,\cdot) &\leftrightarrow \psi^{(n)}(\xi,x) \qquad 0 \leqslant \xi \leqslant C_n - C_{n-1}. \end{aligned} \tag{15}$$

Set

$$h(\xi,x) = \begin{cases} \psi^{(n)}(\xi - C_{n-1}, x) & \text{if } C_{n-1} \leqslant \xi < C_n \\ \psi(\xi - \lim_{n\to\infty} C_n, x) & \text{if } \lim_{n\to\infty} C_n \leqslant \xi \leqslant 1. \end{cases} \tag{16}$$

The function $h(\xi,x)$ is clearly a Borel function of ξ and x. Condition (1) follows from the definition of h in terms of the $\psi^{(n)}$'s and Lemma 4. Conditions (6) and (7) also follow from the definition of h in terms of the $\psi^{(n)}$'s and $T^{(n)}$'s. The proof of Lemma 5 is complete.

A probability space is now set up on which the desired representation will be obtained. Let $\sum$ be the space of points

$$\omega = \begin{pmatrix} \ldots, \xi_{-2}, \xi_{-1}, \xi_0, \xi_1, \xi_2, \ldots \\ \ldots, x_{-2}, x_{-1}, x_0, x_1, x_2, \ldots \\ y_0, y_1, y_2, \ldots \end{pmatrix} \tag{17}$$

with $\xi_i, x_i, y_i \in [0,1]$. Let $\mathscr{C}$ be the smallest Borel field of $\sum$ for which each coordinate projection is measurable. Notationally we shall write $X_i(\omega) = x_i$, $Y_i(\omega) = y_i$, $\xi_i(\omega) = \xi_i$. Let Q be the probability measure on $(\Omega, \mathscr{C})$ with the following properties:

(i) $Q\{\omega : X_0 \in A\} = P(A)$ for Borel sets $A \subset [0,1]$.

(ii) $\{\xi_i\}$, $i = 1, 2, \ldots$ is a sequence of independent random variables uniformly distributed on $[0,1]$ with $\{\xi_i\}$ independent of X_j, $j \leqslant 0$.

(iii) $Q\{\omega : X_i(\omega) = h(\xi_i(\omega), X_{i-1}(\omega))\} = 1$ for $i = 1, 2, \ldots$.

(iv) $\{(\xi_i, X_i)\}$ is a stationary sequence of pairs of random variables, $i = 0, \pm 1, \ldots$.

(v) Y_0 is independent of $\{(\xi_i, X_i)\}$, $i = 0, \pm 1, \ldots$ and

$$Q\{\omega : Y_0(\omega) \in A\} = P(A)$$

for Borel $A \subset [0,1]$.

(vi) $Q\{\omega : Y_i(\omega) = h(\xi_i(\omega), Y_{i-1}(\omega))\} = 1$ for $i = 1, 2, \ldots$

The processes $\{Z_n\}$ and $\{X_n\}$, $n = 0, \pm 1, \ldots$ have the same probability structure. Also, $\{Z_n\}$ and $\{Y_n\}$, $n \geqslant 0$, have the same probability structure and the sequence of pairs $\{(\xi_n, Y_n)\}$, are stationary for $n \geqslant 0$. For the representation it is enough to show that

$$E|X_0 - E(X_0 \mid \xi_0, \ldots, \xi_{-n})| \to 0 \tag{18}$$

or equivalently

$$E|X_n-E(X_n|\xi_n,\ldots,\xi_1)|\to 0 \tag{19}$$

as $n\to\infty$. Let us first show that (19) holds if

$$X_n-Y_n\to 0 \quad \text{in probability} \tag{20}$$

as $n\to\infty$. Since the random variables X_n and Y_n are bounded by 1, $E|X_n-Y_n|\to 0$ is equivalent to (20). Let the mapping T_ξ be given by $T_\xi x=h(\xi,x)$. Then

$$E|X_n-E(X_n|\xi_n,\ldots,\xi_1)|\leqslant E|X_n-Y_n|+E|Y_n-E(X_n|\xi_n,\ldots,\xi_1)| \tag{21}$$

and

$$\begin{aligned}E|Y_n-E(X_n|\xi_n,\ldots,\xi_1)| &=E|T_{\xi_n}\cdots T_{\xi_1}Y_0-E(T_{\xi_n}\cdots T_{\xi_1}X_0|\xi_n,\ldots,\xi_1,Y_0)|\\ &\leqslant E|E(Y_n-X_n|\xi_n,\ldots,\xi_1,Y_0)| \qquad (22)\\ &\leqslant E|Y_n-X_n|.\end{aligned}$$

It is clear that (19) is implied by (20) from inequalities (21) and (22). We now show that $Q\{X_n=Y_n\}\to 1$ as $n\to\infty$. Notice that $X_n=Y_n$ implies that $X_{n+k}=Y_{n+k}$ for integer $k\geqslant 0$ except possibly for a set of measure zero. Set $S_n=\{\omega: X_n=Y_n\}$. The sequence of sets $\{S_n\}$ is nondecreasing up to sets of measure zero. Thus, $\lim_{n\to\infty} Q(S_n)$ exists. Suppose

$$\lim_{n\to\infty} Q(S_n)=1-\gamma, \qquad \gamma>0.$$

Let m be such that $P(A_m)>1-\gamma/4$ where the A_m's are sets of Lemma 5. Set $\alpha=\prod_{i=1}^{m}(C_i-C_{i-1})$ where the C_i's are the numbers of Lemma 5. For sufficiently large n, $Q(S_n)>1-\gamma-(\alpha\gamma)/2$. Except for a set of measure zero

$$S_{n+m}\supset S_n+S_n^c\cap\{\omega: X_n, Y_n\in A_m\}\cap\bigcap_{i=1}^{m}\{C_{m-i}\leqslant\xi_{n+i}<C_{m-i+1}\} \tag{23}$$

so that

$$1-\gamma=\lim_{n\to\infty} Q(S_n)\geqslant Q(S_{n+m})\geqslant Q(S_n)+\left\{\gamma-\frac{\alpha\gamma}{2}\right\}>1-\gamma \tag{24}$$

which is impossible. Therefore $Q(S_n)\to 1$ as $n\to\infty$ and the proof of the theorem is complete.

Corollary 1. *The conclusion of Theorem* 1 *holds if* $\{Z_n\}$ *is a purely nondeterministic stationary Markov process whose state space has an atom.*

Let the set $\{z_0\}$ consisting of the single real number z_0 be the atom. Then for some $\eta>0$ the set $\{z: P(z,\{z_0\})\geqslant\eta\}$ has positive P measure. Condition (ii) of the Theorem 3 is satisfied.

By applying Corollary 1 we immediately get the following extension of Theorem 4.2.

Corollary 2. *Let* $\{Z_n\}$ *be a stationary aperiodic ergodic Markov chain. Then conditions* (i) *and* (ii) *of Theorem* 1 *are satisfied and the conclusion of the Theorem is valid.*

Notes

6.1 Rozanov's book on stationary processes [98] contains an exposition of the main results in the linear prediction problem for weakly stationary processes. An extension of some of these results to the linear prediction problem for vector-valued weakly stationary processes is also considered.

6.2 and 6.3. The uniformizing transformation given by formula 2.6 seems to have first been mentioned by P. Lévy in [66] (see page 71). N. Wiener suggested using this transformation in his book on nonlinear methods in random theory [111] when discussing coding and decoding of random sequences.

The concept of an abstract dynamical system was introduced in section 6.3. Let (Ω, μ, φ) and $(\Omega', \mu', \varphi')$ be two given abstract dynamical systems. The two systems are said to be isomorphic if there is an invertible mapping ψ of the Borel field $\mathscr{F}$ of the first system onto the Borel field $\mathscr{F}'$ of the second (modulo sets of measure zero) that is measure-preserving and such that $\varphi' = \psi^{-1} \varphi \psi$. If the transformation given by formula (2.6) is invertible, then the initially given stationary stochastic process $\{X_k\}$ and the sequence $\{\xi_k\}$ of independent uniformly distributed (on $[0,1]$) random variables are isomorphic when considered as dynamical systems with $\varphi = \varphi' = \tau$ the shift transformation. The isomorphism problem for abstract dynamical systems is considered in Arnold and Avez [2] with respect to the question of specifying invariants under isomorphism. The entropy of an abstract dynamical system as defined by Kolmogorov (see section 12 of Arnold and Avez) is one such invariant. Consider any partition $\mathscr{F}$ of Ω into a finite number of disjoint measurable (with respect to $\mathscr{A}$) sets. Given any finite partition $\mathscr{F}$ let

$$h(\mathscr{F}) = -\sum_{B \in \mathscr{F}} \mu(B) \log \mu(B).$$

Let $\mathscr{F} \vee \varphi \mathscr{F} \vee \cdots \vee \varphi^{n-1} \mathscr{F}$ be the partition generated by $\mathscr{F}, \varphi \mathscr{F}, \ldots, \varphi^{n-1} \mathscr{F}$. This is the smallest finite partition whose sets generate a field containing $\mathscr{F}, \varphi \mathscr{F}, \ldots, \varphi^{n-1} \mathscr{F}$. One can show that

$$h(\mathscr{F}, \varphi) = \lim_{n \to \infty} \frac{h(\mathscr{F} \vee \varphi \mathscr{F} \vee \cdots \vee \varphi^{n-1} \mathscr{F})}{n}$$

exists (see [2]). The entropy $h(\varphi)$ of the abstract dynamical system (Ω, μ, φ) is

$$h(\varphi) = \sup_{\mathscr{F}} h(\mathscr{F}, \varphi)$$

where the supremum is taken over all finite partitions $\mathscr{F}$ of Ω into $\mathscr{A}$ measurable sets.

An abstract dynamical system (Ω, μ, φ) is called a K-system (after Kolmogorov) if there is a sub-Borel field $\mathscr{B}$ of $\mathscr{A}$ such that $\mathscr{B} \subset \varphi \mathscr{B}$, $\bigcap_{n=-\infty}^{\infty} \varphi^n \mathscr{B}$ is the trivial Borel field containing the empty set and Ω, and the Borel fields $\varphi^n \mathscr{B}, n = \ldots, -1, 0, 1, \ldots,$

generate $\mathscr{A}$. It is clear that every purely nondeterministic stationary process $\{X_k, k=\ldots,-1,0,1,\ldots\}$ is a K-system if we set $\mathscr{B}=\mathscr{B}_0=\mathscr{B}\{X_k, k\leq 0\}$ and $\varphi=\tau$ the shift transformation. Of course, μ is simply the probability measure P of the process. However, there are stationary stochastic processes which can be considered as K-systems but are not purely nondeterministic. In such a case one has to take $\mathscr{B}$ as a sub-Borel field of $\mathscr{A}$ that involves values of the process in the future as well as possibly the past. Nonetheless, from the way in which the notion of a K-system is introduced, it is clear that it is motivated by and is a generalization of the concept of a purely nondeterministic process. A stationary process is a K-system (with $\varphi=\tau$) if there is a possible new state specification for the process (possibly depending on the whole history of the process from the past to the future) such that with this new specification the derived process is purely nondeterministic. K-systems are also discussed in Arnold and Avez [2].

6.4 Consider the two state Markov chain $\{X_n\}$ with stationary distribution given by formula (4.10) and transition probability matrix (4.11). The process $\{\xi_n\}$ of independent and identically distributed random variables given by formula (4.13) has

$$P\{\xi_n=1\}=p=1-P\{\xi_n=2\}=1-q.$$

It is given by a one-sided function in terms of the $\{X_n\}$ process. However, as indicated in the text one cannot recover $\{X_n\}$ from $\{\xi_n\}$ by a one-sided representation. A recent remarkable result of Ornstein [83] (see also Smorodinski [106]) indicates that these two processes are isomorphic in the sense described in the notes for section 6.3. In the isomorphism problem two-sided representations depending on the whole history of the process are allowed.

6.5 The derivation given in this section is due to D.L. Hanson (see[34]). He generalized certain results in a paper of M. Rosenblatt [92]. Notice that condition (ii) in Theorem 5.1 is a Doeblin-like condition (see the notes for section 4.3). Intuitively, one feels that a stationary process should have a one-sided representation in terms of a sequence of independent uniformly distributed (on $[0,1]$) random variables if and only if the process is purely nondeterministic. The "only if" part is obvious. However, the "if" part of this conjectured result has not been obtained. The result is valid in the special case of a stationary Markov chain. Hanson has obtained a nice analogue of the Wold representation in the case of a stationary Markov chain [35].

One can pose similar problems for processes with continuous time parameter. These problems are likely to be more difficult because questions of sample function behavior are likely to enter. Nisio [79] and Ito and Nisio [48] have looked at some of these questions. For example, one could ask for the class of continuous time parameter stationary processes that are representable in a sense as one sided functions and their shifts in terms of the Brownian motion process $B(t)$ defined for all real t in section 1.4. More explicitly, let

$$\mathscr{B}_t=\mathscr{B}(B(u)-B(v);\, u,v\leq t),$$

that is $\mathscr{B}_t$ is the Borel field generated by the differences $B(u)-B(v)$ with $u,v\leq t$. We ask for the class of stationary continuous parameter processes $X(t)$, $-\infty<t<\infty$, that can be represented on the probability space of the Brownian motion so that $X(t)$ is measurable with respect to $\mathscr{B}_t$ and such that $X(t)=\tau^t X(0)$ where τ is the shift transformation. Ito and Nisio [48] have shown that a large class of one-dimensional diffusion processes can be represented in this way. It is quite natural to ask the corresponding question concerning the representation of stationary processes in terms of differences $H(u)-H(v)$ of a homogeneous process, that is, a process $H(t)$ whose differences have as their distribution an infinitely divisible law and such that the differences corresponding to nonoverlapping intervals are independent and stationary. Very little is known here. Nisio [79] has shown that a simple stationary process derived from the Poisson process cannot be given such a one-sided representation in terms of the Brownian motion process.

Assuming that one is interested in a one-sided representation of a stationary process in terms of Brownian motion, it is natural to investigate the computational aspects of setting up such a representation. Formally, one would hope that a process $X(t)$, $EX(t)\equiv 0$, with such a representation could be written as

$$X(t)=\sum_{n=1}^{\infty}\int_{-\infty}^{t}\cdots\int k_n(t-u_1,\ldots,t-u_n)\,dB(u_1)\ldots dB(u_n). \tag{1}$$

Of course, appropriate conditions would have to be imposed on the process $X(t)$ and the functions k_n so as to make such a representation meaningful. N. Wiener considered such representations in his book [111]. Ito [47] has noted conditions under which integrals of the type arising in (1) exist and are meaningful. McShane [74] also deals with representations of this type as well as corresponding representations in terms of certain homogeneous processes. A number of people (see [6] and [14]) have tried to use such representations in discussing the problem of turbulence.

Chapter VII

Mixing and the Central Limit Theorem

0. Summary

The object of this chapter is to show how conditions like uniform ergodicity, uniform mixing and related notions allow one to extend certain limit theorems valid for sequences of independent random variables to stationary Markov sequences. In the first section, the case of independence is discussed and the remarkable result of Kolmogorov on the approximation of an nth convolution of a distribution by an infinitely divisible law with error term is obtained. Uniform ergodicity and strong mixing are introduced in section 2. The relation of these to the ordinary concepts of ergodicity and mixing are discussed. Cogburn's interesting result on limit laws for stationary Markov sequences that are uniformly ergodic is derived. An operator formulation of strong mixing and uniform ergodicity is given in the following section. Various L^P norm conditions on the Markov transition operator are also introduced. The L^2 norm condition is shown to be equivalent to the maximal absolute correlation between past and future tending to zero as the distance between past and future tends to zero. The various conditions are examined in the case of random walks on compact groups. A central limit theorem is proved for stationary Markov sequences in the last section and applied to random walks on compact groups.

1. Independence

The random sequences that have been studied most extensively from the earliest days of probability theory are those with no time dependence. Sequences of independent and identically distributed random variables $X_n(\omega)$, $n=0, \pm 1, \ldots$, are the simplest examples of sequences of this type. They are Markov processes with transition function $P(x, A)=P(A)$ independent of x and invariant measure $\mu(A)=P(A)$.

Let $f_{k,n}(\cdot)$, $k=0,1,\ldots,n$, be a sequence of bounded real-valued Borel functions on the range space of the $\{X_n\}$ sequence. The random variables $f_{k,n}(X_k(\omega))$ are independent but not necessarily identically distributed. The study of the possible limiting probability distributions of partial sums of independent random variables

$$\sum_{k=1}^{n} f_{k,n}(X_k(\omega))$$

(when properly centered and normalized) has been of great interest and is now a beautiful and well developed part of probability theory [31]. Interest in questions of this type was motivated in part by statistical problems in the theory of errors as has already been noted in section 3 of Chapter 1.

The main aim in this section is to derive the remarkable theorem of Kolmogorov that was casually referred to at the end of section 3 of Chapter 1. The theorem will be of interest to us for two reasons. First of all, it indicates very clearly and in a constructive manner the basic role that the infinitely divisible laws play as limit laws or, better yet, as approximations to the distribution functions of sums of independent and identically distributed random variables. The Kolmogorov theorem will also play an important role deriving results for infinitely divisible laws as approximations to the distributions of partial sums of instantaneous nonlinear functionals of Markov sequences and their shifts. These results will be obtained in section 2. They constitute a natural generalization of the results obtained in this section for the case of independent summands.

In order to prove the Kolmogorov theorem a number of Lemmas are required. Some of these preliminary results are of independent interest. The theorems of Prohorov and Berry-Esseen derived in section 3 of Chapter 1 will also be used. In many of the arguments one can assume that a distribution function F is continuous since this is always easy to arrange by convoluting with a Gaussian distribution with small variance.

The first Lemma is quite important and makes use of the notion of a concentration function of a distribution, an idea due to P. Lévy [66]. Let F be the distribution function of a random variable. The concentration function $Q_F(l)$ of F is given by

$$Q_F(l)=\sup_x[F(x+l)-F(x)], \qquad l\geqslant 0. \tag{1}$$

From the definition (1) it is clear that $Q_F(l)$ is the maximal probability mass in an interval of size l when F is the distribution function.

Lemma 1. *Let F be the common distribution function of $X_1, \ldots, X_n$ which are independent. Let H be the distribution function of the sum $\sum_{j=1}^{n} X_j$. If*

$$s = n[1 - Q_F(r)] \tag{2}$$

and $R \geqslant r$, then

$$Q_H(R) \leqslant \frac{CR}{r\sqrt{s}} \tag{3}$$

where C is an absolute constant.

The X_k's can be represented as a nondecreasing function

$$X_k = f(\eta_k), \qquad k = 1, \ldots, n \tag{4}$$

of independent random variables η_k uniformly distributed on $[0,1]$. Here

$$f(y) = F^{-1}(y) = \sup\{x \colon F(x) < y\}.$$

Let

$$2\alpha(\theta) = f(\tfrac{1}{2} + \theta) + f(\tfrac{1}{2} - \theta)$$

$$2\beta(\theta) = f(\tfrac{1}{2} + \theta) - f(\tfrac{1}{2} - \theta)$$

for $\theta \in [0,1]$. We can replace the X_k's by

$$X_k' = \alpha(\theta_k) + \xi_k \beta(\theta_k) \tag{5}$$

where the $\xi_k = \pm 1$ each with probability $\frac{1}{2}$ and are independent and the θ_k are uniformly and independently distributed on $[0, \frac{1}{2}]$ and independent of the ξ_k's. By considering the Lemma conditionally for fixed values of the θ_k's, the problem is reduced to the binomial case. Consider now a sum

$$S = \sum_{k=1}^{m} \xi_k x_k, \qquad m \leqslant n, \tag{6}$$

where the independent random variables $\xi_k = \pm 1$ each with probability $\frac{1}{2}$ and the fixed values $x_k \geqslant \gamma > 0$. The object is to get an upper bound for

$$P[a < S < a + 2\gamma]. \tag{7}$$

Let δ be a sequence of ± 1's such that $a < S = \sum \xi_k x_k < a + 2\gamma$. If the number of such distinct sequences can be bounded above, an upper bound for (7) will be available. For the sequence δ let δ^+ be the set of subscripts k for which $\xi_k = +1$ in δ. Call a collection of subsets of the integers $1, 2, \ldots, m$ consistent if no set of the collection is a subset of another set of the collection. The collection of subsets δ^+ is consistent. For if this were not so there would be two sets, say δ_1^+ and δ_2^+ with $\delta_1^+ \subset \delta_2^+$. But if $a < \delta < a + 2\gamma$ for δ_1^+ it is clear that it cannot be so for δ_2^+ since the difference between the S values for δ_1 and δ_2 is a nonzero multiple of 2γ. However, a Theorem of Sperner [104] says that there can be no more than $\binom{m}{[m/2]}$ sets in a consistent collection of subsets of the integers $1, 2, \ldots, m$. But this implies that

$$P[a<S<a+2\gamma] \leqslant \frac{C}{\sqrt{m}} \tag{8}$$

where C is an absolute constant and S the sum in (6). Suppose that Q is the concentration function for

$$S' = \sum_{k=1}^{n} \xi_k x_k$$

where m is the number of x_k's greater than $l>0$. From the very definition it follows that the concentration function of the sum of two independent random variables is less than or equal to the concentration function of each summand. This remark coupled with inequality (8) implies that under these conditions

$$Q(L) \leqslant 2\frac{CL}{l\sqrt{m}}$$

if $L \geqslant l$. Our object is to now use the results in the conditional situation to get the desired conclusion. Let m be the number of $\beta(\theta_k)$'s in (5) greater than $l>0$. Then m is a binomial variable of sample size n with mean

$$nP\{\beta(\theta)>l\} = n\gamma$$

and variance

$$n\alpha(1-\alpha).$$

By the Chebyshev inequality

$$P\left[m \leqslant \frac{n}{K}\right] = P\left[m - n\alpha \leqslant -n\left(\alpha - \frac{1}{K}\right)\right] \leqslant \frac{\alpha(1-\alpha)}{n\left[\alpha - \frac{1}{K}\right]^2}$$

if $1/K < \alpha$. This implies that

$$Q_H(L) \leqslant P\left[m \leqslant \frac{n}{K}\right] + \frac{2C\sqrt{K}L}{l\sqrt{n}} \leqslant \frac{\alpha(1-\alpha)}{n\left[\alpha - \frac{1}{K}\right]^2} + \frac{2C\sqrt{K}L}{l\sqrt{n}}.$$

If we set $1/K = \alpha/2$, the estimate

$$Q_H(L) \leqslant \frac{C'L}{l}\frac{1}{\sqrt{n\alpha}} \leqslant \frac{C'L}{l}\frac{1}{\sqrt{n(1-Q_F(l))}}$$

is obtained with C' an absolute constant. The proof is complete.

Φ has already been used to represent the standard Gaussian distribution with mean zero and variance one. Let Φ_{σ^2} denote the Gaussian distribution with mean zero and variance σ^2 so that $\Phi = \Phi_1$. The following Lemma concerns a simple inequality for Gaussian distributions.

Lemma 2. *If* $\sigma, \sigma_1 > 0$ *then there is a constant* C *such that*

$$|\Phi_{\sigma^2}(x) - \Phi_{\sigma_1^2}(x)| < C \left| \frac{\sigma^2}{\sigma_1^2} - 1 \right| \tag{9}$$

for all x.

It is clearly sufficient to prove this lemma for $\sigma_1^2 = 1$ and $\sigma^2 > \sigma_1^2$. Let $\delta = \sigma^2 - 1$. Then

$$|\Phi_{\sigma^2}(x) - \Phi(x)| = |\int \{\Phi(x-y) - \Phi(x)\} d\Phi_\delta(y)| \leqslant \int Q_\Phi(y) d\Phi_\delta(y).$$

However, if $k\delta \leqslant |y| < (k+1)\delta$, then $Q_\Phi(x) \leqslant (k+1) Q_\Phi(\delta)$. Thus

$$\int Q_\Phi(y) d\Phi_\delta(y) \leqslant Q_\Phi(\delta) \left\{ \int d\Phi_\delta(y) + \frac{1}{\delta} \int |y| d\Phi_\delta(y) \right\} \leqslant C Q_\Phi(\delta) \leqslant C' |\delta|.$$

The inequality (9) follows immediately from this last result.

Let Δ denote the degenerate distribution function with a unit jump at the origin.

Lemma 3. *Let* F *be a distribution function with mean zero and variance* σ^2. *If* $h \geqslant \sigma > 0$, *then*

$$\sum_{r=-\infty}^{\infty} \sup_{rh \leqslant x \leqslant (r+1)h} |F(x) - \Delta(x)| \leqslant C \tag{10}$$

where C *is an absolute constant.*

If ξ is a random variable with distribution function F, then by the Chebyshev inequality

$$|F(x) - \Delta(x)| \leqslant P(|\xi| \geqslant |x|) \leqslant \frac{\sigma^2}{x^2}.$$

This inequality implies (10).

In Lemma 4 a bound is obtained for a similar measure applied to the difference between the distribution of the sum of bounded, identically distributed independent random variables and an approximating Gaussian distribution.

Lemma 4. *Let* ξ_k, $k = 1, \ldots, n$, *be independent identically distributed random variables that are bounded,* $|\xi_k| \leqslant l$. *Let the variance of the sum* $\xi = \sum_1^n \xi_k$ *be* $\sigma^2 > 0$. *Then if* $h = \sqrt{n}\sigma > 0$

$$\sum_{r=-\infty}^{\infty} \sup_{rh \leqslant x \leqslant (r+1)h} |H(x) - \Phi_{\sigma^2}(x)| < C \frac{l}{\sigma} \tag{11}$$

where H *is the distribution of* ξ *and* C *is an absolute constant.*

Now

$$\sup_{|x|\leqslant h} |H(x)-\Phi_{\sigma^2}(x)| < C_1 \frac{l}{\sigma} \tag{12}$$

where C_1 is an absolute constant by the Berry-Esseen theorem (Theorem 2 of section 1.3). Since the ξ_k are identically distributed with $|\xi_k|\leqslant l$, it follows that $l^2>\sigma^2/n$ or $l/\sigma>1/\sqrt{n}$. The Chebyshev inequality implies that

$$\sup_{|x|\geqslant kh} |H(x)-\Delta(x)|, \quad \sup_{|x|\geqslant kh} |\Phi_{\sigma^2}(x)-\Delta(x)| \leqslant \frac{\sigma^2}{k^2 n\sigma^2} \leqslant \frac{1}{k^2}\frac{l}{\sigma} \tag{13}$$

for $k=1,2,\ldots$. Inequalities (12) and (13) yield the desired result.

The last tool required to prove Kolmogorov's theorem is the theorem of Prohorov (Theorem 1 of section 1.3) which we restate here and shall refer to as Lemma 5.

Lemma 5. *If* $0\leqslant p\leqslant 1$, *then*

$$\sum_{k=0}^{\infty} \left| b_k(p) - \frac{e^{-np}(np)^k}{k!} \right| < Cp$$

where

$$b_k(p) = \begin{cases} \binom{n}{k} p^k(1-p)^{n-k} & 0\leqslant k\leqslant n \\ 0 & \text{otherwise} \end{cases}$$

and C *is an absolute constant independent of* n.

We are now prepared to prove the following theorem of Kolmogorov.

Theorem. *Let* F *be any given distribution function. Given any fixed integer* $n\geqslant 1$ *there is an infinitely divisible distribution function* L *(depending on* n *and* F*) such that*

$$\sup_x |F^{(n)}(x)-L(x)| < Cn^{-\frac{1}{3}} \tag{14}$$

where C *is a universal constant independent of* F *and* n.

An infinitely divisible distribution function L providing the desired approximation (14) is actually constructed in the course of the proof. Let ξ_k, $k=1,\ldots,n$, be independent identically distributed random variables with common distribution F. We can assume that

$$\xi_k = F^{-1}(\eta_k), \qquad k=1,\ldots,n,$$

with F^{-1} the generalized nondecreasing inverse of F (as given in (4)) and the η_k are independent random variables uniformly distributed on $[0,1]$. Set

$$p=p(n)=n^{-\frac{1}{3}}$$

with

$$\mu_k = \begin{cases} 0 & \text{if } \frac{p}{2} < \eta_k < 1 - \frac{p}{2} \\ 1 & \text{otherwise,} \end{cases}$$

$k=1, \ldots, n$. Consider the conditional mean and variance

$$a = E(\xi_k \mid \mu_k = 0) \quad \text{and} \quad \sigma^2 = E((\xi_k - a)^2 \mid \mu_k = 0),$$

respectively, and the conditional distribution functions

$$A(x) = P\{\xi_k \leqslant x \mid \mu_k = 0\}$$
$$B(x) = P\{\xi_k \leqslant x \mid \mu_k = 1\}.$$

Notice that a and σ^2 are then the mean and variance of the distribution A. There will be no real loss of generality in assuming $a=0$ (consider $\xi_k - a$ instead of ξ_k) and we shall assume that this is the case. Loosely speaking, A could be called the midrange distribution and B the tail distribution of the ξ_k's. The distribution

$$F(x) = p B(x) + (1-p) A(x).$$

Also, the distribution A has its mass concentrated in $[x^-, x^+]$ with

$$x^- = F^{-1}\left(\frac{p}{2}\right), \qquad x^+ = F^{-1}\left(1 - \frac{p}{2}\right),$$

while the distribution B has its mass on $(-\infty, x^-] \cup [x^+, \infty)$. Since B has mass 1/2 in each of $(-\infty, x^-]$ and $[x^+, \infty)$, Lemma 1 implies that

$$Q_{B^{(m)}}(\lambda) \leqslant C m^{-\frac{1}{2}}$$

where

$$\lambda = x^+ - x^-$$

with C an absolute constant. Now

$$F^{(n)} = [pB + (1-p)A]^{(n)} = \sum_{m=0}^{n} \binom{n}{m} p^m (1-p)^{n-m} B^{(m)} * A^{(n-m)}.$$

Different infinitely divisible approximations will be proposed in the two cases (a) $\lambda \geqslant \sqrt{n}\,\sigma$ and (b) $\lambda < \sqrt{n}\,\sigma$.

First consider the case (a). The infinitely divisible approximation is given by the formal convolution exponential

$$L = \exp^{(*)}[np(B - \Delta)] = \sum_{m=0}^{\infty} \frac{(np)^m}{m!} e^{-np} B^{(m)}. \tag{15}$$

Let

$$H = \sum_{m=0}^{n} \binom{n}{m} p^m (1-p)^{n-m} B^{(m)}.$$

By Lemma 5

$$|L-H|<C'p=C'n^{-\frac{1}{3}}. \tag{16}$$

with C' an absolute constant. We now estimate the difference between $F^{(n)}$ and H. By Lemma 3 (here $\lambda \geqslant \sqrt{n}\,\sigma$ is used) and (3)

$$\begin{aligned}|B^{(m)}*A^{(n-m)}-B^{(m)}| &\leqslant \int |A^{(n-m)}(x-z)-\Delta(x-z)|\,B^{(m)}(dz)\\ &\leqslant Q_{B^{(m)}}(\lambda)\sum_r \sup_{r\lambda\leqslant y\leqslant (r+1)\lambda} |A^{(n-m)}(y)-\Delta(y)|\\ &\leqslant C''m^{-\frac{1}{2}}\end{aligned} \tag{17}$$

with C'' an absolute constant. However,

$$\begin{aligned}|F^{(n)}-H| &\leqslant \sum_{m=0}^{n}\binom{n}{m}p^m(1-p)^{n-m}|B^{(m)}*A^{(n-m)}-B^{(m)}|\\ &\leqslant C_3 n^{-\frac{1}{3}}+2\Sigma',\end{aligned} \tag{18}$$

where

$$\Sigma'=\sum_{m<\frac{1}{2}n^{\frac{2}{3}}}\binom{n}{m}p^m(1-p)^{n-m}=P(X<\tfrac{1}{2}n^{\frac{2}{3}})$$

by (17). Here X is a binomial variable of sample size n with mean $EX=np<n^{\frac{2}{3}}$ and variance $\sigma^2(X)=np(1-p)<n^{\frac{2}{3}}$. Thus,

$$\Sigma'\leqslant P\{|X-n^{\frac{2}{3}}|>\tfrac{1}{2}n^{\frac{2}{3}}\}\leqslant 4n^{-\frac{2}{3}}. \tag{19}$$

The inequalities (16), (18) and (19) indicate that the infinitely divisible distribution L as given by (15) satisfies (14).

The discussion of the case (b) proceeds in a similar manner but requires a more detailed estimation. Here L is given by

$$L=\exp^{(*)}[np(B-\Delta)]*\Phi_{n(1-p)\sigma^2}=\sum_{m=0}^{\infty}\frac{(np)^m}{m!}e^{-np}B^{(m)}*\Phi_{n(1-p)\sigma^2}. \tag{20}$$

Let

$$H_1=\sum_{m=0}^{n}\binom{n}{m}p^m(1-p)^{n-m}B^{(m)}*\Phi_{(n-m)\sigma^2}$$

and

$$H_2=\sum_{m=0}^{n}\binom{n}{m}p^m(1-p)^{n-m}B^{(m)}*\Phi_{n(1-p)\sigma^2}.$$

By Lemma 5

$$|L-H_2|\leqslant Cn^{-\frac{1}{3}} \tag{21}$$

with C an absolute constant. Consider estimating a typical term in the series for $F^{(n)}-H_1$. Lemmas 4 and 1 imply that

$$
\begin{aligned}
&|B^{(m)} * A^{(n-m)} - B^{(m)} * \Phi_{(n-m)\sigma^2}| \\
&\leqslant \int |A^{(n-m)}(x-z) - \Phi_{(n-m)\sigma^2}(x-z)|\, B^{(m)}(dz) \qquad (22)\\
&\leqslant Q_{B^{(m)}}(\sqrt{n}\,\sigma) \sum_r \sup_{r\sqrt{n}\sigma \leqslant y \leqslant (r+1)\sqrt{n}\sigma} |A^{(n-m)}(y) - \Phi_{(n-m)\sigma^2}(y)| \\
&\leqslant C_1 \frac{\sqrt{n}\,\sigma}{\lambda} m^{-\frac{1}{2}} C_2 \frac{\lambda}{\sqrt{n-m}\,\sigma} = C_3 m^{-\frac{1}{2}} \left(1 - \frac{m}{n}\right)^{-\frac{1}{2}}.
\end{aligned}
$$

Just as in (18) we find by using (22) that

$$|F^{(n)} - H_1| \leqslant C_4 n^{-\frac{1}{3}} + 2\Sigma' \leqslant C_5 n^{-\frac{1}{3}}. \qquad (23)$$

Use Lemma 2 to obtain

$$|H_1 - H_2| \leqslant C_6 n^{-\frac{1}{3}} + \Sigma''$$

with

$$\Sigma'' = \sum_{\left|\frac{n-m}{n(1-p)} - 1\right| \geqslant \frac{1}{2} n^{-\frac{1}{3}}} \binom{n}{m} p^m (1-p)^{n-m}.$$

By Chebyshev it is easily seen that

$$\Sigma'' \leqslant C\, n^{-\frac{1}{3}}$$

and so

$$|H_1 - H_2| \leqslant C_7 n^{-\frac{1}{3}}. \qquad (24)$$

The inequalities (21), (23) and (24) show that L as given by (20) satisfies (14). The proof is complete.

Corollary: *The possible limiting distributions of normalized and properly centered partial sums of a sequence of independent and identically distributed random variables are all infinitely divisible.*

Of course, this result follows immediately from Kolmogorov's theorem if one makes use of the closure of the class of infinitely divisible laws under limit operations.

From section 1.3 it is clear that the distribution function F of a random variable is infinitely divisible if and only if for each integer $n \geqslant 2$ there is a distribution function F_n such that $F_n^{(n)} = F$ where $F_n^{(n)}$ is the nth convolution of F_n with itself. A distribution function F is *symmetric* if $F(x) = 1 - F(-x)$ at every continuity point x of F. The following two Lemmas will be helpful in developing results on limit laws for partial sums of simple functions of Markov sequences. The first Lemma provides a bound on the probability mass in the tail of a symmetric infinitely divisible distribution.

Lemma 6. *Given* $0<a<b$ *and* $\varepsilon>0$ *there is a* $\delta(\varepsilon)>0$ *such that* $P[Y\geqslant a]<\varepsilon$ *for every real-valued random variable* Y *having a symmetric infinitely divisible distribution and satisfying* $P[Y\geqslant b]<\delta$.

Let $Z_1,\ldots,Z_n$ be independent, identically distributed random variables with $\sum_{k=1}^{n} Z_k=Y$. The random variables Z_k can be taken to be symmetrically distributed and we assume this to be the case. Since $P[Z_k\geqslant b/n]\leqslant\{P[Y\geqslant b]\}^{1/n}$ it follows that

$$P\left[\max_{1\leqslant k\leqslant n}|Z_k|\geqslant\frac{b}{n}\right]\leqslant 2n\{P[Y\geqslant b]\}^{\frac{1}{n}}.$$

The conditional probability

$$P[Y\geqslant a\,|\,|Z_1|,\ldots,|Z_n|]\leqslant\frac{1}{a^2}E(Y^2\,|\,|Z_1|,\ldots,|Z_n|)\leqslant\frac{1}{a^2}\sum_{k=1}^{n}|Z_k|^2$$

almost surely since $E(Z_jZ_k\,|\,|Z_j|,|Z_k|)=0$ for $j\neq k$ by the symmetry of the random variables Z_j. This implies that

$$P[Y\geqslant a]\leqslant P\left[\max_{1\leqslant k\leqslant n}|Z_k|\geqslant\frac{b}{n}\right]+\frac{b^2}{na^2}\leqslant 2n\{P[Y\geqslant b]\}^{\frac{1}{n}}+\frac{b^2}{na^2}.$$

The proof is completed by first selecting n and then δ appropriately.

The second Lemma describes a condition sufficient for the convergence of a sequence of infinitely divisible distribution functions F_n to the distribution function Δ with total mass one at zero, that is, $F_n(x)\to\Delta(x)$ for all $x\neq 0$.

Lemma 7. *Let* F_n *be a sequence of infinitely divisible distribution functions. Assume that* $\int g(x)\,dF_n(x)\to 0,\ n\to\infty$, *for some continuous function* $g\geqslant 0$ *with limit* $\lim_{|x|\to\infty} g(x)>0$ *and that* $\int h(x)\,dF_n(x)\to 0,\ n\to\infty$, *for some continuous bounded function* h *such that* $h(x)\neq h(0)$ *for* $x\neq 0$. *Then it follows that* $F_n\to\Delta$ *as* $n\to\infty$.

Let $\bar{F}_n(x)=1-F_n(-x)$ at every continuity point x of F_n. Then $F_n^s=F_n*\bar{F}_n$ ($*$ denotes the convolution operation) is a symmetric distribution function generated from F_n. The hypothesis implies that $F_n(b)-F_n(-b)\to 1$ for some finite b. Therefore, $F_n^s(2b)-F_n^s(-2b)\to 1$ as $n\to\infty$. Lemma 6 implies that $F_n^s(a)-F_n^s(-a)\to 1$ as $n\to\infty$ for every finite $a>0$. Let m_n be a median of F_n, that is,

$$F_n(m_n-),\quad 1-F_n(m_n)\leqslant\tfrac{1}{2}.$$

Then $\int f(x)\,dF_n(x)-f(m_n)\to 0$ as $n\to\infty$ for any continuous bounded f. Thus $g(m_n)\to 0$ and $h(m_n)\to h(0)$ as $n\to\infty$. The first remark implies that the m_n are bounded and the second remark that $m_n\to 0$. Therefore $F_n\to\Delta$.

2. Uniform Ergodicity, Strong Mixing and the Central Limit Problem

The theorem of Kolmogorov derived in the preceding section indicates what a central role is played by the infinitely divisible laws as limit laws of properly centered and normalized partial sums of independent and identically distributed random variables. In fact it does more. The infinitely divisible laws are shown to provide good approximations to the distributions of sums of independent and identically distributed random variables even when one can not normalize and center so as to get limiting distributions. Our object in this section is to obtain similar results for infinitely divisible laws as approximations to the distributions of partial sums (suitably normalized) of the shifts of instantaneous real-valued functions of Markov processes. In this way an extension of the types of results obtained in section 1 for independence will be obtained for an interesting class of dependent processes. As we shall see, certain conditions will have to be imposed on the Markov processes in order to get global results of the type desired.

Let $\{X_n\}$, $n=0, \pm 1, \ldots$, be a stationary Markov process. Suppose $Y_n(\omega)$, $n=0, \pm 1, \ldots$ is a series of real-valued random variables on the probability space of the Markov process $\{X_n\}$. Call the series $\{Y_n\}$ *time consistent* if $Y_n(\omega)$ is measurable with respect to $\mathscr{A}_n^n$, $n=0, \pm 1, \ldots$ and stationary if $\tau Y_n(\omega) = Y_{n+1}(\omega)$, $n=0, \pm 1, \ldots$ where τ is the shift transformation. Clearly, a time consistent series is one in which $Y_n(\omega)$ is a function of $X_n(\omega)$ so that $Y_n(\omega) = f_n(X_n(\omega))$ generated by a series of Borel functions f_n is time consistent. If $f_n = f$ is independent of n, the series is also stationary but not necessarily Markovian as was noted in detail in Chapter 3. Our concern will be with the approximation (in an appropriate sense) of the distribution of normalized partial sums of time consistent stationary series by infinitely divisible laws.

We now consider the condition to be imposed on the stationary Markov sequences. Recall that in Corollary 4 of section 4.2, ergodicity for a measure-preserving mapping φ of a Borel field $\mathscr{A}$ was found to be equivalent to

$$\lim_{n\to\infty} \frac{1}{n} \sum_{j=1}^{n} P(A \cap \varphi^{-j} B) = P(A) P(B)$$

for each pair of sets $A, B \in \mathscr{A}$. The probability measure P is preserved by φ. In the infinite sequence space representation of a stationary Markov process, the shift transformation τ is an invertible measure-preserving transformation. Let

$$a(n) = \sup_{\substack{B \in \mathscr{B}_0 \\ F \in \mathscr{F}_0}} \left| \frac{1}{n} \sum_{k=1}^{n} P(B \cap \tau^k F) - P(B) P(F) \right|. \tag{1}$$

where $\mathscr{B}_0=\mathscr{B}\{X_j, j\leqslant 0\}$ is the Borel field generated by X_j, $j\leqslant 0$ (the past and present relative to time 0) and $\mathscr{F}_0=\mathscr{B}\{X_j, j\geqslant 0\}$ is the Borel field generated by $X_j, j\geqslant 0$ (the future and present relative to time 0). It is natural to call the stationary Markov sequence $(X_n\}$ *uniformly ergodic* if $a(n)\to 0$ as $n\to\infty$. This condition of uniform ergodicity will be used to get the results desired. It is simple to give examples of stationary Markov sequences that are ergodic but not uniformly ergodic. One such example is the Markov process with state space the real numbers modulo one, transition functions determined by the transformation $\varphi x=x+\alpha$ modulo one with α a fixed irrational number and invariant instantaneous measure the uniform distribution on the unit interval $0\leqslant x<1$ (the Borel field on the state space is just the Borel field of Borel sets of real numbers on $[0,1)$). This Markov process is certainly ergodic. Some simple computations given in section 4 show that it is not uniformly ergodic.

In passing, let us also mention a more stringent condition. Set

$$d(n)=\sup_{\substack{B\in\mathscr{B}_0\\ F\in\mathscr{F}_n}}|P(B\cap F)-P(B)P(F)|. \tag{2}$$

The Markov process $\{X_n\}$ is called *strongly mixing* if $d(n)\to 0$ as $n\to\infty$. Notice that the use of the term strongly mixing given here is different and somewhat stronger than that in standard use in ergodic theory [2]. What we have called mixing (see section 2 of Chapter 4) is often called strong mixing in ergodic theory. A better term for what we call strong mixing would be uniform mixing. By a small change in the casting of the strong mixing condition, it can be seen that one could equally well refer to the process possessing it as being *uniformly purely nondeterministic*. Recall (see section 6.2) that a stationary process is purely nondeterministic if for each measurable set F

$$P(F\,|\,\mathscr{B}_{-k})\to P(F) \tag{3}$$

almost everywhere as $k\to\infty$. Actually, $P(F\,|\,\mathscr{B}_{-k})$ is the best predictor of the indicator function I_F of F that is $\mathscr{B}_{-k}$-measurable in the sense of minimizing the mean square error of prediction. The mean square error of prediction is

$$E|I_F-P(F\,|\,\mathscr{B}_{-k})|^2.$$

If there were no information from the past we would simply approximate I_F by $P(F)$ and obtain mean square error

$$E|I_F-P(F)|^2=P(F)\{1-P(F)\}.$$

Notice that

$$\begin{aligned} E|I_F - P(F)|^2 &= E|\, I_F - P(F|\mathscr{B}_{-k}) + P(F|\mathscr{B}_{-k}) - P(F)|^2 \\ &= E|I_F - P(F|\mathscr{B}_{-k})|^2 + E|P(F|\mathscr{B}_{-k}) - P(F)|^2. \end{aligned} \tag{4}$$

Relations (3) and (4) imply that one ought to call a stationary process uniformly purely nondeterministic if

$$\sup_{F \in \mathscr{F}_n} E|P(F|\mathscr{B}_0) - P(F)|^2 \to 0$$

as $n \to \infty$. Since

$$P(B \cap F) - P(B)P(F) = \int_B \{P(F|\mathscr{B}_0)(x) - P(F)\}\, dP$$

it follows that

$$\begin{aligned} \tfrac{1}{2} \int |P(F|\mathscr{B}_0)(x) - P(F)|\, dP &\leqslant \sup_{B \in \mathscr{B}_0} |P(B \cap F) - P(B)P(F)| \\ &\leqslant \int |P(F|\mathscr{B}_0) - P(F)|\, dP. \end{aligned} \tag{5}$$

Furthermore,

$$\begin{aligned} \{\int |P(F|\mathscr{B}_0)(x) - P(F)|^2\, dP\}^{\frac{1}{2}} &\geqslant \int |P(F|\mathscr{B}_0) - P(F)|\, dP \\ &\geqslant \int |P(F|\mathscr{B}_0) - P(F)|^2\, dP. \end{aligned} \tag{6}$$

The inequalities (5) and (6) lead directly to the following simple result.

Lemma 1. *A stationary Markov sequence is strongly mixing if and only if it is uniformly purely nondeterministic.*

We now continue with our discussion of infinitely divisible laws as approximations to the distributions of partial sums of time consistent stationary series. Recall that in dealing with sums of independent random variables, the Kolmogorov theorem showed that the maximal absolute deviation between the distribution of the sums and the approximating infinitely divisible law distribution could be made small. Because the maximal absolute deviation was used as a measure of closeness, there was no need to use a normalization of the partial sums. In the situation we are now dealing with, a weaker notion of approximation and something like normalization will both have to be used. The following simple example will illustrate the reasons for this. Let ε_n, $n = \ldots, -1, 0, 1, \ldots$, be a sequence of independent and identically distributed random variables. The sequence of two-vectors $X_n = (\varepsilon_n, \varepsilon_{n-1})$, $n = \ldots, -1, 0, 1, \ldots$, is then certainly a stationary Markov process which is very well behaved. It is uniformly ergodic, even strongly mixing. Consider the time consistent series $Y_n = \varepsilon_n - \varepsilon_{n-1}$. The partial sum

$$\sum_{j=1}^{n} Y_j = \varepsilon_n - \varepsilon_0 \tag{7}$$

so that the partial sums (7) have a limiting distribution which is that of $\varepsilon_1 - \varepsilon_0$. However, the class of distributions of differences of independent and identically distributed random variables contains many distributions that are not infinitely divisible. If (7) is normalized by anything that diverges as $n \to \infty$, then it is clear that the limiting distribution is the degenerate distribution with all its mass at 0, which is infinitely divisible.

The example just discussed suggests the introduction of the concept of uniform asymptotic negligibility. This concept has been extensively used in studying sums of independent random variables (see [31] and [70]). Let $\{Y_k^{(n)}, k=0, \pm 1, \ldots\}$ be a sequence of time consistent series. Such a sequence is called uniformly asymptotically negligible if for each $\varepsilon > 0$

$$\sup_k P[|Y_k^{(n)}(\omega)| > \varepsilon] \to 0 \tag{8}$$

as $n \to \infty$. Set

$$Y^{(n)}(s,t) = \sum_{k=s}^{t} Y_k^{(n)}.$$

Consider a sequence of partial sums $Y^{(n)}(s_n, t_n)$, $s_n < t_n$, and the corresponding distribution functions F_n. The sequence of distribution functions is *well approximated by infinitely divisible laws* if there is a corresponding sequence of infinitely divisible laws G_n such that

$$\int g(x)\, dF_n(x) - \int g(x)\, dG_n(x) \to 0 \tag{9}$$

as $n \to \infty$ for any fixed bounded uniformly continuous function g. The stationary Markov process $\{X_k\}$ is said to have *central structure* if for any uniformly asymptotically negligible stationary sequence $\{Y_k^{(n)}\}$ of time consistent series, any partial sums $Y^{(n)}(s_n, t_n)$, $-\infty < s_n < t_n < \infty$, are well approximated in distribution by infinitely divisible laws as $n \to \infty$. Kolmogorov's theorem (see section 1) implies that a stationary Markov process of independent random variables has central structure. The main result of this section is given by the following theorem which is a natural generalization of the remark just made about series of independent identically distributed random variables.

Theorem 1. *Let $\{X_n\}$ be a stationary Markov process. The process has central structure if and only if it is uniformly ergodic.*

The following Lemmas are required for a proof of Theorem 1. The symbol τ has already been introduced to represent the shift transformation. Let

$$\bar{\tau}^k = \frac{1}{k} \sum_{j=1}^{k} \tau^j.$$

Lemma 2. *Let $\{X_n\}$ be stationary and uniformly ergodic. Assume that Y, Z are real or complex-valued random variables measurable on $\mathscr{B}_n$, $\mathscr{F}_n$ respectively with $|Y|, |Z| \leqslant 1$. Then for integer $m > 0$*

$$|E(Y \tau^m Z) - E\,Y\,E\,Z| \leqslant 4a(m).$$

Assume that Z is real-valued first and let $A = \{\omega : E(\tau^m Z \mid \mathscr{B}_n) - E\,Z \geqslant 0\}$. Then

$$\begin{aligned}|E(Y\tau^m Z) - E\,Y\,E\,Z| &\leqslant E\,|E(\tau^m Z \mid \mathscr{B}_n) - E\,Z| \\ &= 2E\,I_A\{E(\tau^m Z \mid \mathscr{B}_n) - E\,Z\} = 2E(Z\tau^{-m} I_A - P(A)E\,Z) \\ &= 2E\,Z\{E(\tau^{-m} I_A \mid \mathscr{F}_n) - P(A)\} \\ &\leqslant 2\sup_{B \in \mathscr{F}_n} E\,I_B\{E(\tau^{-m} I_A \mid \mathscr{B}_n) - P(A)\} \leqslant 2a(m).\end{aligned}$$

In the inequalities given above I_A denotes the indicator function of the set A. The case for complex Z follows by applying the bound obtained separately to the real and imaginary parts of Z.

Lemma 3. *Let $\{X_n\}$ be a stationary uniformly ergodic Markov process with $\{Y_n\}$ a stationary time consistent series on the probability space of the process. Assume that for some ε, $\delta > 0$*

$$\sup_{0 < n-m \leqslant l} P[\,|Y(m,n)| > \varepsilon] \leqslant \delta. \tag{10}$$

Let $g_u(x) = e^{iux}$, u real. Then

$$|E\,g_u(Y(r,s))\,E\,g_u(Y(s,t)) - E\,g_u(Y(r,t))| \leqslant 2(\delta + |u|\varepsilon + 2a(l)), \quad r < s < t.$$

If $|x - y| \leqslant \varepsilon$, then

$$|g_u(x) - g_u(y)| \leqslant |u|\varepsilon. \tag{11}$$

For $0 < h \leqslant l$, the inequalities (10) and (11) imply that

$$E\,|\tau^h g_u(Y(s,t)) - g_u(Y(s,t))| = E\,|g_u(\tau^h Y(s,t)) - g_u(Y(s,t))| \leqslant 2(\delta + |u|\varepsilon).$$

Therefore,

$$\begin{aligned}&|E\,g_u(Y(r,s))\{\tau^h g_u(Y(s,t)) - g_u(Y(s,t))\}| \\ &\leqslant E\,|g_u(\tau^h Y(s,t)) - g_u(Y(s,t))| \leqslant 2(\delta + |u|\varepsilon).\end{aligned}$$

Applying Lemma 2 yields the desired conclusion.

Proof of Theorem 1*:* Assume that $\{X_n\}$ is a stationary Markov process that is uniformly ergodic. We shall show that $\{X_n\}$ has central structure. Now $a(n) \to 0$ as $n \to \infty$. Let $\{Y_k^{(n)}\}$ be a sequence of uniformly asymptotically negligible stationary time consistent series. Consider the sequence of partial sums $Y^{(n)}(s_n, t_n)$, $-\infty < s_n < t_n < \infty$. Assume that

$t_n - s_n \to \infty$ as $n \to \infty$ since the distribution of $Y^{(n)}(s_n, t_n)$ will converge to the trivial infinitely divisible distribution Δ with all mass at zero for any subsequence (s_n, t_n) with $t_n - s_n$ bounded as $n \to \infty$ because of the asymptotic negligibility of the summands. Let ${}_j s = s_n + j b_n$ where b_n is the greatest integer less than or equal to $(t_n - s_n)/k$. The explicit dependence of ${}_j s$ on n is not indicated to avoid complicated notation. Given ε, $l > 0$, choose n large enough so that

$$\sup_{0 < t-s \leqslant l} P[|Y^{(n)}(s,t)| > \varepsilon] \leqslant \varepsilon.$$

By Lemma 3

$$\begin{aligned}
&|E g_u(Y^{(n)}({}_0 s, {}_k s)) - [E g_u(Y^{(n)}({}_0 s, {}_1 s))]^k| \\
&\leqslant \sum_{j=2}^{n} E|g_u(Y^{(n)}({}_0 s, {}_j s)) - g_u(Y^{(n)}({}_0 s, {}_{j-1} s)) g_u(Y^{(n)}({}_0 s, {}_1 s))| \qquad (12) \\
&\leqslant 2(k-1)(\varepsilon + |u|\varepsilon + 2a(l)).
\end{aligned}$$

Let $n \to \infty$, $\varepsilon \to 0$, $l(=o(n)) \to \infty$ and $k = k(n) \to \infty$ sufficiently slowly so that (12) tends to zero and $Y^{(n)}({}_k s, t_n) \to 0$ in probability. By Kolmogorov's theorem the distribution of the sum of k independent, identically distributed random variables with common distribution that of $Y^{(n)}({}_0 s, {}_1 s)$ is well approximated by infinitely divisible laws as $n \to \infty$. Thus $Y^{(n)}({}_0 s, {}_k s)$ and $Y^{(n)}(s_n, t_n)$ (because of the uniform asymptotic negligibility of summands) are well approximated in distribution by infinitely divisible laws.

We now show that central structure for the stationary Markov process $\{X_n\}$ implies uniform ergodicity. Assume that the process is not uniformly ergodic. Then there is a sequence of events $A_n \in \mathscr{F}_0$ and integers $t_n \to \infty$ such that for some $\varepsilon > 0$

$$E|E(\tau^{t_n} I_{A_n} | \mathscr{B}_0) - P(A_n)| \geqslant \varepsilon. \qquad (13)$$

Let

$$Y_k^{(n)} = \tau^k (P(A_n | \mathscr{B}_0) - P(A_n)) t_n^{-1}.$$

The sequence of series $\{Y_k^{(n)}\}$ is stationary time consistent and uniformly asymptotically negligible as $n \to \infty$. Let the distribution functions of the sums $Y^{(n)}(0, t_n) = \sum_{k=0}^{t_n} Y_k^{(n)}$ be F_n. Since the process $\{X_k\}$ has central structure, there are infinitely divisible distributions G_n such that $\int g(x)\,dF_n(x) - \int g(x)\,dG_n(x) \to 0$ as $n \to \infty$ for any fixed bounded uniformly continuous function g. Let g be continuous, nonnegative and such that $g(\infty) = \lim_{|x| \to \infty} g(x) > 0$ with $g(x) = 0$ for $|x| \leqslant 1$. Also let h be continuous with $h(x) = x$ for $|x| \leqslant 1$, $h(x) \neq 0$ for $x \neq 0$ and

$$h(\infty) = \lim_{|x| \to \infty} h(x)$$

well defined. Now $|Y(0,t_n)|\leqslant 1$ so that $\int g(x)\,dF_n(x)=0$. Also,

$$\int h(x)\,dF_n(x)=E\,Y(0,t_n)=P(A_n)-P(A_n)=0.$$

Therefore, $\int g(x)\,dG_n(x)$, $\int h(x)\,dG_n(x)\to 0$ as $n\to\infty$. By Lemma 1.7 it follows that $G_n\to\Delta$ and therefore $F_n\to\Delta$ as $n\to\infty$. However, for continuous f with $f(\infty)=\lim_{|x|\to\infty} f(x)$ well defined and $f(x)=|x|$ for $|x|\leqslant 2$

$$E|E(\overline{\tau}^{t_n} I_{A_n})-P(A_n)|\leqslant \int f(x)\,dF_n(x)\to 0$$

as $n\to\infty$. Since this contradicts (13), we must have uniform ergodicity.

3. An Operator Formulation of Strong Mixing and Uniform Ergodicity

Let $\{X_k;\, k=0,\pm 1,\dots\}$ be a stationary Markov process with transition probability function $P(x,A)$, $x\in\Omega$, $A\in\mathscr{A}$, and probability measure μ invariant with respect to $P(\cdot,\cdot)$

$$\int \mu(dx)\,P(x,A)=\mu(A).$$

The process $\{X_k\}$ is assumed to be defined on the infinite product space $\dots\times\Omega\times\Omega\times\dots$ with $\mathscr{B}_n$ the backward Borel field generated by X_k, $k\leqslant n$, and $\mathscr{F}_n$ the forward Borel field generated by X_k, $k\geqslant n$. It will be convenient to assume that $\mathscr{A}$ is the Borel field generated by any single random variable X_k, $k=0,\pm 1,\dots$. Let T be the operator induced by the transition probability function $P(\cdot,\cdot)$. A simple argument will be used to give equivalent formulations of strong mixing and uniform ergodicity for stationary Markov processes in terms of the operator T.

We first consider strong mixing. For any $B\in\mathscr{B}_m$ and $F\in\mathscr{F}_{m+n}$, $n>0$,

$$\begin{aligned}&|P(B\cap F)-P(B)P(F)|\\ &=\left|\int P(B|x_m)\,\mu(dx_m)\int [P_n(x_m,dx_{m+n})-\mu(dx_{m+n})]\,P(F|x_{m+n})\right|\end{aligned}\tag{1}$$

so that $\{X_k\}$ is strongly mixing if and only if

$$\sup_{F\in\mathscr{F}_n}\int \mu(dx)\left|\int [P_n(x,dy)-\mu(dy)]\,P(F|y)\right|\to 0\tag{2}$$

as $n\to\infty$. Condition (2) is equivalent to

$$\sup_{\|f\|_\infty\leqslant 1}\int \mu(dx)\left|\int [P_n(x,dy)-\mu(dy)]\,f(y)\right|\to 0\tag{3}$$

as $n\to\infty$ where $\|f\|_\infty$ is the norm of $L^\infty(d\mu)$. It is clear that (3) implies (2). The converse follows on approximating f by step functions. The proof of Lemma 1 is complete.

Lemma 1. *A stationary Markov process with transition operator T and invariant probability measure μ is strongly mixing if and only if*

$$\sup_{f\perp 1} \frac{\|T^n f\|_1}{\|f\|_\infty} \to 0 \tag{4}$$

as $n\to\infty$, where $\|\cdot\|_p$ is the L^p norm with respect to μ and $f\perp 1$ means that $\int f(x)\mu(dx)=0$.

The corresponding result for uniform ergodicity is given in Lemma 2 and is obtained by a similar argument.

Lemma 2. *A stationary Markov process with transition operator T and invariant probability measure μ is uniformly ergodic if and only if*

$$\sup_{f\perp 1} \frac{\left\| \frac{1}{n} \sum_{j=1}^{n} T^j f \right\|_1}{\|f\|_\infty} \to 0$$

as $n\to\infty$.

The simple norm inequality

$$\|f\|_1 \leqslant \|f\|_p \leqslant \|f\|_\infty, \quad 1\leqslant p\leqslant\infty$$

leads to the simple and occasionally useful results that follow.

Lemma 3. *Consider a stationary Markov process with transition operator T and invariant probability measure μ. If*

$$\sup_{f\perp 1} \frac{\|T^n f\|_p}{\|f\|_p} \to 0 \tag{5}$$

as $n\to\infty$ for some p, $1\leqslant p\leqslant\infty$, the process is strongly mixing. If

$$\sup_{f\perp 1} \frac{\left\| \frac{1}{n} \sum_{j=1}^{n} T^j f \right\|_p}{\|f\|_p} \to 0$$

as $n\to\infty$ for some p, $1\leqslant p\leqslant\infty$, the process is uniformly ergodic.

Notice that if (5) is satisfied there are constants K and α, $0<\alpha<1$, such that

$$\sup_{f\perp 1} \frac{\|T^n f\|_p}{\|f\|_p} \leqslant K\alpha^n .$$

Any positive recurrent stationary Markov chain with no periodic states is strongly mixing (see Lemma 2 of section 4). There are simple examples of such Markov chains which indicate that apparently stronger (and now verifiably stronger) conditions of the type (5) need not be satisfied.

Let us now see what strong mixing and uniform ergodicity mean in the case of a special class of stationary Markov processes taking on values in a compact topological group G. Let ν be a regular probability measure on G whose support is not contained in any proper closed normal subgroup of G. This condition is assumed to insure that the process constructed is ergodic. The Borel field $\mathscr{A}$ is taken to be the Borel field given by the topology on G. Consider the transition probability function (as in Chapter 5)

$$P(g, A) = \nu(g^{-1} A), \qquad A \in \mathscr{A}, \qquad g \in G. \tag{6}$$

The n-step transition probability function $P_n(g, A)$ corresponds to the product of n independent identically distributed elements of G with common distribution ν

$$P_n(g, A) = \nu^{(n)}(g^{-1} A). \tag{7}$$

Here $\nu^{(n)}$ is the nth convolution of ν with itself. The transition function $P(\cdot, \cdot)$ is set up here in terms of a right product but a construction in terms of left products is completely analogous. In the case of a commutative group the two constructions in terms of right and left products are the same. The results obtained in the commutative case are the most natural ones and it is good to keep the circle group in mind as a simple example. In any case, the unique invariant measure for transition function (6), under the assumptions made, is the uniform or Haar measure μ on the group.

We first consider a commutative group G. Let $\gamma(g) = (g, \gamma)$ denote the value of a character γ at the point $g \in G$. The set of all continuous characters of G forms the dual group Γ. If η is a set function on G of bounded variation, the Fourier-Stieltjes transform $\hat{\eta}$ (on Γ) of η is given by

$$\hat{\eta}(\gamma) = \int_G (-g, \gamma) \eta(dg).$$

If η is absolutely continuous with respect to μ, $d\eta = f d\mu$, then the Fourier-Stieltjes transform of η is the Fourier transform of f and we write $\hat{\eta} = \hat{f}$. Γ is given the weak topology induced by the Fourier transform $\hat{f}$ and since G is compact, this implies that Γ is discrete (see [101]). Consider now how the operator T^n induced by (7) acts on a function $f \in L^2(d\mu)$, $f \perp 1$,

$$\begin{aligned}(T^n f)(g) &= \int_G f(k) \nu^{(n)}(g^{-1} dk) = \int_G \sum_{\gamma \neq 0} \hat{f}(\gamma)(k, \gamma) \nu^{(n)}(g^{-1} dk)\\ &= \sum_{\gamma \neq 0} \hat{f}(\gamma) \int_G (gk, \gamma) \nu^{(n)}(dk) = \sum_{\gamma \neq 0} \hat{f}(\gamma) \overline{[\hat{\nu}(\gamma)]^n (g, \gamma)}.\end{aligned}$$

Here by $\gamma=0$ we mean the character γ with $\gamma(g)\equiv 1$ on G. Notice that

$$\|T^n f\|_2^2 = \sum_{\gamma\neq 0} |\hat{f}(\gamma)|^2 |\hat{\nu}(\gamma)|^{2n}$$

because of the orthonormality of the characters. It is clear that (5) is satisfied with $p=2$ if

$$\sup_{\gamma\neq 0} |\hat{\nu}(\gamma)| < 1. \tag{8}$$

However, if (8) is not satisfied one can find a sequence of characters $\gamma_n\neq 0$ such that $|\hat{\nu}(\gamma_n)|\to 1$ as $n\to\infty$. As characters they are orthogonal to one and of absolute value one as functions on G. On setting $f_k(g)=\gamma_k(g)$ we see that

$$1 \geqslant \sup_{f\perp 1} \frac{\|T^n f\|_1}{\|f\|_\infty} \geqslant \sup_k \frac{\|T^n f_k\|_1}{\|f_k\|_\infty} = \sup_k |\hat{\nu}(\gamma_k)|^n = 1, \qquad n=1,2,\dots,$$

so that the process is not strongly mixing. A similar argument shows that the process is uniformly ergodic if and only if

$$\inf_{\gamma\neq 0} |\hat{\nu}(\gamma)-1| > 0.$$

Theorem 1. *Let G be a compact Abelian group and ν a regular probability measure on G with support contained in no proper closed subgroup. The stationary Markov process generated by transition function*

$$P(g,A)=\nu(g^{-1}A)$$

and invariant measure the Haar measure is strongly mixing if and only if

$$\sup_{\gamma\neq 0} |\hat{\nu}(\gamma)| < 1.$$

The process is uniformly ergodic if and only if

$$\inf_{\gamma\neq 0} |\hat{\nu}(\gamma)-1| > 0.$$

It is easy to give a trivial example of a Markov process of this type on a compact group which is uniformly ergodic but not strongly mixing. Consider the finite commutative group of complex numbers $\exp\{2\pi i j/n\}$, $j=0,1,\dots,n-1$, under multiplication or equivalently the group of integers modulo n under addition. Using the additive representation, it is clear that the characters on the group are $\exp\{2\pi i jk/n\}$, $k=0,1,\dots,n-1$, so that the character group in this case is the original group. Let ν be the measure with mass one at $j=1$ and mass zero elsewhere. Then

$$\hat{\nu}(k) = \sum_j \exp\left\{\frac{2\pi i k j}{n}\right\} \nu(\{j\}) = \exp\left\{\frac{2\pi i k}{n}\right\}. \tag{9}$$

It follows that the Markov process with transition function determined by ν and the uniform invariant measure on the integers modulo n is uniformly ergodic but not strongly mixing. All this can be immediately verified without referring to Theorem 1. This simple example is not too interesting because the mass of ν is concentrated on a coset of a proper closed subgroup of G.

Let us now consider the relationship between uniform ergodicity and strong mixing on the circle group (or on a single torus). The circle group can be represented as the set of real numbers $0 \leqslant x < 1$ under addition modulo one. The continuous characters on the circle group are $\exp\{2\pi i n x\}$, $n=0, \pm 1, \ldots$. The character group is thus the set of integers under addition. Strong mixing implies uniform ergodicity. Let ν be a regular measure on the circle group whose support is not contained in any proper closed subgroup. Assume that the stationary Markov process of Theorem 1 determined by ν is uniformly ergodic but not strongly mixing. Then $\inf_{n \neq 0} |\hat{\nu}(n) - 1| > 0$ but $\sup_{n \neq 0} |\hat{\nu}(n)| = 1$ where

$$\hat{\nu}(n) = \int_0^1 \exp\{2\pi i n x\}\, \nu(dx).$$

Then there is a pair of sequences $\{\alpha_j\}$, $0 \leqslant \alpha_j < 1$, and $\{\varepsilon_j\}$ with $\varepsilon_j \downarrow 0$ as $j \to \infty$ such that for some integer sequence $n_j \to \infty$

$$|\hat{\nu}(n_j) - \exp\{2\pi i \alpha_j\}| \leqslant \varepsilon_j.$$

There is a limit point $\alpha (\neq 0)$ of $\{\alpha_j\}$ such that for an integer subsequence $m_j \to \infty$ of $\{n_j\}$

$$|\hat{\nu}(m_j) - \exp(2\pi i \alpha)| \to 0.$$

But then

$$\nu\left\{\bigcup_{k=-\infty}^{\infty} (x : |m_j x - \alpha + k| < \varepsilon_j')\right\} > 1 - \varepsilon_j'.$$

for some sequence $\varepsilon_j' \to 0$. However, for some integer sequence k_j, $k_j \alpha \to 0$ modulo one. This implies that for an integer sequence $r_j \to \infty$, $\hat{\nu}(r_j) \to 1$ and so we have a contradiction. Uniform ergodicity and strong mixing are equivalent on the circle group for stationary processes with translation transition operator given in terms of such a measure ν. A simple modification of this argument holds equally well in the case of a k-torus, the product of k circle groups.

Corollary 1. *Consider a stationary Markov process on the k-torus (k finite) with transition function*

$$P(g, A) = \nu(A - g)$$

and invariant measure the Haar measure, where the support of the regular measure ν is not contained in any proper closed subgroup. The process is strongly mixing if and only if it is uniformly ergodic.

In the case of a noncommutative compact group it is not clear how to obtain a natural and meaningful necessary and sufficient condition for a stationary Markov process generated by transition function (6) to be strongly mixing. A sufficient condition can be obtained by making use of Lemma 3.

Let $\{M^{(r)}(g), r \in R\}$ be a maximal collection of finite dimensional, non-equivalent, continuous irreducible unitary representations of the compact group G. The matrix elements of $M^{(r)}(g)$ are

$$m_{i,j}^{(r)}(g), \qquad i,j=1,\ldots,n(r)$$

with $n(r)$ the order of $M^{(r)}(g)$. The set of all matrix elements $m_{i,j}^{(r)}(g)$ obtained from such a maximal collection are a complete orthogonal system in L^2 on G with respect to the Haar measure of G by the Peter-Weyl theorem (see Appendix 5 on topological groups)

$$\int m_{i,j}^{(r)}(g)\overline{m_{k,l}^{(s)}(g)}\,d\mu(g)=\delta_{r,s}\delta_{i,k}\delta_{j,l}n(r)^{-1}.$$

Let ν be a regular probability measure on G with $M^{(r)}(\nu)$ the Fourier-Stieltjes coefficient of ν relative to $M^{(r)}(g)$, that is,

$$M^{(r)}(\nu)=\int M^{(r)}(g)\,\nu(dg).$$

For any square $n\times n$ matrix $A=(a_{i,j})$ let A^* denote the conjugated transpose of A and

$$|A|=\sup_x \frac{\|Ax\|_2}{\|x\|_2},$$

the norm of A as a linear operator on n-vectors x with $\|x\|_2^2=\sum |x_j|^2$, $x=(x_j)$. Notice that $|A|$ is the square root of the maximal eigenvalue of AA^* or A^*A.

Theorem 2. *Let ν be a regular probability measure on the compact group G whose support is not contained in any proper closed normal subgroup of G. Let T be the transition operator generated by the transition function $\nu(g^{-1}A)$. Then*

$$\sup_{f\perp 1}\frac{\|T^n f\|_2}{\|f\|_2}\to 0 \tag{10}$$

as $n\to\infty$ if and only if for some integer $k\geq 1$

$$\sup_{r\neq 0}|\{M^{(r)}(\nu)\}^k|\leq\rho<1 \tag{11}$$

where $M^{(0)}(g)$ is the representation consisting of the function identically equal to one. The L^2 norms are understood to be taken with respect to the Haar measure of G.

For any function $f \perp 1$ in L^2 with respect to Haar measure on G one can write the expansion

$$f = \sum_{r \neq 0} \sum_{i,j} f_{i,j}^{(r)} m_{j,i}^{(r)}(g),$$

where

$$\|f\|_2^2 = \sum_{r \neq 0} \sum_{i,j} \frac{|f_{i,j}^{(r)}|^2}{n(r)} < \infty .$$

Notice that

$$(T^k f)(g) = \int f(g h) \nu^{(k)}(dh) = \sum_{r \neq 0} \operatorname{tr}(f^{(r)} M^{(r)}(g) \{M^{(r)}(\nu)\}^k] \tag{12}$$

where $\operatorname{tr}(A)$ denotes the trace of the matrix A and the matrix

$$f^{(r)} = (f_{i,j}^{(r)}; i, j) .$$

$\nu^{(k)}$ stands for the kth convolution of ν with itself. Equation (12) follows from the multiplicative property of a group representation. Now

$$\|T^k f\|_2^2 = \sum_{r \neq 0} n(r)^{-1} \operatorname{tr}[f^{(r)*} f^{(r)} \{M^{(r)}(\nu)^*\}^k \{M^{(r)}(\nu)\}^k]$$

while

$$\|f\|_2^2 = \sum_{r \neq 0} \operatorname{tr}[f^{(r)*} f^{(r)}] n(r)^{-1} .$$

From condition (11) it follows that

$$\sup_{f \perp 1} \frac{\|T^k f\|_2}{\|f\|_2} \leqslant \rho < 1$$

and therefore (10) is valid. Suppose (11) is not satisfied. There is then an integer sequence $k_n \to \infty$ such that for corresponding $r_n \neq 0$

$$|\{M^{(r_n)}(\nu)\}^{k_n}| \to 1$$

as $n \to \infty$. Let $v^{(n)}$ be the eigenvector of $\{M^{(r_n)}(\nu)^*\}^{k_n} \{M^{(r_n)}(\nu)\}^{k_n}$ with maximal eigenvalue. Let $f^{(r_n)}$ be the matrix with all rows the vector $v^{(n)}$. The sequence of functions $f_n \perp 1$

$$f_n = \operatorname{tr}\{f^{(r_n)} M^{(r_n)}(g)\}$$

will then be such that

$$\|T^{k_n} f_n\|_2 / \|f_n\|_2 \to 1$$

as $n\to\infty$ and therefore

$$\sup_{f\perp 1}\frac{\|T^k f\|_2}{\|f\|_2}\equiv 1$$

for all k. The proof is complete.

Notice that in the case of a commutative group, if (10) is satisfied, then

$$\sup_{f\perp 1}\frac{\|Tf\|_2}{\|f\|_2}<1.$$

However, this needn't be the case for noncommutative groups.

4. L^p Norm Conditions and a Central Limit Theorem

We have already mentioned L^p norm conditions implicitly in Lemma 3.1. Let $\{X_k; k=0,\pm 1,\ldots\}$ as before be a stationary Markov process with transition probability function $P(\cdot,\cdot)$ and probability measure μ invariant with respect to $P(\cdot,\cdot)$. The Markov process $\{X_k\}$ *is said to satisfy an* L^p *norm condition*, $1\leqslant p\leqslant\infty$, *for* T *if*

$$\sup_{f\perp 1}\frac{\|T^n f\|_p}{\|f\|_p}\to 0 \tag{1}$$

as $n\to\infty$, where T is the transition operator on $L^p(d\mu)$ induced by $P(\cdot,\cdot)$. The L^2 norm condition will be reinterpreted from an interesting point of view. Furthermore, a clearer understanding of the relationship between the different L^p conditions will be given.

Suppose $\{X_k, k=0,\pm 1,\pm 2,\ldots\}$ is a stationary process. Let $f,g\neq 0$ be any two random variables with finite second moments on the probability space of the process. With no loss of generality f and g can be assumed to have first moment zero. The correlation between f and g is then

$$\frac{\mathrm{E}(fg)}{\|f\|_2\,\|g\|_2}.$$

Call the set of $\mathscr{B}_0$-measurable functions on the probability space with mean zero and finite second moment $L^2(\mathscr{B}_0)$. Let $L^2(\mathscr{F}_n)$ be the set of $\mathscr{F}_n$-measurable functions on the probability space with mean zero and finite second moment. The maximal correlation between $L^2(\mathscr{B}_0)$ and $L^2(\mathscr{F}_n)$ is given by

$$\sup_{\substack{f\in L^2(\mathscr{B}_0)\\ g\in L^2(\mathscr{F}_n)}}\frac{E(fg)}{\|f\|_2\,\|g\|_2}=c(n). \tag{2}$$

Call the stationary process $\{X_k\}$ *asymptotically uncorrelated if* $c(n) \to 0$ *as* $n \to \infty$.

Lemma 1. *A stationary Markov process* $\{X_k; k = \ldots, -1, 0, 1, \ldots\}$ *is asymptotically uncorrelated if and only if it satisfies the* L^2 *norm condition.*

If $\{X_k\}$ is asymptotically uncorrelated, it obviously satisfies the L^2 norm condition. Let us prove the converse. Consider any $f \in L^2(\mathscr{B}_0)$, $g \in L^2(\mathscr{F}_n)$ with $\|f\|_2 = \|g\|_2 = 1$. Because of the Markov property

$$h = E(g \mid \mathscr{A}_n) = E(g \mid \mathscr{B}_n)$$

with

$$Eh = 0, \qquad \|h\|_2 \leq \|g\|_2 .$$

Moreover,

$$|E(fg)| = |E(f E(h \mid \mathscr{A}_0))| \leq \|f\|_2 \, \|T^n h\|_2 .$$

Since the L^2 norm condition is assumed to be satisfied

$$c(n) \leq \sup_{h \perp 1} \frac{\|T^n h\|_2}{\|h\|_2} \to 0$$

as $n \to \infty$ and the Markov process is seen to be asymptotically uncorrelated.

Notice that the proof of Theorem 3.1 indicates that a random walk on a compact Abelian group is strongly mixing if and only if it satisfies the L^2 norm condition. However, the L^2 norm condition is more restrictive than strong mixing in general. This can easily be seen by looking at stationary aperiodic Markov chains with an irreducible set of states.

Lemma 2. *A stationary Markov chain is aperiodic and ergodic if and only if it is strongly mixing.*

It is immediately clear that if a stationary Markov chain is strongly mixing, it must be aperiodic with an irreducible set of states. The converse requires a small argument. Suppose that the chain is aperiodic with an irreducible set of states. It follows from Theorem 1.2.2 that

$$p_{j,k}^{(n)} \to \frac{1}{\mu_k} = \alpha_k > 0 \tag{3}$$

as $n \to \infty$ for all j, k where $\{\alpha_k\}$ is the stationary instantaneous distribution of the Markov chain. The relation (3) together with the fact that $\sum \alpha_j = 1$ implies that for each k

$$\sum_j \alpha_j \, | p_{j,k}^{(n)} - \alpha_k | \to 0 \tag{4}$$

as $n\to\infty$. Consider now any vector $f=(f_j)$ with

$$\sup_j |f_j|=1 \tag{5}$$

and $f\perp 1$, *that is*

$$\sum_j f_j\alpha_j=0.$$

Now

$$\|T^n f\|_1=\sum_j \alpha_j\left|\sum_k p_{j,k}^{(n)} f_k\right|=\sum_j \alpha_j\left|\sum_k \{p_{j,k}^{(n)}-\alpha_k\} f_k\right|.$$

Given any $\varepsilon>0$ there is an integer $N=N(\varepsilon)$ such that

$$\sum_{k>N(\varepsilon)} \alpha_k<\varepsilon. \tag{6}$$

From (4) it follows that

$$\sum_j \alpha_j \sum_{k=1}^{N} |p_{j,k}^{(n)}-\alpha_k|\to 0 \tag{7}$$

as $n\to\infty$. However,

$$\sum_j \alpha_j\left|\sum_{k>N(\varepsilon)} \{p_{j,k}^{(n)}-\alpha_k\} f_k\right|\leqslant 2\sum_{k>N(\varepsilon)} \alpha_k|f_k|<2\varepsilon \tag{8}$$

because of (5) and (6). The strong mixing follows from (7) and (8) since they are valid for each $\varepsilon>0$. We could equally well say that strong mixing is equivalent to mixing in the case of a stationary Markov chain.

Now for a simple example of a stationary mixing Markov chain that does not satisfy the L^2 norm condition (or any L^p norm condition with $1\leqslant p\leqslant\infty$). Let the states of the chain be 0 and (j,k) with $j=1,2,\ldots$ and $k=0,1,\ldots,j-1$. The transition probabilities are

$$\begin{aligned} &p_{0,(j,0)}=q_j>0, \quad \sum q_j=1\\ &p_{(j,k),(j,k+1)}=1 \quad \text{for } k=0,\ldots,j-2\\ &p_{(j,j-1),0}=1 \end{aligned} \tag{9}$$

with all other transition probabilities zero. Assume that $\sum j q_j<\infty$. Then the stationary probability distribution is given by

$$\begin{aligned} \alpha_0 &=\left(1+\sum_1^\infty j q_j\right)^{-1}\\ \alpha_{(j,0)}&=q_j\alpha_0\\ \alpha_{(j,k)}&=\alpha_{(j,k+1)}, \qquad k=0,1,\ldots,j-2. \end{aligned} \tag{10}$$

The stationary Markov chain given by the transition probabilities (9) and the stationary probability distribution (10) is irreducible and aperiodic. Let ${}_jf$ be the vector with ${}_jf_s=0$ unless $s=(j,j-1)$, $(j,j-2)$ when

$${}_jf_{(j,j-1)}=(2\alpha_0 q_j)^{-\frac{1}{p}}=-{}_jf_{(j,j-2)}.$$

Then

$$\|{}_jf\|_p\equiv 1, \qquad \|T^{j-2}{}_jf\|_p\equiv 1$$

so that

$$\sup_{f\perp 1}\frac{\|T^n f\|_p}{\|f\|_p}\equiv 1, \qquad 1\leqslant p\leqslant\infty.$$

In determining the relationship between the different L^p conditions, it is helpful to specify a certain version of the Doeblin condition. A broader version of the Doeblin condition was already mentioned in the notes for section 4.3. We shall briefly describe some useful results assuming this version of the condition without giving any derivation. However, a derivation of the results can be found, for example, in Doob's book [18]. *Let $P(\cdot,\cdot)$ be a transition probability function with invariant probability measure μ. If the whole state space is a minimal ergodic set, there are no cyclically moving sets and there is an integer m and an ε, $0<\varepsilon<1$, such that*

$$P_m(x,A)\leqslant 1-\varepsilon$$

for all x if $\mu(A)<\varepsilon$ we shall say that the Doeblin condition D_0 is satisfied.

A derivation of the following Lemma can be found in Doob ([18]).

Lemma 3. *If condition D_0 is satisfied, then there are constants γ, ρ with $0<\rho<1$ such that*

$$|P_n(x,A)-\mu(A)|\leqslant\gamma\rho^n$$

for all x and all $A\in\mathscr{A}$.

As we shall see by making use of Lemma 3, an essentially equivalent formulation of D_0 is given by

$$\sup_{A\in\mathscr{A}}|P_n(x,A)-\mu(A)|\leqslant\alpha(n)\to 0 \tag{11}$$

for almost all x (with respect to μ) as $n\to\infty$. The only difficulty will be in taking care of the almost everywhere statement. It is clear from Lemma 3 that if D_0 is satisfied, then (11) holds. We should like to show that if (11) is satisfied, then D_0 holds on an appropriate set of μ measure one. Let E_0 be the set of points of μ-measure zero for which (11) is not valid. Let

$$E_j=\{x: P_j(x,E_0)>0\}.$$

Then $E=\bigcup_{j=0}^{\infty} E_j$ has μ measure zero and for $x\in E^c$

$$P_n(x,A)=P_n(x,A\,E^c) \tag{12}$$

and this implies one can restrict oneself to E^c as the new state space and on this state space by virtue of (12), condition (11) implies D^0. For this reason, we shall also refer to condition (11) as condition D_0.

If a backward transition probability function Q exists, the operator induced by Q (just as T is induced by P) is the adjoint operator T^* of T. However, even if a backward transition probability function doesn't exist, we can still define T^* on L^p, $1\leqslant p\leqslant\infty$. It will be enough to indicate how this is done for L^1 and L^∞. First consider $L^\infty(d\mu)$. Let $h\in L^\infty(d\mu)$. Since μ is invariant with respect to $P(\cdot\,,\cdot)$, it follows that the set function

$$v(A;h)=\int\mu(dx)h(x)P(x,A), \qquad A\in\mathscr{A}, \tag{13}$$

is absolutely continuous with respect to μ. The Radon-Nikodym derivative of $v(\cdot\,;h)$ with respect to $\mu(\cdot)$ exists. Set

$$T^*h=\frac{v(dx;h)}{\mu(dx)}. \tag{14}$$

Then it is clear that T^* is a bounded linear operator taking L^∞ into L^∞ that is nonnegative with

$$T^*1=1$$

and hence

$$|T^*|_\infty=\sup_{\substack{f\in L^\infty(d\mu)\\ f\neq 0}}\frac{\|T^*f\|_\infty}{\|f\|_\infty}=1\,.$$

In order to extend T^* to an operator on $L^1(d\mu)$ it is enough to define it for $h\geqslant 0$, $h\in L^1(d\mu)$. Let

$$h_N(x)=\begin{cases}h(x) & \text{if } h(x)\leqslant N\\ 0 & \text{otherwise}.\end{cases}$$

Then $h_N\in L^\infty(d\mu)$ and so T^*h_N is well-defined. Further, for $g\geqslant 0$, $g\in L^\infty(d\mu)$

$$\int(T^*g)(x)\mu(dx)=\int g(x)\mu(dx)$$

by (13) and (14). But then T^*h_N is a nondecreasing sequence of nonnegative functions with

$$\int(T^*h_N)(x)\mu(dx)\leqslant\int h(x)\mu(dx).$$

It follows that $\lim_{N\to\infty} T^* h_N$ is well-defined almost everywhere with respect to μ. Set

$$T^* h = \lim_{N\to\infty} T^* h_N. \tag{15}$$

From (15) it follows that T^* is well-defined as a nonnegative bounded operator taking $L^1(d\mu)$ into $L^1(d\mu)$ with

$$T^* 1 = 1$$

and

$$|T^*|_1 = \sup_{\substack{f\in L^1(d\mu)\\ f\neq 0}} \frac{\|T^* f\|_1}{\|f\|}.$$

Let us now consider the relationship between the different L^p norm conditions as given in the following theorem.

Theorem 1. *Consider a transition probability function* $P(\cdot,\cdot)$ *with invariant probability measure* μ. *The* L^p *norm conditions,* $1<p<\infty$, *for* T *(the operator induced by* $P(\cdot,\cdot)$*) are all equivalent to each other. Further, the* L^∞ *norm condition for* T *and the* L^1 *norm condition for* T^* *are equivalent to each other and to the condition* D_0 *as given by* (11), *but are stronger than the* L^2 *norm condition.*

Consider any operator U defined for $L^p(d\mu)$, $1\leqslant p\leqslant\infty$, and taking $L^p(d\mu)$ into $L^p(d\mu)$. The L^p norm of U is then

$$|U|_p = \sup_{\substack{f\in L^p(d\mu)\\ f\neq 0}} \frac{\|U f\|_p}{\|f\|_p}.$$

The Riesz convexity theorem (see Zygmund [113], pp. 93–100 for a derivation and discussion of this important result) states that $\log|U|_p$ is a convex function of p^{-1}. We shall use the Riesz convexity theorem to establish the equivalence of the L^p norm conditions, $1<p<\infty$. The L^p norm condition for T is given by

$$\langle T^n\rangle_p = \sup_{f\perp 1} \frac{\|T^n f\|_p}{\|f\|_p} \to 0$$

as $n\to\infty$. Let us set $Uh = Th - Eh$ for $h\in L^p(d\mu)$, $1\leqslant p\leqslant\infty$. Now

$$\|h - Eh\|_p \leqslant \|h\|_p + |Eh| \leqslant 2\|h\|_p$$

and

$$U^n h = U^{n-1}(Th - Eh) = T^{n-1}(Th - Eh)$$

for $n\geqslant 1$. Thus,

$$|U^n|_p \leqslant 2\langle T^n\rangle_p \leqslant 2, \qquad 1\leqslant p\leqslant\infty,$$

for $n\geqslant 1$. However,

$$\langle T^n\rangle_p \leqslant |U^n|_p.$$

The Riesz convexity theorem tells us that if $\infty > q > p$

$$\log|U^n|_q \leqslant \frac{p}{q}\log|U^n|_p + \left(1-\frac{p}{q}\right)\log|U^n|_\infty \tag{16}$$

while if $1<q<p$

$$\log|U^n|_q \leqslant \left(\frac{1-\frac{1}{q}}{1-\frac{1}{p}}\right)\log|U^n|_p + \left(1-\frac{1-\frac{1}{q}}{1-\frac{1}{p}}\right)\log|U^n|_1\,. \tag{17}$$

We already know from section 3 of the geometric decay of $\langle T^n\rangle_p$ if the L^p norm condition is satisfied. Thus if

$$\langle T^n\rangle_p \leqslant \alpha < 1 \quad \text{then} \quad \langle T^{nk}\rangle_p \leqslant \alpha^k$$

for $k=1,2,\ldots$. Now $|U^n|_\infty, |U^n|_1 \leqslant 2$. If the L^p norm condition for T is satisfied for some p with $1<p<\infty$, it follows from (16) and (17) that the L^q norm condition is satisfied for all q with $1<q<\infty$. Since the L^p norm conditions are all equivalent for $1<p<\infty$, it is natural to single out the one that is easiest to deal with computationally, the L^2 norm condition. Also, (16) and (17) indicate that the L^1 and L^∞ norm conditions are at least as strong as the L^2 norm condition.

Let us now show that the Doeblin condition (11) is equivalent to the L^1 norm condition for T^*. As a first step it is shown to be equivalent to

$$\sup_{B\in\mathscr{A}_n} |P(AB)-\mu(A)\mu(B)| \leqslant \varphi(n)\mu(A) \tag{18}$$

for all $A\in\mathscr{A}_0$ with φ some function such that $\varphi(n)\to 0$ as $n\to\infty$. Now (11) certainly implies (18) as one can readily see by integrating

$$|P_n(x,B)-\mu(B)|$$

with respect to μ over the set A. The function $\varphi(n)$ can simply be set equal to $\alpha(n)$. Conversely, if (18) is not satisfied for any such function α, then there is an $\varepsilon>0$ and there are sequences of sets $A_n\in\mathscr{A}_0$, $B_n\in\mathscr{A}_n$ with $\mu(A_n)>0$ such that

$$P_n(x,B)-\mu(B_n)\geqslant\varepsilon \qquad (\leqslant -\varepsilon) \tag{19}$$

for $x\in A_n$. We carry out the argument for only the first alternative ($\geqslant$) since it is exactly the same in the case of the second alternative. From (19) we see that

$$P(A_nB_n)-\mu(A_n)\mu(B_n)\geqslant\varepsilon\mu(A_n)$$

contradicting (11). The equivalence is established. Let us now show that (18) is equivalent to the L^1 norm condition for T^*. Assume (18). Consider any $f \perp 1$ with $\int |f(x)|\mu(dx)=1$. For fixed $\varepsilon>0$ let

$$B_j(\varepsilon)=\{x: j\varepsilon \leqslant f(x)<(j+1)\varepsilon\}.$$

Define f_ε by setting

$$f_\varepsilon(x)=j\varepsilon \quad \text{on } B_j(\varepsilon).$$

Then $|f_\varepsilon(x)-f(x)|\leqslant\varepsilon$ so that $|T^{*n}(f-f_\varepsilon)(x)|\leqslant\varepsilon$ for all x and

$$\left|\int f_\varepsilon(x)\mu(dx)\right|\leqslant\varepsilon, \left|\int|f_\varepsilon(x)|\mu(dx)-1\right|\leqslant\varepsilon.$$

Now

$$\begin{aligned}\int |T^{*n}f_\varepsilon - Ef_\varepsilon|\mu(dx) &= \int \Big|\sum_j (j\varepsilon)\{T^{*n}I_{B_j}-\mu(B_j)\}\Big|\mu(dx)\\ &\leqslant 2\varphi(n)\sum_j |j\varepsilon|\mu(B_j)\leqslant 2\varphi(n)(1+\varepsilon).\end{aligned}$$

Thus,

$$\int |(T^{*n}f)(x)|\mu(dx)\leqslant 2\varphi(n)(1+\varepsilon)+\varepsilon$$

and we get the L^1 norm condition for T^* on letting $n\to\infty$, $\varepsilon\to 0$. It is easy to show that the L^1 norm condition for T^* implies (18). Just set $f(x)=I_A(x)-\mu(A)$. Then

$$\int |f(x)|\mu(dx)=2\mu(A)\{1-\mu(A)\}$$

and for $A\in\mathscr{A}_0$, $B\in\mathscr{A}_n$

$$\begin{aligned}|P(AB)-\mu(A)\mu(B)| &= \Big|\int_B\{T^{*n}I_A-\mu(A)\}\mu(dx)\Big|\leqslant\int |T^{*n}I_A-\mu(A)|\mu(dx)\\ &=\|T^{*n}f\|_1\leqslant K\alpha^n\mu(A)\{1-\mu(A)\}\end{aligned}$$

for an appropriate constant K and some α, $0<\alpha<1$. We can also see directly that the Doeblin condition D_0 as given by (11) is equivalent to the L^∞ norm condition for T. The proof of Theorem 1 is complete.

The following simple example of a stationary Markov chain shows that either one of the L^1 or L^∞ norm conditions for T can hold without the other one being valid. Let $0,1,2,\dots$ be the states of the chain with transition probabilities

$$p_{j,k}=\begin{cases} q_j>0 & \text{if } k=j+1\\ 1-q_j>0 & \text{if } k=0\\ 0 & \text{otherwise}\end{cases} \tag{20}$$

$j,k=0,1,\ldots$. If $a=\sum_{j=1}^{\infty}\prod_{k=0}^{j=1} q_k<\infty$, there is an invariant probability distribution μ with

$$\mu_k=\left(\prod_{j=0}^{k-1} q_j\right)(1+a)^{-1}, \qquad k=0,1,\ldots. \tag{21}$$

Consider the stationary Markov chain with one step transition probabilities (20) and invariant measure (21). The L^∞ norm condition for T (the Doeblin condition D_0) is satisfied if, for example, $q_j\equiv\varepsilon$ with $0<\varepsilon<1$. However, the chain with time reversed does not satisfy the Doeblin condition D_0 and thus by Theorem 1 neither the L^∞ norm condition for T^* or the L^1 norm condition for T is satisfied by the original chain. Notice that for a stationary reversible Markov process, Theorem 1 implies that the L^1 norm condition and L^∞ norm conditions for T are equivalent.

It is also easy to give an example which satisfies the L^2 norm condition but neither the L^1 or L^∞ norm conditions for T which are equivalent since the process is reversible. Take the stationary Gaussian Markov process with correlation sequence $E(X_0 X_k)=\rho^{|k|}$, $|\rho|<1$, with $\rho\neq 0$. Assume the process has mean $E(X_j)\equiv 0$. The invariant probability density is then

$$f(x)=\frac{1}{\sqrt{2\pi}}\exp\left(-\frac{x^2}{2}\right)$$

and the transition probability density is

$$p(y|x)=\frac{1}{\sqrt{2\pi(1-\rho^2)}}\exp\left(-\frac{(y-\rho x)^2}{2(1-\rho^2)}\right). \tag{22}$$

Mehler's formula indicates that (22) can be written as

$$p(y|x)=\sum_{\nu=0}^{\infty}\rho^\nu h_\nu(y)h_\nu(x)\frac{e^{-\frac{y^2}{2}}}{\sqrt{2\pi}} \tag{23}$$

where the h_ν are the Hermite polynomials orthonormal with respect to the standard Gaussian density $f(x)$. By using (23) one can see that $\langle T^n\rangle_2=|\rho|^{2n}$ so that the L^2 norm condition is satisfied. By inspection, however, one can see that the Doeblin condition D_0 is not satisfied. One can also give examples satisfying the L^2 norm condition with totally singular transition functions of all orders. Consider the context discussed in Theorem 3.1 with G the circle group (addition of real numbers modulo one). We can identify the group with the real numbers of the interval $(0,1]$. The random walk on the group is generated by transition function

$$P(x,A)=\nu(A-x) \tag{24}$$

where ν is a probability measure on the Borel sets of $(0,1]$ and the invariant measure is the uniform measure (Lebesgue measure) on $(0,1]$. The random walk satisfies the L^2 norm condition if and only if

$$\sup_{n\neq 0} |\hat{\nu}(n)|>1$$

where

$$\hat{\nu}(n)=\int_0^1 \exp(2\pi i n x)\,\nu(dx).$$

However, it is easy to find measures ν such that

$$1>\limsup_{|n|\to\infty} |\hat{\nu}(n)|>0$$

with all transition functions $P_n(\cdot,\cdot)$ generated from (24) totally singular with respect to Lebesgue measure.

Our aim is now to derive a central limit theorem for stationary Markov processes satisfying an L^2 norm condition. Let

$$f_{k,n}(\cdot), \qquad k=1,\ldots,k_n, \qquad n=1,2,\ldots$$

with $k_n\to\infty$ as $n\to\infty$ be a triangular array of real-valued $\mathscr{A}$-measurable functions on the state space of the Markov process. Asymptotic normality for the sums

$$S_n=\sum_{k=1}^{k_n} f_{k,n}(X_k)$$

when properly normalized is obtained as $n\to\infty$ under appropriate conditions on the triangular array. For convenience, it is assumed that

$$E f_{k,n}(X)\equiv 0.$$

Lemma 4. *Let $\{X_k, k=0,\pm 1,\ldots\}$ be the stationary Markov process with transition probability function $P(\cdot,\cdot)$ and invariant probability measure μ. Assume that the functions $f_{k,n}$ are uniformly bounded in k,n with*

$$E|S_n|^2\cong k_n\sigma^2, \qquad \sigma^2>0 \tag{25}$$

as $n\to\infty$. If the process satisfies an L^2 norm condition, then

$$\lim_{n\to\infty} P\{k_n^{-\frac{1}{2}}\sigma^{-1}S_n\leqslant x\}=\int_{-\infty}^{x}\frac{1}{\sqrt{2\pi}}e^{-\frac{u^2}{2}}du.$$

The L^2 norm condition is satisfied if and only if there are constants K and α, $0<\alpha<1$, such that

$$\sup_{f\perp 1} \frac{\|T^n f\|_2}{\|f\|_2} \leqslant K\alpha^n. \tag{26}$$

Our object is to obtain certain estimates for fourth order moments so that Liapounov's conditions for a central limit theorem for independent random variables can finally be applied. Let f_0, f_1, f_2, f_3 be bounded functions with mean zero. Let $k_1, k_2, k_3 \geqslant 0$ be integers. A bound for $|E(f_0 T^{k_1} f_1 T^{k_2} f_2 T^{k_3} f_3)|$ will be obtained. Now

$$\begin{aligned}|E(f_0 T^{k_1} f_1 T^{k_2} f_2 T^{k_3} f_3)| &= |(T^{*k_1} f_0, f_1 T^{k_2} f_2 T^{k_3} f_3)| \\ &\leqslant K\alpha^{k_1} \|f_0\|_2 \|f_1\|_\infty \|T^{k_2} f_2 T^{k_3} f_3\|_2\end{aligned} \tag{27}$$

using (26). However,

$$\begin{aligned}\|T^{k_2} f_2 T^{k_3} f_3\|_2 &\leqslant \|T^{k_2}\{f_2 T^{k_3} f_3 - E(f_2 T^{k_3} f_3)\}\|_2 + |E(f_2 T^{k}\ f_3)| \\ &\leqslant K\alpha^{k_2} \|f_2 T^{k_3} f_3\|_2 + (1+K\alpha^{k_2})|E(f_2 T^{k_3} f_3)| \\ &\leqslant K\alpha^{k_3} \|f_2\|_\infty \|f_3\|_2 (1+K\alpha^{k_2}).\end{aligned} \tag{28}$$

Inequalities (27) and (28) imply that

$$\begin{aligned}&|E(f_0 T^{k_1} f_1 T^{k_2} f_2 T^{k_3} f_3)| \\ &\leqslant \|f_0\|_2 \|f_1\|_\infty \|f_2\|_\infty \|f_3\|_2 K^2 \alpha^{k_1+k_3} (1+K\alpha^{k_2}).\end{aligned} \tag{29}$$

Let

$$S_n(a,b) = \sum_{k=a+1}^{b} f_{k,n}(X_k).$$

Denote the uniform upper bound of $|f_{k,n}|$ by A. The variance of $S_n(a,b)$ is smaller than $A'(b-a)$ where A' is a number independent of a, b, and n (but depending on α). Then

$$\sup_{0<b-a\leqslant l} P[|k_n^{-\frac{1}{2}} S_n(a,b)| > \varepsilon] \leqslant \varepsilon$$

as long as $k_n^{-1} A' l \leqslant \varepsilon^3$ by the Chebyshev inequality (see Appendix 1). Consider a positive integer r and let a_n be the greatest integer less than or equal to $k_n r^{-1}$. Set $n_j = j a_n$, $j=0, \dots, r$. It is clear that $k_n^{-\frac{1}{2}} S_n(0, n_r)$ and $k_n^{-\frac{1}{2}} S_n$ have the same limiting distribution (if one exists) as $n \to \infty$. The argument leading to inequality (12) in section 2 can be modified in a straightforward way so as to obtain the inequality

$$\begin{aligned}&\left|E\exp\{iuk_n^{-\frac{1}{2}} S_n(0,n_r)\} - \prod_{j=0}^{r-1} E\exp\{iuk_n^{-\frac{1}{2}} S_n(n_j, n_{j+1})\}\right| \\ &\leqslant 2(r-1)(\varepsilon + |u|\varepsilon + 2K\alpha^l).\end{aligned} \tag{30}$$

If we now let $\varepsilon \downarrow 0$, $n \to \infty$, $l(=o(n)) \to \infty$ and $r = r(n) \to \infty$ sufficiently slowly so that the right side of inequality (30) tends to zero, the problem of determining the limiting distribution is reduced to one in which the summands are r independent random variables whose marginal distributions are those of $k_n^{-\frac{1}{2}} S_n(n_j, n_{j+1})$, $j = 0, \ldots, r-1$. By (25) and (26) the sum of the variances

$$\sum_{j=0}^{r-1} \sigma^2(k_n^{-\frac{1}{2}} S_n(n_j, n_{j+1})) \cong \sigma^2 \quad \text{as } r \to \infty .$$

The bound (29) for fourth order moments implies that

$$\sum_{j=0}^{r-1} E|k_n^{-\frac{1}{2}} S_n(n_j, n_{j+1})|^4 \leq r \frac{C\left(\frac{k_n}{r}\right)^2}{k_n^2} = \frac{C}{r} \to 0 \tag{31}$$

as $r \to \infty$, where C is a constant. Liapounov's conditions (see section 1.3) for the asymptotic normality of the sum of independent random variables is satisfied so that by (31) $k_n^{-\frac{1}{2}} S_n(0, n_r)$ is asymptotically normal with mean zero and variance σ^2 as $n \to \infty$. This implies that $k_n^{-\frac{1}{2}} \sigma^{-1} S_n$ is asymptotically normal with mean zero and variance one.

Theorem 2. *Let* $\{X_k, k = 0, \pm 1, \ldots\}$ *be a stationary Markov process with transition probability function* $P(\cdot, \cdot)$ *and invariant probability measure* μ. *If the functions* $f_{k,n}$ *are uniformly square integrable in* k, n *with*

$$E|S_n|^2 \cong k_n \sigma^2, \quad \sigma^2 > 0,$$

as $n \to \infty$ *and if the process* $\{X_k\}$ *satisfies an* L^2 *norm condition, then*

$$\lim_{n \to \infty} P\{k_n^{-\frac{1}{2}} \sigma^{-1} S_n \leq x\} = \int_{-\infty}^{x} \frac{1}{\sqrt{2\pi}} e^{-\frac{u^2}{2}} du .$$

The uniform square integrability of the functions $f_{k,n}$ implies that

$$E|f_{k,n}(X)|^2 < C < \infty$$

for all k, n. Given the number $\varepsilon > 0$ let

$$A_{k,n}(\varepsilon) = \{x : -N_{k,n}(\varepsilon) \leq f_{k,n}(x) \leq M_{k,n}(\varepsilon)\}$$

with $N_{k,n}(\varepsilon)$, $M_{k,n}(\varepsilon)$ determined so that if

$${}_{\varepsilon}f_{k,n}(x) = \begin{cases} f_{k,n}(x) & \text{if } x \in A_{k,n}(\varepsilon), \\ 0 & \text{otherwise} \end{cases}$$

then

$$\int {}_{\varepsilon}f_{k,n}(x)\, \mu(dx) = 0, \quad \int \mu(dx) |f_{k,n}(x) - {}_{\varepsilon}f_{k,n}(x)|^2 < \varepsilon .$$

The uniform square integrability of the $f_{k,n}$'s implies that the numbers $N_{k,n}(\varepsilon)$, $M_{k,n}(\varepsilon)$ for fixed $\varepsilon>0$ can be assumed uniformly bounded in k, n. Let

$$
\begin{aligned}
{}_\varepsilon r_{k,n}(x) &= f_{k,n}(x) - {}_\varepsilon f_{k,n}(x) \\
{}_\varepsilon S_n &= \sum_{j=1}^{k_n} {}_\varepsilon f_{j,n}(X_j), \qquad {}_\varepsilon a_n^2 = E|{}_\varepsilon S_n|^2. \\
{}_\varepsilon S_n' &= S_n - {}_\varepsilon S_n.
\end{aligned}
$$

Now

$$E|{}_\varepsilon S_n'|^2 = \sum_{j,l=1}^{k_n} E\{{}_\varepsilon r_{j,n}(X_j)\, {}_\varepsilon r_{l,n}(X_l)\}.$$

Since

$$\left|E\{{}_\varepsilon r_{j,n}(X_j)\, {}_\varepsilon r_{l,n}(X_l)\}\right| \leqslant K\alpha^{|l-j|}\{E|{}_\varepsilon r_{j,n}(X)|^2\, E|{}_\varepsilon r_{l,n}(X)|^2\}^{\frac{1}{2}} \leqslant K\alpha^{|l-j|}\varepsilon$$

it is clear that

$$\lim_{\substack{n\to\infty \\ \varepsilon\to 0}} \frac{{}_\varepsilon a_n^2}{k_n} = \sigma^2.$$

By Lemma 4 we know that ${}_\varepsilon a_n^{-1}\, {}_\varepsilon S_n$ is asymptotically normally distributed with mean zero and variance one. The desired conclusion is obtained by a standard approximation argument on letting $\varepsilon\to 0$.

Corollary 1. *Let the assumptions of Theorem 2 be satisfied with the Markov process $\{X_k\}$ the stationary random walk on a compact group G generated by the regular probability measure ν. The random walk satisfies the L^2 norm condition if and only if the Fourier-Stieltjes coefficients of ν relative to a complete irreducible set of unitary representations satisfy for some integer $s\geqslant 1$*

$$\sup_{r\neq 0} \left|\{M^{(r)}(\nu)\}^s\right| \leqslant \rho < 1.$$

Then if $E|S_n|^2 \cong k_n\sigma^2$, $\sigma^2>0$, the normalized sums $k_n^{-\frac{1}{2}}\sigma^{-1}S_n$ are asymptotically normally distributed with mean zero and variance one.

This follows immediately from Theorem 2 and Theorem 3.2.

Notes

7.1 and 7.2 See Prohorov [86] for remarks on Kolmogorov's theorem. The derivation given in these two sections of Theorem 2.1 is an adaptation of one given by Cogburn [12]. Cogburn discussed uniform ergodicity in [11] and [12]. The notion of strong or uniform mixing is due to M. Rosenblatt [91].

7.3 Some work has been done in obtaining conditions under which certain classes of stationary processes satisfy the strong mixing condition. Unfortunately, very little has been done with respect to uniform ergodicity. Kolmogorov and Rozanov [59] considered strong mixing for stationary Gaussian processes. If

$\{X_k; k=\cdots,-1,0,1,\ldots\}$ is a Gaussian stationary process with $EX_k\equiv 0$, the covariances $r_k=EX_nX_{n+k}$ have representation

$$r_k=\int_{-\pi}^{\pi} e^{ik\lambda}dF(\lambda)$$

in terms of a bounded nondecreasing function F, the spectral distribution function of the process (see [18] page 474). The spectral distribution function F must be absolutely continuous with spectral density $f(\lambda)=F'(\lambda)$ if $\{X_k\}$ is strongly mixing since it is then purely nondeterministic in the sense of the linear prediction problem (see [98]). Kolmogorov and Rozanov show that $\{X_k\}$ is strongly mixing if and only if

$$\sup_p \left\| \{f(\lambda)-e^{-in\lambda}p(e^{i\lambda})\}\frac{1}{f(\lambda)}\right\|_\infty \to 0$$

as $n\to\infty$ where the sup is taken over all one-sided trigonometric polynomials $p(e^{i\lambda})=\sum_{j=0}^{m} c_j e^{ij\lambda} (m\geq 0)$ and $\|\cdot\|_\infty$ is the essential supremum norm. This implies that if $f(\lambda)$ is continuous with $f(-\pi)=f(\pi)$ and bounded away from zero, then $\{X_k\}$ is strongly mixing. Ibragimov [45] has obtained further results on the form of f for strongly mixing Gaussian stationary processes. Helson and Sarason [39] have completely recast this problem as a question in harmonic analysis. Generalizations to the case of matrix-valued functions have also been considered (see [77]).

In the case of a stationary random walk on a compact Abelian group Theorem 1 indicates that the stationary process is strongly mixing if and only if

$$\sup_{f\perp 1}\frac{|T^n f\|_2}{\|f\|_2}\to 0$$

as $n\to\infty$, where T is the transition operator of the random walk. The norm of the operators T^n on the space of square integrable functions $f\perp 1$ decreases to zero geometrically as $n\to\infty$. This geometric rate of decay of the norm of the powers of T is reminiscent of the "geometric ergodicity" that D. Kendall [55] mentions in his analysis of a special class of Markov chains. A paper of E. Hopf [42] is interesting in this connection since it provides eigenvalue estimates that may be useful in determining such a geometric decay for certain types of Markov processes.

7.4 Uniform mixing and strong mixing were introduced as conditions under which limit theorems like a central limit theorem would hold for stochastic processes satisfying auxiliary moment conditions. A variety of papers such as those of M. Rosenblatt [91], Rozanov and Volkonski [97], and Ibragimov [44] apply strong mixing to get such limit theorems. For more recent discussions of central limit theorems see [22] and [96].

Appendix 1

Probability Theory

Let ν be a probability measure on the Borel field $\mathscr{B}$ on the space of points Ω, that is, $\nu(\Omega)=1$ (see [18], [70], and [100] for a discussion of measure theory). The triple $(\Omega, \mathscr{B}, \nu)$ is often called a *probability space*. A function $X(\omega)$ measurable with respect to $\mathscr{B}$ is sometimes called a *random variable* because it can be thought of as a possible observable in an experiment whose outcome is governed by the measure ν. The integral

$$E(X)=\int X(\omega)\,\nu(d\omega)$$

(if it exists) is called the *mean* or *expectation* of X.

Consider any random variable X with a finite absolute moment $E|X|<\infty$. The following simple but useful *Chebyshev inequality*

$$P(|X|\geqslant c)=\int\limits_{|X|\geqslant c} dP \leqslant \frac{1}{c}\int\limits_{|X|\geqslant c} |X(\omega)|\,dP \leqslant \frac{1}{c}E|X|$$

is valid for $c>0$. This inequality is helpful in getting bounds on the probability mass in the tail of a distribution. Of course, if $E|X|^{\alpha}<\infty$ for some $\alpha>0$ one finds in the same manner that

$$P(|X|\geqslant c)\leqslant \frac{1}{c^{\alpha}}E|X|^{\alpha}$$

for $c>0$.

Another useful inequality for moments is *Jensen's inequality*. Let u be a real-valued continuous convex function of a real variable, that is, for each point z there is a number $\lambda=\lambda(z)$ such that

$$u(y)\geqslant u(z)+\lambda(y-z) \tag{1}$$

for all y. This states that for each point $(z,u(z))$ there is a straight line with slope λ passing through $(z,u(z))$ lying below curve $u(y)$. Let X be a random variable with EX, $Eu(X)$ finite. Set $z=EX$ and $y=X$ in (1). By taking expectations one obtains *Jensen's inequality*

$$Eu(X)\geqslant u(EX).$$

Suppose $A_1, A_2, \ldots, A_n$ are a finite number of events or measurable sets. Let $A_i^{(0)} = A_i$ and $A_i^{(1)} = A_i^c$, $i=1,\ldots,n$. The events are said to be independent if

$$P\left(\bigcap_{i=1}^{n} A_i^{(m_i)}\right) = \prod_{i=1}^{n} P(A_i^{(m_i)})$$

for $m_i = 0, 1$, $i=1,\ldots,n$. An infinite sequence of events $A_1, A_2, \ldots$ is said to be independent if every finite subset is independent.

Now let $A_1, A_2, \ldots$ be an infinite sequence of events. Then $\{A_i \text{ i.o.}\} = \bigcap_{n=1}^{\infty} \bigcup_{j=n}^{\infty} A_j$ is the set of points lying in an infinite number of the events A_j, $j=1,2,\ldots$. The Borel-Cantelli Lemma determines the probability $P(A_i \text{ i.o.})$ in terms of the probabilities $P(A_i)$ of the individual events A_i under particularly simple circumstances.

Borel-Cantelli Lemma. Let $P(A_i)$ be the probabilities of the events A_i, $i=1,2,\ldots$. If $\sum_{i=1}^{\infty} P(A_i) < \infty$ then

$$P(A_i \text{ i.o.}) = 0. \tag{2}$$

If the events A_i, $i=1,2,\ldots$, are independent then $\sum_{i=1}^{\infty} P(A_i) = \infty$ implies that

$$P(A_i \text{ i.o.}) = 1. \tag{3}$$

Assume that $\sum_{i=1}^{\infty} P(A_i) < \infty$. Then

$$P(A_i \text{ i.o.}) \leq P\left(\bigcup_{j=n}^{\infty} A_j\right) \leq \sum_{j=n}^{\infty} P(A_j)$$

and on letting $n \to \infty$ formula (2) follows. Suppose the events A_i, $i=1,2,\ldots$ are independent and that $\sum_{i=1}^{\infty} P(A_i) = \infty$. Then

$$P(A_i \text{ i.o.}) \geq 1 - P\left(\bigcap_{j=n}^{\infty} A_j^c\right) = 1 - \prod_{j=n}^{\infty} (1 - P(A_j)) \geq 1 - \exp\left(-\sum_{j=n}^{\infty} P(A_j)\right)$$

and (3) follows on letting $n \to \infty$.

We now make an observation that is used often in computations. Let $X_1(\omega), \ldots, X_n(\omega)$ be a finite number of random variables on a probability space $(\Omega, \mathscr{B}, P)$. Let

$$F(x_1, \ldots, x_n) = P[X_1(\omega) \leq x_1, \ldots, X_n(\omega) \leq x_n]$$

be the distribution function (defined for real x_i) of the random variables $X_1(\omega), \ldots, X_n(\omega)$. Then F determines a probability measure on the n-dimensional Borel sets (see [18]) and we also loosely refer to this measure as F. Assume that Φ is a real-valued function measurable with respect to the n-dimensional Borel sets. If

$$\int |\Phi(X_1(\omega), \ldots, X_n(\omega))| \, dP < \infty,$$

then

$$\int \Phi(X_1(\omega), \ldots, X_n(\omega)) \, dP = \int \Phi(x_1, \ldots, x_n) \, dF$$

where the integrals are understood to be Lebesgue integrals.

Suppose X is integrable with respect to v. Then

$$\varphi(B) = \int_B X(\omega) \, v(d\omega) \tag{4}$$

is a completely additive but not necessarily positive set function. The completely additive set function φ is *absolutely continuous with respect to* v, that is, given any $B \in \mathscr{B}$ for which $v(B) = 0$ it follows that then $\varphi(B) = 0$. The Radon-Nikodym theorem characterizes finite completely additive set functions φ on $\mathscr{B}$ absolutely continuous with respect to v as integrals (4).

Theorem 1. *Let v be a probability measure on the Borel field $\mathscr{B}$. A finite completely additive set function φ on $\mathscr{B}$ is absolutely continuous with respect to v if and only if it is an integral*

$$\varphi(B) = \int_B X(\omega) \, v(d\omega), \qquad B \in \mathscr{B},$$

of an integrable $\mathscr{B}$ measurable function X. The function X is uniquely defined up to a set of v measure zero.

The function X can be thought of as the derivative $\varphi(d\omega)/v(d\omega)$ of φ with respect to v relative to the Borel field $\mathscr{B}$. Actually, Theorem 1 is still valid even if the measure v is σ-finite.

As already indicated, the Borel field $\mathscr{B}$ can be thought of as representing the information available potentially in an experiment. Suppose another, "cruder" experiment is also considered with a corresponding Borel field $\mathscr{A} \subseteq \mathscr{B}$. We might try to define a conditional mean of the random variable X given the cruder experiment, that is, a *conditional expectation of X given the Borel field $\mathscr{A}$*. This suggests looking at φ and v on the sets A of the smaller Borel field $\mathscr{A}$ and trying to define a derivative of φ with respect to v relative to the Borel field $\mathscr{A}$

$$\frac{\varphi(d\omega)}{v(d\omega)} = f(\omega),$$

where the derivative f is measurable with respect to the Borel field $\mathscr{A}$. Since φ is absolutely continuous with respect to ν on $\mathscr{B}$, it is automatically absolutely continuous with respect to ν on the smaller Borel field $\mathscr{A}$. The following is an immediate corollary of the Radon-Nikodym theorem.

Corollary 1. *Let $X(\omega)$ be a random variable on the probability space $(\Omega, \mathscr{B}, P)$ integrable with respect to P. Assume that $\mathscr{A}$ is a sub-Borel field of $\mathscr{B}$. Then there is an $\mathscr{A}$ measurable function f integrable with respect to P such that*

$$\int_A X(\omega) P(d\omega) = \int_A f(\omega) P(d\omega) \tag{5}$$

for all $A \in \mathscr{A}$. The $\mathscr{A}$ measurability of f determines f uniquely almost everywhere, that is, up to an exceptional set of P-measure zero.

The $\mathscr{A}$-measurable function f obtained in (5) is called the conditional expectation of X given the Borel field $\mathscr{A}$

$$f(\omega) = E(X|\mathscr{A})(\omega).$$

There are basic properties of the conditional expectation that follow almost immediately from the definition (5):

(i) $E(1|\mathscr{A}) = 1$;

(ii) $E(X|\mathscr{A}) \geqslant 0$ if $X \geqslant 0$ and $E(X) < \infty$;

(iii) If $E(|X|), E(|Y|) < \infty$ and α, β are constants, then

$$E(\alpha X + \beta Y|\mathscr{A}) = \alpha E(X|\mathscr{A}) + \beta E(Y|\mathscr{A});$$

(iv) If $\mathscr{C} \subset \mathscr{A} \subset \mathscr{B}$ and $E(|X|) < \infty$, then

$$E(E(X|\mathscr{A})|\mathscr{C}) = E(X|\mathscr{C}).$$

The conditional expectation also provides the solution to an interesting but simple minimum problem.

Theorem 2. *Let X be a random variable with finite second moment $EX^2 < \infty$. If $\mathscr{A}$ is a Borel subfield of $\mathscr{B}$, the best $\mathscr{A}$-measurable predictor or approximator of X in the sense of minimizing the mean square error of approximation is given by $E(X|\mathscr{A})$.*

Since $EX^2 < \infty$, $E|X| < \infty$ and this implies that $E(X|\mathscr{A})$ is well-defined. Further, $E\{(\alpha X + 1)^2|\mathscr{A}\} \geqslant 0$ for all real α and from this it follows that

$$\{E(X|\mathscr{A})\}^2 \leqslant E(X^2|\mathscr{A}).$$

But this implies that $E(X|\mathscr{A})$ has a finite second moment. Now consider any $\mathscr{A}$-measurable random variable Y with finite second moment.

It can be considered a possible predictor of X and the mean square error of approximation is

$$E(X-Y)^2 = E(\{X-E(X|\mathscr{A})\}+\{E(X|\mathscr{A})-Y\})^2 \tag{6}$$
$$= E(X-E(X|\mathscr{A}))^2 + 2E\{(X-E(X|\mathscr{A}))(E(X|\mathscr{A})-Y)\}$$
$$+ E(E(X|\mathscr{A})-Y)^2 .$$

However, (5) implies that

$$E\{(X-E(X|\mathscr{A}))Z\}=0$$

for any $\mathscr{A}$-measurable random variable Z with finite second moment. The second term on the right of (6) is seen to be zero on setting $Z=E(X|\mathscr{A})-Y$. It is now clear that $E(X|\mathscr{A})$ is the best predictor.

If $X(\omega)$ is the indicator function $I_B(\omega)$ of the set $B\in\mathscr{B}$, relation (5) becomes

$$P(B\cap A) = \int_A I_B(\omega)P(d\omega) = \int_A f(\omega)P(d\omega)$$

and the $\mathscr{A}$ measurable function f is called the conditional probability of the set (or event) B given the Borel field $\mathscr{A}\subseteq\mathscr{B}$

$$f(\omega) = P(B|\ \mathscr{A})(\omega).$$

The defining relation can be written

$$P(B\cap A) = \int_A P(B|\mathscr{A})(\omega)P(d\omega) \tag{7}$$

for all $A\in\mathscr{A}$. This is a natural extension of the way in which conditional probabilities are defined on a space with at most a countable set of points. Notice that (7) implies that for a countable sequence B_1, B_2, ... of disjoint sets of $\mathscr{B}$

$$\sum_{i=1}^{\infty} P(B_i|\mathscr{A})(\omega) = P(\bigcup B_i|\mathscr{A})(\omega) \tag{8}$$

for almost all ω. However, the exceptional set of measure zero for which (8) does not hold may depend on the sequence B_i, $i=1,2,\ldots$. One would like to be able to arrange things so that the exceptional set of measure zero is independent of the sequence B_i. For then $P(B|\mathscr{A})(\omega)$ would be a probability measure in $B\in\mathscr{B}$ for almost all ω. This can be done under conditions of a mixed topological and measure theoretic character. For example, this is possible if the space Ω is a complete separable metric space with $\mathscr{B}$ the Borel field generated by the topology of the space (see Appendix 2 for a definition of the topological terms used).

There is an interesting and useful class of processes called martingales that arise in a number of situations. Let X_n, $n=\ldots, -1,0,1,\ldots$ be a sequence of random variables on a probability space with an increasing family of Borel fields $\mathscr{B}_n$ such that X_n is $\mathscr{B}_n$ measurable. One may often have

$$\mathscr{B}_n = \mathscr{B}(X_k, k \leqslant n).$$

If $E|X_n| < \infty$ for all n and

$$E(X_n | \mathscr{B}_m) = X_m, \qquad m < n, \tag{9}$$

the sequence of random variables (and the corresponding Borel fields) is called a *martingale*. If the equality (9) is replaced by

$$E(X_n | \mathscr{B}_m) \geqslant X_m, \qquad m < n, \tag{10}$$

the sequence is called a *submartingale*. The parameter set needn't be the full set of integers. It could be the set of nonnegative (nonpositive) integers or any totally ordered set.

Let $X_1, X_2, \ldots$ be a sequence of independent identically distributed random variables with $EX_i \equiv 0$. A simplest example of a martingale is given by the partial sums

$$S_n = \sum_{j=1}^{n} X_j$$

with $\mathscr{B}_j = \mathscr{B}(X_k, k \leqslant j)$ since

$$E(S_n | \mathscr{B}_m) = S_m, \qquad m < n.$$

If X_j is regarded as the gain (or loss) at a jth trial in a "fair" game of chance, S_n represents the net gain after n independent trials. For a discussion of the use of martingales in analyzing games of chance see [18].

Let Z be a real-valued random variable on a probability space with an increasing family of Borel fields. If $E|Z| < \infty$ we can set

$$Y_n = E(Z | \mathscr{B}_n).$$

The sequence $\{Y_n\}$ with the Borel fields $\mathscr{B}_n$ is a martingale since

$$E(Y_n | \mathscr{B}_m) = E(E(Z|B_n) | \mathscr{B}_m) = E(Z | \mathscr{B}_m) = Y_m \quad \text{if } m < n.$$

A third example can be given in terms of a Markov process $\{X_n\}$ with transition probability function $P(\cdot, \cdot)$. Let f be an invariant function with respect to the transition function, that is

$$\int P(x, dy) f(y) = f(x).$$

Notice that the idea of such an invariant function is dual to the notion of an invariant measure. Suppose that $E|f(X_n)| < \infty$ for all n. Let

$$Y_n = f(X_n).$$

The sequence $\{Y_n\}$ with the Borel fields $\mathscr{B}_n=\mathscr{B}(X_k, k\leqslant n)$ is a martingale since

$$E(Y_n|\mathscr{B}_m)=E(f(X_n)|\mathscr{B}_m)=\int P_{n-m}(X_m|dy)\,f(y)=f(X_m)=Y_m \quad \text{if } m<n.$$

Lemma 1. *Let $\{X_n\}$ be a martingale. Consider a convex function u such that $E|u(X_n)|<\infty$ for all n. Then*

$$Y_n=u(X_n)$$

is a submartingale.

Let $\mathscr{B}_n$ be the associated increasing family of Borel fields. Then

$$E(Y_n|\mathscr{B}_m)=E(u(X_n)|\mathscr{B}_m)\geqslant u(E(X_n|\mathscr{B}_m))=u(X_m)$$

by Jensen's inequality (which still holds for conditional probabilities) and the martingale condition.

Kolmogorov's inequality for submartingales. Let Y_k, $k=1,2,\ldots$, be a nonnegative sequence of random variables with corresponding Borel fields $\mathscr{B}_k=\mathscr{B}(Y_j, j\leqslant k)$. Then

$$P(\max(Y_1,\ldots,Y_n)>c)\leqslant\frac{E(Y_n)}{c} \tag{11}$$

for $c>0$.

Let N be the smallest subscript $j\leqslant n$ such that $Y_j>c$. Set $N=0$ if there is not such a subscript. Thus,

$$P(\max(Y_1,\ldots,Y_n)>c)=P(N\neq 0)=\sum_{j=1}^{N}P(N=j)$$

while

$$E(Y_n)=\sum_{j=0}^{n}E(Y_n|N=j)P(N=j).$$

However, if $j\geqslant 1$, it follows that

$$E(Y_n|N=j)P(N=j)=\int_{N=j}Y_n\,dP\geqslant c\,P(N=j).$$

The Kolmogorov inequality (11) is an immediate consequence.

The Kolmogorov inequality can be used to prove an interesting convergence theorem for martingales. Let us first remark that (10) indicates that the moments EY_n of a submartingale sequence do not decrease as n increases.

Martingale convergence theorem. Let $\{X_n;\, n=\ldots,-1,0\}$ be a one-sided martingale sequence with $E|X_0|^2<\infty$. Then X_n converges to a random variable Y as $n\to-\infty$. If $\mathscr{B}_j=\mathscr{B}(X_k, k\leqslant j)$, then

$$Y=E(X_0|\mathscr{B}_{-\infty})$$

where $\mathscr{B}_{-\infty}=\bigcap_j\mathscr{B}_j$.

As already remarked. EX_n^2 is a nondecreasing sequence. It follows that

$$E(X_n^2)\to\mu\geqslant 0 \tag{12}$$

as $n\to-\infty$. By the Kolmogorov inequality

$$\begin{aligned} P(|X_{-n-k}-X_{-n}|>c \text{ for some } 1\leqslant k\leqslant m) &\leqslant \frac{E(X_{-m-n}-X_{-n})^2}{c^2} \\ &= \frac{E(X_{-n})^2-E(X_{-n-m})^2}{c^2}. \end{aligned} \tag{13}$$

The right hand side of (13) tends to zero as $n\to\infty$ by (12). Thus (13) implies that the sequence X_n converges as $n\to-\infty$. The limit Y is clearly $\mathscr{B}_{-\infty}$ measurable. Since

$$\Big|\int\limits_{|Y|\leqslant a} YdP\Big| = \Big|\lim_{n\to-\infty}\int\limits_{|X_n|\leqslant a} X_n dP\Big| \leqslant \{EX_n^2\}^{\frac{1}{2}}$$

it follows that Y is integrable. However, we must then have

$$\int\limits_B YdP = \int\limits_B E(X_0|\mathscr{B}_{-\infty})dP$$

for any $\mathscr{B}_{-\infty}$ measurable set. This implies that $Y=E(X_0|\mathscr{B}_{-\infty})$.

As an immediate corollary we have the following result on convergence of conditional probabilities.

Corollary 2. *If A is an event and $\mathscr{B}_{-n}\subseteq\mathscr{B}_{-n+1}$, $n=2,3,\ldots$ is a sequence of Borel fields of events, the conditional probabilities $P(A|\mathscr{B}_{-n})$ converge boundedly almost everywhere to $P(A|\mathscr{B}_{-\infty})$ as $n\to\infty$, where $\mathscr{B}_{-\infty}=\bigcap_{n=1}^{\infty}\mathscr{B}_{-n}$.*

This corollary is obtained immediately from the martingale convergence theorem by setting $X_n=E(I_A|\mathscr{B}_{-n})=P(A|\mathscr{B}_{-n})$ and noticing that $\{X_{-n}\}$ is a martingale.

Appendix 2

Topological Spaces

In a number of applications and examples topological spaces will naturally enter. For this reason some of the basic properties of these spaces will be outlined and motivated together with references to more extensive treatments. A number of simple illustrations will be given.

A topological space is a nonempty set Ω of points with a family τ of subsets (called open sets) having the following properties:

1. $\Omega \in \tau$, $\Omega^c = \Phi \in \tau$;
2. If $O_1, O_2 \in \tau$, then $O_1 \cap O_2 \in \tau$;
3. If $O_\alpha \in \tau$, then $\bigcup_\alpha O_\alpha \in \tau$.

The family τ is called a topology for Ω. One can always define several topologies on a space Ω. There is always the trivial topology in which Φ and Ω are the only open sets. At the other extreme, there is the discrete topology in which every subset of Ω is an open set. An open set O containing a point $x \in \Omega$ is called a *neighborhood* of x. The family of *closed* subsets of Ω consists precisely of the complements of the open subsets of Ω. The characterizing properties of closed sets are immediately read off from the properties 1–3 of open sets (a topology). It is sometimes convenient to refer to the space Ω and its topology by the doublet (Ω, τ).

A mapping f of a topological space (Ω_1, τ_1) into a topological space (Ω_2, τ_2) is called *continuous* if for every open set $O \in \tau_2$ the inverse image $f^{-1}O$ is open, that is, $f^{-1}O \in \tau_1$. A one-to-one mapping f of Ω_1 onto Ω_2 is called a *homeomorphism* if f and f^{-1} are continuous.

A collection of open subsets of a topological space Ω is called a *base* for the topology τ of Ω if for every open set $O \in \tau$ and every point $x \in O$ there is a set B in the base such that $x \in B \subset O$. It is clear that a topology is generated by a base and often it is convenient to speak in terms of the base. For example, the ordinary topology (the linear open sets) for the set of real numbers has as a base the collection of open intervals $(a,b) = \{a < x < b\}$ for all pairs of real numbers $a < b$.

A topological space is called a *Hausdorff* space if for every pair of distinct points x, y of the space there are disjoint open sets O_1, O_2 with

$x \in O_1$ and $y \in O_2$. The Hausdorff space is defined by means of separation properties. Different types of separation properties are discussed and examined at some length in various texts (see for example Royden [100]).

A metric space (Ω, d) is a nonempty set of elements (or points) with a nonnegative distance function $d(x,y)$ defined for all pairs of elements $x, y \in \Omega$ satisfying the following conditions for all points x, y, $z \in \Omega$:

1. $d(x,y) \geqslant 0$
2. $d(x,y) = 0$ if and only if $x = y$
3. $d(x,y) = d(y,x)$
4. $d(x,y) \leqslant d(x,z) + d(y,z)$.

The distance function d is often called a *metric*. The set of all real n-tuples $(x_1, \ldots, x_n)$ with the distance function

$$d(x,y) = \left\{ \sum_i (x_i - y_i)^2 \right\}^{\frac{1}{2}}$$

is an example of a metric space. If the open sets of a metric space taken to be those generated by the base

$$\{y : d(x,y) < \varepsilon\}$$

for all $x \in \Omega$ and $\varepsilon > 0$, the metric space can be seen to be Hausdorff. The metric space (Ω, d) is typically taken with this topology.

Given two topological spaces (Ω_1, τ_1) and (Ω_2, τ_2) a topology is defined on the product space $\Omega_1 \times \Omega_2 = \{(x,y) : x \in \Omega_1, y \in \Omega_2\}$ by taking as a base the collection of all sets $O_1 \times O_2$ where $O_1 \in \tau_1$, $O_2 \in \tau_2$. The topology generated in this way is called the *product topology* for $\Omega_1 \times \Omega_2$. This notion can be extended to any indexed family $(\Omega_\alpha, \tau_\alpha)$ of topological spaces as follows. Define the product topology on $\prod_\alpha \Omega_\alpha$ of points $x = (x_\alpha)$ with α^{th} coordinate $x_\alpha \in \Omega_\alpha$ by taking as a base all sets of the form $\prod_\alpha O_\alpha$ where $O_\alpha \in \tau_\alpha$ and $\Omega_\alpha = O_\alpha$ except for a finite number of α. When the Ω_α are all the same space Ω and are indexed by the index set A, it is usual to write Ω^A for $\prod_\alpha \Omega_\alpha$. Denote the discrete space with n elements (n finite) by n and a countable (nonfinite) set by ω. Then n^ω is the space of n-ary sequences. The reader can compare n^ω with the discrete topology and n^ω with the product topology. It is also easy to show that the product of Hausdorff spaces is a Hausdorff space.

The notion of compactness is introduced to characterize a class of spaces for which the conclusion of the Heine-Borel theorem holds. A collection $\mathscr{C}$ of open sets in a topological space is said to be an open covering for a set A if A is contained in the union of sets in $\mathscr{C}$. A topological space Ω is *compact* if every open covering $\mathscr{C}$ of Ω has a finite subcovering, that is, if there are a finite number of sets $O_1, \ldots, O_n \in \mathscr{C}$

such that $\Omega = \bigcup_{\alpha=1}^{n} O_\alpha$. A subset A of Ω can be looked upon as a topological subspace by taking the sets $O \cap A$ for all $O \in \tau$ (the topology on Ω) as the topology on A. The set A is called compact if it is compact as a subspace of Ω.

Theorem 1. *A closed subset of a compact space is compact. A compact subset of a Hausdorff space is closed.*

Let Ω be compact with A a closed subset of Ω and $\mathscr{C}$ an open covering for A. The open covering $\mathscr{C} \cup \{A^c\}$ for Ω must then have a finite subcovering $\{O_1, \ldots, O_n, A^c\}$. Since the sets $O_1, \ldots, O_n$ cover A, $\mathscr{C}$ has a finite subcovering.

Assume now that Ω is Hausdorff with C a compact subset of Ω. We show that C^c is open. Let $y \in C^c$. Since Ω is Hausdorff, for each $x \in C$ there are disjoint open sets O_x, Q_x with $x \in O_x, y \in Q_x$. The sets $\{O_x : x \in C\}$ are an open covering of C so that there is by the compactness of C a finite subcollection $(O_{x_1}, \ldots, O_{x_n})$ which covers C. Let

$$Q = \bigcap_{i=1}^{n} Q_{x_i}.$$

The set Q is open and is disjoint from $\bigcup_{i=1}^{n} O_{x_i}$. Since $C \subset \bigcup_{i=1}^{n} O_{x_i}$ it follows that $Q \subset C^c$. C^c is open and hence C is closed.

A point $x \in \Omega$ is called a *point of closure* of a set $A \subseteq \Omega$ if every open set $O \in \tau$ containing x has nonvacuous intersection $O \cap A \neq \Phi$. One can show that the set $\bar{A}$ of points of closure of A is the intersection of all the closed sets containing A. The set $\bar{A}$ is called the closure of A.

A metric space (Ω, d) with a countable set of points $A = \{x_i, i = 1, 2, \ldots\}$ whose closure $\bar{A} = \Omega$ is said to be *separable*. A Cauchy sequence in (Ω, d) is a sequence of points x_i, $i = 1, 2, \ldots$, such that $\lim_{i,j \to \infty} d(x_i, x_j) = 0$. The metric space (Ω, d) is said to be complete if each Cauchy sequence x_i has a limit point, that is, a point $x \in \Omega$ such that $\lim_{i \to \infty} d(x_i, x) = 0$.

A topological space (Ω, τ) is *locally compact* if each point $x \in \Omega$ has a neighborhood O_x whose closure $\bar{O}_x$ is compact. Every topological space with the discrete topology is locally compact. A more restrictive but sometimes more useful concept is σ-compactness. The space Ω is *σ-compact* if it is the union of a countable number of compact sets. The set of real numbers with the usual topology is σ-compact since it is the union of the compact sets $[n, n+1] = \{x : n \leqslant x \leqslant n+1\}$, $n = 0, \pm 1, \pm 2, \ldots$.

We would like to consider a generalization of the notion of a sequence which is sometimes convenient. It is the notion of a net. For this we will first have to consider the idea of a directed system. Let B

be a set with a relation $<$. The set B and the relation $<$ are called a directed system if the following conditions are satisfied:
(i) if $\alpha<\beta$ and $\beta<\gamma$, then $\alpha<\gamma$;
(ii) if $\alpha, \beta \in B$ there is a $\gamma \in B$ with $\alpha<\gamma$, $\beta<\gamma$.
The set N of positive integers with $<$ given by $\leqslant$ are a directed system. Another example is given by the set of all open sets containing a point x where $O_1 < O_2$ means that $O_1 \supset O_2$.

A net is a mapping of a directed system into a topological space Ω. Notice that a net is a sequence if the directed system is the set N of positive integers. Let x_α be the value of the net at α and let $\{x_\alpha\}$ denote the net itself. A point $x \in \Omega$ is the *limit* of a net $\{x_\alpha\}$ if for each open set O containing x there is an $\alpha_0 \in B$ such that $x_\alpha \in O$ for all $\alpha > \alpha_0$.

Lemma 1. *A point x belongs to the closure of a set A if and only if it is the limit of a net $\{x_\alpha\}$ in A.*

If x is the limit of a net $\{x_\alpha\}$ in A it follows from the definition of the closure $\bar{A}$ of A that $x \in \bar{A}$. Assume that $x \in \bar{A}$. Take as the directed system the collection of open sets containing x with $O_1 < O_2$ if $O_1 \supset O_2$. Then for each $O \subset A$ there is a point x_O in $O \cap B$. One can then verify that $\{x_O\}$ is a net in A converging to x.

If a topological space is compact Hausdorff, it actually has a strong separation property.

Lemma 2. *If Ω is a compact Hausdorff space and C_1, C_2 are two disjoint closed sets, there are disjoint open sets O_1, O_2 with $C_1 \subset O_1$, $C_2 \subset O_2$.*

We note briefly that this result is obtained by first showing that a closed set C and a point $x \notin C$ can be separated by disjoint open sets. It is the compactness coupled with the Hausdorff property that leads to this strengthening of the Hausdorff separation property.

Lemma 2 can be used to obtain the following useful result which is usually called Urysohn's Lemma.

Lemma 3. *Let C_1, C_2 be disjoint closed subsets of a compact Hausdorff space. There is then a real-valued continuous function f defined on the space with $0 \leqslant f \leqslant 1$ everywhere and such that $f=0$ on C_1 and $f=1$ on C_2.*

The following theorem due to Tychonoff tells us that the product space of compact Hausdorff spaces is itself compact in the product topology. This is a very useful result to have since product spaces occur very often.

Theorem 2. *Let $\{\Omega_\alpha\}$ be an indexed family of compact Hausdorff spaces. Then the product space $\prod_\alpha \Omega_\alpha$ is compact Hausdorff in the product topology.*

The classical Weierstrass approximation theorem (see [93]) states that every real-valued continuous function on a finite closed interval $[a,b]=\{x:a\leqslant x\leqslant b\}$ on the real line can be uniformly approximated by polynomials. There is an interesting generalization of this result on compact Hausdorff spaces called the Stone-Weierstrass theorem. Consider the family $\mathscr{F}$ of real-valued continuous functions on a compact Hausdorff space Ω. They are an algebra, that is, given any two functions $f,g\in\mathscr{F}$

1. the product $fg\in\mathscr{F}$ and
2. the linear combination $\alpha f+\beta g\in\mathscr{F}$ for any two real constants α, β. $\mathscr{F}$ contains the constant functions. Further, given any two distinct points $x,y\in\Omega$ there is a function $f\in\mathscr{F}$ with $f(x)\neq f(y)$, that is, $\mathscr{F}$ *distinguishes between the points of* Ω. *Also* $\mathscr{F}$ *is closed in the absolute supremum norm*

$$\|f\|=\sup_{x\in\Omega}|f(x)|\,.$$

The Stone-Weierstrass theorem says that these properties characterize the algebra of continuous functions on Ω.

Theorem 3. *Let* $\mathscr{F}$ *be a closed algebra of real continuous functions on the compact Hausdorff space* Ω. *If* $\mathscr{F}$ *contains the constant functions and distinguishes between points of* Ω, *it is the algebra of continuous functions on* Ω.

We shall briefly outline the proof of this theorem. Let $f\in\mathscr{F}$. First we show that $|f|\in\mathscr{F}$. Since f is continuous on the compact space Ω, there is a sufficiently large integer n such that $|f|\leqslant n$ on Ω. By the classical Weierstrass theorem, given any $\varepsilon>0$ there is a polynomial $p(\lambda)$ on $[-n,n]$ such that $\big||\lambda|-p(\lambda)\big|<\varepsilon$ on $[-n,n]$. But then $\big||f(\lambda)|-p(f(\lambda))\big|<\varepsilon$ on Ω. Thus $|f|\in\mathscr{F}$. From this, it follows that for any two functions $f,g\in\mathscr{F}$ the functions $\min(f,g)$ and $\max(f,g)$ are in $\mathscr{F}$ since

$$\min(f,g)=\tfrac{1}{2}\{f+g-|f-g|\}$$
$$\max(f,g)=-\min(-f,-g)\,.$$

Given any two points $x,y\in\Omega$ and any two real numbers α, β one can find a function $f\in\mathscr{F}$ such that $f(x)=\alpha$, $f(y)=\beta$. This follows from the assumption that $\mathscr{F}$ contains the constant functions and distinguishes between the points of Ω. Let us now consider any given continuous function h on Ω. Consider any fixed $\varepsilon>0$. A function $f\in\mathscr{F}$ such that $\sup_x|f(x)-h(x)|<\varepsilon$ will be determined. Let $f_{y,z}\in\mathscr{F}$ be a function with $f_{y,z}(y)=h(y)$, $f_{y,z}(z)=h(z)$. There is a neighborhood U_z of z such that $f_{y,z}(x)>h(x)-\varepsilon$ for $x\in U_z$ By the compactness of Ω there is a finite collection $U_{z_1},\ldots,U_{z_p}$ of neighborhoods covering Ω. Let $f_y=\min_i f_{y,z_i}$.

Then $f_y(x) > h(x) - \varepsilon$ for all $x \in \Omega$. Notice that $f_y(y) = h(y)$. There is a neighborhood V_y of y such that $f_y(x) < h(x) + \varepsilon$ for $x \in V_y$. Again there is a finite collection $V_{y_1}, \ldots, V_{y_q}$ of neighborhoods covering Ω. Let $f = \max_i f_{y_i}$. The function f approximates h uniformly to within ε and is an element of $\mathscr{F}$.

Let (Ω, τ) be a compact Hausdorff space. The smallest Borel field $\mathscr{B}_c$ with respect to which all the continuous functions on Ω are measurable is called the Borel field of *Baire sets*. The smallest Borel field $\mathscr{B}_\tau$ with respect to which all open sets are measurable is called the Borel field of *Borel sets*. $\mathscr{B}_\tau$ is the Borel field generated by the topology τ. Generally the family of Baire sets $\mathscr{B}_c$ is a proper subset of $\mathscr{B}_\tau$. However, if (Ω, τ) is a separable metric space the two Borel fields $\mathscr{B}_c$ and $\mathscr{B}_\tau$ coincide. A measure ν on $\mathscr{B}_\tau$ is said to be *regular* if for any set $B \in \mathscr{B}_\tau$ and any given $\varepsilon > 0$ there are closed and open sets C and O respectively such that $C \subset B \subset O$ and

$$\begin{aligned} \nu(B - C) &< \varepsilon \\ \nu(O - B) &< \varepsilon . \end{aligned}$$

The regularity of ν is a consistency of ν with the given topology τ.

Let ν be a finite regular measure on the Borel sets of the compact Hausdorff space (Ω, τ). Let $L^p(d\nu)$ be the set of Borel measurable functions f on Ω such that

$$\int |f(x)|^p \, \nu(dx) < \infty .$$

Given any $\varepsilon > 0$, we can find a continuous function f_ε on Ω such that

$$\int |f(x) - f_\varepsilon(x)|^p \, \nu(dx) < \varepsilon .$$

This follows from the regularity of the measure ν and an application of the Urysohn Lemma.

Appendix 3

The Kolmogorov Extension Theorem

A stochastic process. It often happens that the existence of a stochastic (or random) process $\{X_t(\omega), t \in \tau\}$, an indexed family of random variables on a probability space, is taken for granted. The usual model is that of a *space* Λ *with points* $\omega \in \Lambda$. *A Borel field* $\mathcal{B}$ *of subsets of* Λ *is given with a probability measure* μ *on the Borel field* $\mathcal{B}$. Here τ is a parameter set and the function $X_t(\omega)$ is to be thought of as an observable at location $t \in \tau$. In this book τ will typically be assumed to be the set (or a subset) of the integers or the real numbers because we are basically interested in Markov processes. The Markov property is most natural for a time-like parameter set (a well-ordered set) though limited investigations of the Markov property have been carried out for a multidimensional parameter t (see Dobrushin [16], Rozanov [99]). Nonetheless there are many (non-Markovian) stochastic processes of interest with multidimensional time. Thus, τ might be the set of lattice points in k space ($k=1,2,\ldots$), the set of real (or complex) k-tuples, or even the set of elements of a Borel field on some space. *The range space* Ω *of values of the functions* $X_t(\omega)$ may be the real (or complex) numbers, real (or complex) k-vectors, or a group. *There is a Borel field* $\mathcal{A}$ *of subsets of* Ω *and it is only natural to call the functions* $X_t(\omega)$ observables (or *random variables*) *if they are measurable with respect to* $\mathcal{B}$, that is, if for each $A \in \mathcal{A}$ and each $t \in \tau$

$$\{\omega : X_t(\omega) \in A\} \in \mathcal{B}. \tag{1}$$

In that case the collection of measurable functions $\{X_t(\omega), t \in \tau\}$ on the probability space Λ *is called a stochastic process.* The Borel field $\mathcal{B}$ is often taken to be the Borel field generated by the functions $\{X_t(\omega), t \in \tau\}$, that is, the smallest Borel field containing the sets (1).

Let $\tau_n = (t_1, \ldots, t_n)$, $n=1,2,\ldots$, be a finite subset of τ with $\mathcal{B}_{\tau_n}$ the Borel field generated by

$$\{\omega : X_t(\omega) \in A\}, \qquad t \in \tau_n, \qquad A \in \mathcal{A}.$$

Let μ_{τ_n} be the probability measure μ on $\mathscr{B}_{\tau_n}$. μ_{τ_n} is the projection of μ on $\mathscr{B}_{\tau_n}$ (the restriction of μ to $\mathscr{B}_{\tau_n}$) and describes the joint probability structure of the finite set of random variables $X_t(\omega)$, $t \in \tau_n$. In many cases, it is only the finite dimensional joint probability measures μ_{τ_n} that are given for all finite subsets τ_n of τ and *it is* then *an open question as to whether there is a stochastic process* $\{X_t(\omega), t \in \tau\}$ *on a probability space realizing these finite dimensional measures* μ_{τ_n}. The Kolmogorov extension theorem describes conditions under which this occurs. We shall give a version of the Kolmogorov theorem in this appendix which is adequate for our purposes. Let us first note that a basic consistency condition must be satisfied in order that there be a stochastic process realizing the finite dimensional probability measures μ_{τ_n}.

Consistency condition: Assume that finite dimensional joint probability measures μ_{τ_n} *are given for all finite subsets* τ_n *of* τ. *Every pair of probability measures* $\mu_{\tau_n}, \mu_{\tau_m}$ *corresponding to any two finite subsets* τ_n, τ_m *of* τ *must agree on the Borel field* $\mathscr{B}_{\tau_n \cap \tau_m}$ ($\mathscr{B}_\Phi$ with Φ the null set is understood to be the trivial Borel field consisting only of the null set and Λ the whole space).

A basic extension theorem called the Carathéodory extension theorem will be required in our derivation of the Kolmogorov extension theorem. We shall briefly sketch the background of this result and give a statement of it. A proof of the Carathéodory theorem can be found in Loéve [70] or Royden [100]. Let $\mathscr{F}$ be a field of subsets of the space Λ with η a nonnegative additive set function on $\mathscr{F}$ and $\eta(\Lambda)=1$. The set function η is nonnegative and additive if

(i) $\eta(B) \geqslant 0$ for $B \in \mathscr{F}$,
(ii) $\eta(B^c) = 1 - \eta(B)$ for $B \in \mathscr{F}$,
(iii) $\eta(B \cup B') = \eta(B) + \eta(B')$ for disjoint sets $B, B' \in \mathscr{F}$.

Let $\mathscr{B}$ be the Borel field generated by the field $\mathscr{F}$. The Carathéodory extension theorem tells us when η can be extended to a measure on the Borel field $\mathscr{B}$. The set function η is said to be continuous at the null set Φ if $\eta(B_n) \downarrow 0$ as $n \to \infty$ for any decreasing family $B_1 \supset B_2 \supset \cdots$ of sets $B_k \in \mathscr{F}$, $k=1,2,\ldots$, with $\bigcap B_k = \Phi$.

Carathéodory extension theorem: Let η *be a nonnegative additive set function on a field* $\mathscr{F}$ *of sets of* Λ $(\eta(\Lambda)=1)$. *If* η *is continuous at the null set* Φ, *there is a uniquely determined extension* μ *of* η $(\mu(B)=\eta(B)$ *for* $B \in \mathscr{F})$ *on the Borel field* $\mathscr{B}$ *generated by the field* $\mathscr{F}$.

To say that η is *continuous at the null set* Φ *is equivalent to* saying that η is *σ-additive on* $\mathscr{F}$. It will be enough to briefly show how continuity of η at Φ implies that η is σ-additive on $\mathscr{F}$ since the converse follows by a similar argument. Let $F_n \in \mathscr{F}$, $n=1,2,\ldots$, be a disjoint sequence of sets with $F=\bigcup_n F_n \in \mathscr{F}$. Then $B_n = F - \bigcup_{k=1}^{n} F_k$ is a decreasing family of sets of $\mathscr{F}$ with $\bigcap_k B_k = \Phi$. However,

$$\mu(F)=\mu\left(\bigcup_{k=1}^{n} F_k\right)+\mu(B_n)=\sum_{k=1}^{n}\mu(F_k)+\mu(B_n)$$

and since $\mu(B_n)\to 0$ as $n\to\infty$ it follows that

$$\mu(F)=\sum_{k=1}^{\infty}\mu(F_k).$$

The assumptions required for the version of the Kolmogorov extension theorem given here are introduced now. Let Ω be a σ-compact Hausdorff space with $\mathscr{A}$ the Borel field generated by the topology (the smallest Borel field containing the closed sets). Let $\Omega_t = \Omega$ be the range space of an observable $X_t(\omega)$, $t\in\tau$, with $\mathscr{A}_t = \mathscr{A}$ the corresponding Borel field on Ω_t. Let the Borel field $\mathscr{B}_{\tau_n}$ be the product Borel field $\prod_{t\in\tau_n}\mathscr{A}_t$ on $\prod_{t\in\tau_n}\Omega_t$ where $\tau_n=(t_1,\dots,t_n)$ is a finite subset of τ. The finite dimensional measures μ_{τ_n} are assumed to be regular on $\mathscr{B}_{\tau_n}$ with respect to the product topology on $\prod_{t\in\tau_n}\Omega_t$.

Kolmogorov extension theorem: Let the range space Ω of an observable be a σ-compact Hausdorff space with $\mathscr{A}$ the Borel field generated by the topology on the space. Assume that for each finite subset τ_n of τ there is given a probability measure μ_{τ_n} on $\mathscr{B}_{\tau_n}$ regular with respect to the product topology on $\prod_{t\in\tau_n}\Omega_t$. If the family of measures $\{\mu_{\tau_n}\}$ satisfy the consistency condition there is a stochastic process $\{X_t(\omega), t\in\tau\}$ on a probability space realizing these finite dimensional measures.

Let $\Omega^\tau=\prod_{t\in\tau}\Omega_t$, that is, Ω^τ is the product space of points $\omega=(\omega_t, t\in\tau)$ whose t^{th} coordinate is $\omega_t\in\Omega_t$. We identify a finite dimensional set $B\in\mathscr{B}_{\tau_n}$ on $\Omega^{\tau_n}=\prod_{t\in\tau_n}\Omega_t$ with the set $B\times\prod_{t\in\tau-\tau_n}\Omega_t = B\times\Omega^{\tau-\tau_n}$ in Ω^τ (we have already implicity done this in an earlier discussion in this appendix for convenience but hope it has not caused any confusion) and set

$$\mu(B\times\Omega^{\tau-\tau_n})=\mu_{\tau_n}(B),\qquad B\in\mathscr{B}_{\tau_n}. \tag{2}$$

In this way $\mathscr{B}_{\tau_n}$ is identified with the Borel field of sets $B\times\Omega^{\tau-\tau_n}$, $B\in\mathscr{B}_{\tau_n}$, on Ω^τ which we shall still call $\mathscr{B}_{\tau_n}$ so as to minimize additional notation. The definition of μ by (2) on all sets $B\times\Omega^{\tau-\tau_n}$, $B\in\mathscr{B}_{\tau_n}$, for all finite subsets τ_n of τ can be carried out because of the consistency of the measures $\{\mu_{\tau_n}\}$. Then μ is a nonnegative additive set function (with $\mu(\Omega^\tau)=1$)

on the field $\mathscr{F} = \bigcup_{\tau_n \subset \tau} \mathscr{B}_{\tau_n}$ obtained by taking the union of the Borel fields $\mathscr{B}_{\tau_n}$ for all finite subsets τ_n of τ. Our construction of the desired stochastic process will be complete if we can show that μ can be extended to the Borel field $\mathscr{B}$ generated by $\mathscr{F}$ as a probability measure—for then one can simply take

$$X_t(\omega) = \omega_t, \qquad t \in \tau,$$

as the random variables (measurable functions) of the stochastic process. The Carathéodory extension theorem tells us that this can be done if μ is continuous at the empty set Φ on the field $\mathscr{F}$, that is, for any decreasing sequence $C_n \in \mathscr{F}$ with null intersection $\bigcap_n C_n = \Phi$ it follows that $\mu(C_n) \to 0$ as $n \to \infty$. Clearly it is enough to show that if $C_n \in \mathscr{F}$ is a decreasing sequence with

$$\mu(C_n) > \varepsilon > 0, \qquad n = 1, 2, \ldots,$$

then $\bigcap_n C_n \neq \Phi$. The sets in $\mathscr{F}$ are determined by conditions on a finite number of coordinates ω_t. Thus, for any given sequence $C_n \in \mathscr{F}$ there is a countable subset $\tau_\infty = (t_1, t_2, \ldots) \subset \tau$ with $\tau_n = (t_1, t_2, \ldots, t_n)$, $n = 1, 2, \ldots$, such that

$$C_n = B_n \times \Omega^{\tau - \tau_n}, \qquad B_n \in \mathscr{B}_{\tau_n},$$

with $B_n \subset \Omega^{\tau_n}$. By the σ-compactness of Ω and the regularity of μ_{τ_n} on Ω^{τ_n} there is a nondecreasing sequence of compacts $A_n \in \mathscr{A}$ such that

$$\mu_{\tau_n}\left(\prod_{r=1}^{n} A_n\right) > 1 - \frac{\varepsilon}{2^{n+2}}, \qquad n = 1, 2, \ldots .$$

The set $K = \prod_{n=1}^{\infty} A_n$ is a nonvacuous compact in Ω^{τ_∞} (with the product topology) by the Tychonov theorem. Also, by the regularity of μ_{τ_n} on the sets $\mathscr{B}_{\tau_n}$ of Ω^{τ_n} there is a compact set $S_n \in \mathscr{B}^{\tau_n}$ (on Ω^{τ_n}) with $S_n \subset B_n$ and

$$\mu_{\tau_n}(B_n - S_n) < \frac{\varepsilon}{2^{n+2}}. \tag{3}$$

Then (3) implies that

$$\mu_{\tau_n}(B_n - S_n') > \frac{\varepsilon}{2}$$

for the compact sets

$$S_n' = \bigcap_{k=1}^{n} (S_k \times \Omega^{\tau_n - \tau_k}).$$

Notice that the sets

$$(S_n' \times \Omega^{\tau_\infty - \tau_n}) \cap K \tag{4}$$

are a decreasing sequence of nonvacuous compact subsets of K. The sets (4) must therefore have a nonvacuous intersection. But that in turn implies that the sets C_n have a nonvacuous intersection. The proof of the theorem is complete.

The assumptions of the Kolmogorov extension theorem are of a mixed topological and measure theoretic character. However, this result is of especial interest because it is valid for non-Markovian as well as Markovian processes and also because it is applicable in both the discrete and continuous parameter case. If Ω is σ-compact Hausdorff and the Borel field generated by the topology is the σ-field of Baire sets, the assumptions of the Kolmogorov extension theorem are satisfied. In particular, this is the case if Ω is a complete separable metric space. If Ω is a compact Hausdorff space and $P(\cdot,\cdot)$ is a transition function taking continuous functions into continuous functions, that is,

$$(Tf)(x) = \int P(x, dy)\, f(y)$$

is continuous if f is, then the Kolmogorov extension theorem can be applied to construct the Markov process with the initial distribution the regular probability measure μ and transition probability function $P(\cdot,\cdot)$. Notice that if μ is regular then μT is regular.

In the discrete time parameter Markovian case a result of Ionescu-Tulcea which we shall briefly state without proof indicates that one can set up the appropriate probability space without assumptions of a topological character. Let Ω be the state space with the Borel field of subsets $\mathscr{A}$. Let $P(x, A)$, $x \in \Omega$, $A \in \mathscr{A}$, be a given *transition probability function*, that is, $P(x, A)$ is an $\mathscr{A}$-measurable function of x for each $A \in \mathscr{A}$ and for each $x \in \Omega$ is a probability measure on $A \in \mathscr{A}$. Given an initial probability measure μ on $\mathscr{A}$, define the set function

$$\begin{aligned} P_\mu(A_0 \times A_1 \times \cdots \times A_n) &= P_\mu(x_0 \in A_0, \ldots, x_n \in A_n) \\ &= \int_{A_0} \mu(dx_0) \int P(x_0, dx_1) \ldots \int_{A_{n-1}} P(x_{n-2}, dx_{n-1}) P(x_{n-1}, A_n) \end{aligned} \tag{5}$$

on the product sets $A_0 \times A_1 \times \cdots \times A_n$ (n a positive integer) where $A_0, A_1, \ldots, A_n \in \mathscr{A}$.

Theorem of Ionescu-Tulcea: Let Ω be a state space with Borel field $\mathscr{A}$ and $P(\cdot,\cdot)$ a transition probability function relative to Ω, $\mathscr{A}$. Consider a probability measure μ on $\mathscr{A}$. The set function (5) *can then be extended to a probability measure P_μ on the Borel field $\mathscr{A}_\infty$ generated by the product* *sets of the form* $A_0 \times A_1 \times \cdots \times A_n = \{x_0 \in A_0, \ldots, x_n \in A_n\}$, $A_i \in \mathscr{A}$, *on the space of points* $(x_0, x_1, x_2, \ldots) = \omega$. *The stochastic process*

$$\{X_n(\omega), n=0,1,2,\ldots\}, \qquad X_n(\omega) = x_n,$$

is a Markov process.

The theorem just stated is conveniently general in that there is no topological condition of type assumed in the Kolmogorov theorem on the finite dimensional set functions (5). However, it is restricted to Markov processes with a discrete time parameter. A proof for a somewhat more general context can be found in example 2.6 of the supplement in Doob [18]. The usual proof again makes use of the Carathéodory extension theorem.

Appendix 4

Spaces and Operators

In this appendix we briefly state and try to motivate relevant results on linear spaces and linear operators on these spaces. Detailed proofs of the results can be found in books like Dunford and Schwartz [21, 22], Lorch [71], and Royden [100].

Let M be a set of elements that is a commutative group under addition. Assume that the product αx is well defined for any real (complex) number α and element $x \in M$ and that $\alpha x \in M$. Further, let

$$\begin{aligned} \alpha(x+y) &= \alpha x + \beta y, \\ (\alpha+\beta)x &= \alpha x + \beta y, \\ (\alpha\beta)x &= \alpha(\beta x), \\ 1x &= x \end{aligned}$$

for all elements $x, y \in M$ and real (complex) numbers α, β. If M satisfies these conditions it is called a *real (complex) linear vector space*. The vector space M is said to be *normed* if there is a nonnegative function $\|\cdot\|$ (the norm) defined on M with

$$\begin{aligned} \|x\| \geqslant 0 \quad \text{and } \|x\| = 0 \quad &\text{if and only if } x=0, \\ \|\alpha x\| &= |\alpha| \cdot \|x\|, \\ \|x+y\| &\leqslant \|x\| + \|y\| \end{aligned} \tag{1}$$

where $x, y \in M$ and α is a real (complex number). Condition (1) is usually referred to as the triangle inequality. A sequence $x_n \in M$, $n=1,2,\ldots$, is said to be a Cauchy sequence if

$$\|x_n - x_m\| \to 0 \tag{2}$$

as $n, m \to \infty$. The normed vector space M is called a *Banach space* if it is complete, that is, if every Cauchy sequence (2) has a limit $x \in M$ so that

$$\|x_n - x\| \to 0$$

as $n \to \infty$.

Let l^p be the set of sequences of real (complex) numbers

$$x = (x_j;\ j=1,2,\ldots)$$

with

$$\|x\| = \left(\sum_{j=1}^{\infty} |x_j|^p\right)^{\frac{1}{p}} < \infty \tag{3}$$

for a fixed $p>0$. It is understood that $x+y=(x_j+y_j; j=1,2,\ldots)$ and $\alpha x=(\alpha x_j; j=1,2,\ldots)$ for α a real (complex) number. If $1 \leqslant p < \infty$, l^p is a Banach space with norm $\|\cdot\|$ given by (3). The triangle inequality is just the Minkowski inequality for sequences (see [61]). l^∞ is the Banach space of bounded sequences $x=(x_j; j=1,2,\ldots)$ with norm

$$\|x\| = \sup_j |x_j| < \infty .$$

Let μ be a σ-finite measure on a σ-field $\mathscr{B}$ of subsets of Ω. Call the set of $\mathscr{B}$ measurable functions f on Ω with

$$\|f\| = \left(\int |f(\omega)|^p \mu(d\omega)\right)^{\frac{1}{p}} < \infty, \qquad 1 \leqslant p < \infty, \tag{4}$$

the space $L^p(d\mu)$. Addition and multiplication are understood in the usual sense. Furthermore, functions that differ only on a set of μ measure zero are identified so that one actually deals with the corresponding equivalence classes of functions as elements of $L^p(d\mu)$. With this understanding and with norm (4) $L^p(d\mu)$, $1 \leqslant p < \infty$, is a Banach space. The triangle inequality is now the Minkowski inequality for functions (see [61]). $L^\infty(d\mu)$ is the set of functions f for which there is a number $c=c(f)$, $0<c<\infty$, such that

$$\mu\{(\omega : |f(\omega)| > c\} = 0 .$$

Functions that differ on a set of μ measure zero are again identified. The set $L^\infty(d\mu)$ is a Banach space with norm

$$\|f\| = \inf_c \mu\{(\omega : |f(\omega)| > c)\} = 0 = \operatorname*{ess\,sup}_{\omega\in\Omega} |f(\omega)| .$$

Still another example of a Banach space is given by the bounded continuous functions f on a locally compact Hausdorff space with norm

$$\|f\| = \sup_\omega |f(\omega)| . \tag{5}$$

A Hilbert space M is a special type of Banach space with the norm generated by a complex-valued inner product $(\cdot,\cdot)$ which is defined for pairs of elements. The norm of an element x is given by

$$\|x\| = \{(x,x)\}^{\frac{1}{2}}$$

and the inner product is assumed to satisfy

$$(x,x) \geqslant 0 \quad \text{with } (x,x)=0 \quad \text{only for } x=0,$$

$$(\alpha x+\beta y, z)=\alpha(x,z)+\beta(y,z), \tag{6}$$

$$(x,y)=\overline{(y,x)}, \tag{7}$$

$$|(x,y)| \leqslant \|x\| \cdot \|y\| \tag{8}$$

for elements $x, y, z \in M$ and complex numbers α, β. The conditions given above are for a complex Hilbert space. In the case of a real Hilbert space the inner product is real-valued, the scalars α, β allowed are real and (7) is replaced by $(x,y)=(y,x)$. Condition (6) is a bilinearity for the inner product and (8) implies the triangle inequality for the norm since

$$\begin{aligned}\|x+y\|^2 &= (x+y, x+y) = \|x\|^2 + 2\operatorname{Re}(x,y) + \|y\|^2 \\ &\leqslant \|x\|^2 + 2\|x\| \cdot \|y\| + \|y\|^2 = (\|x\| + \|y\|)^2 .\end{aligned}$$

Notice that l^2 and $L^2(d\mu)$ are examples of Hilbert spaces with the inner product given by

$$(x,y)=\sum_{j=1}^{\infty} x_j \overline{y_j} \tag{9}$$

and

$$(f,g)=\int f(\omega)\overline{g(\omega)}\,\mu(d\omega) \tag{10}$$

respectively. Condition (8) for the inner products (9) and (10) can be seen to be the Schwarz inequality.

A linear functional L defined on a Banach space M is a real (complex) valued function defined for each element of M and satisfying

$$L(\alpha x+\beta y)=\alpha L(x)+\beta L(y)$$

for each pair of elements $x, y \in M$ and each pair of real (complex) numbers. The functional is said to be *bounded* if

$$|L| = \sup_{\substack{x \in M \\ x \neq 0}} \frac{|L(x)|}{\|x\|} < \infty . \tag{11}$$

Since a linear functional on M is just a function on M into the real (complex) numbers, it is natural to consider continuity of functionals.

Theorem 1. *A linear functional L on a Banach space is bounded if and only if it is continuous.*

The set of bounded linear functionals L on a Banach space M is itself a Banach space with norm $|L|$ given by (11). This derived Banach space of bounded linear functionals on M is called the conjugate space M^*. A basic result in functional analysis (see Royden [100]) characterizes the bounded linear functionals on l^p and $L^p(d\mu)$ for $1 \leqslant p < \infty$. We shall just state the result for $L^p(d\mu)$.

Theorem 2. *Let μ be a σ-finite measure. Then there is a one-to-one correspondence between the bounded linear functionals L on $L^p(d\mu)$, $1 \leqslant p < \infty$, and elements g of $L^q(d\mu)$, where $p^{-1} + q^{-1} = 1$, such that*

$$L(f) = \int f(\omega) \overline{g(\omega)} \mu(d\omega)$$

for $f \in L^p(d\mu)$, and the norm $|L|$ of L is equal to the norm of g as an element of $L^q(d\mu)$.

The special case of Theorem 2 in which $p = q = 2$ suggests a general result for Hilbert spaces M. If L is a bounded linear functional on M then there is an element $y \in M$ such that

$$L(x) = (x, y)$$

for all $x \in M$. Thus, if M is a Hilbert space the conjugate space $M^* = M$.

The case of a bounded linear functional on the continuous functions of a compact Hausdorff space is of especial interest. Here M is the Banach space of continuous functions with the norm (5). The Riesz representation theorem characterizes such linear functionals as integrals with respect to a completely additive set function consistent with the topology.

Theorem 3. *Let L be a bounded real-valued linear functional on the continuous functions of a compact Hausdorff space (Ω, τ). Then*

$$L(f) = \int f(\omega) \nu(d\omega)$$

for all continuous f where ν is a completely additive set function on the Borel sets $\mathscr{B}_\tau$ of Ω. Further, ν is uniquely determined as the difference $\nu = \eta_1 - \eta_2$ of two regular measures η_1, η_2 on $\mathscr{B}_\tau$ singular with respect to each other, that is, there is a set $B \in \mathscr{B}_\tau$ such that $\eta_1(A) = 0$ for $A \subseteq B^c$, $A \in \mathscr{B}_\tau$, and $\eta_2(A) = 0$ if $A \subseteq B$, $A \in \mathscr{B}_\tau$. If L is a positive functional ($L(f) \geqslant 0$ for $f \geqslant 0$) then $\nu = \eta_1$. The norm $\|L\| = \eta_1(\Omega) + \eta_2(\Omega)$.

If L is positive one can show that

$$\nu(B) = \sup_{\substack{0 \leqslant f \leqslant 1 \\ f = 0 \text{ on } B^c}} L(f) \tag{12}$$

where the supremum is taken over all continuous functions satisfying the conditions noted on the right side of (12).

Let T be a linear operator on a Banach space M into itself so that if $x, y \in M$ and α, β are real (complex) scalars than

$$T(\alpha x + \beta y) = a T x + \beta T y.$$

Just as in the case of a linear functional, the linear operator T is said to be bounded if

$$|T| = \sup_{\substack{x \neq 0 \\ x \in M}} \frac{\|Tx\|}{\|x\|} < \infty .$$

The following result parallels Theorem 1.

Theorem 4. *A linear operator T mapping the Banach space M into itself is bounded if and only if it is continuous.*

A bounded linear operator S on the Banach space M is called a *projection* if $S^2 = S$. Notice that S determines the subspaces M_1, M_2 of M with $M_1 = \{x \in M : Sx = x\}$, $M_2 = \{x \in M : Sx = 0\}$. Given any $x \in M$ one can write $x = Sx + (x - Sx)$ with $Sx \in M_1$ and $(x - Sx) \in M_2$. One commonly speaks of S as projecting M on M_1.

A linear operator T is called a *contraction operator* if $|T| \leqslant 1$. Given bounded linear operators T and T' on Banach spaces M and M' respectively with $M \subset M'$, the operator T' is called a dilation of T if there is a projection operator S projecting M' on M and such that

$$T = ST'S .$$

If T is a bounded linear operator on a Hilbert space M, there is a bounded linear operator T^* on M such that

$$(T^* x, y) = (x, Ty)$$

for all elements $x, y \in M$. The operator T^* is called the adjoint of T and $|T| = |T^*|$. The operator T is called a *self-adjoint operator* if $T = T^*$. As an example, consider the probability measure P on the Borel field $\mathscr{B}$ of subsets of Ω. Let $\mathscr{A}$ be a sub-Borel field of $\mathscr{B}$. Consider the Hilbert space $L^2(dP)$ of square integrable random variables $X(\omega)$. It can be shown that the conditional expectation with respect to $\mathscr{A}$

$$SX = E(X \mid \mathscr{A})$$

is uniquely determined by the property that it is a self-adjoint projection on $L^2(dP)$ leaving the $\mathscr{A}$-measurable functions in $L^2(dP)$ invariant.

We have assumed a linear operator is defined on all of a Banach space M. Let us refer to a linear mapping defined on a linear subspace of M and mapping it into M as a linear transformation. Let T be a bounded linear operator on M. If T is one-to-one, then T^{-1} is a linear transformation defined on the range $R = \{Tx : x \in M\}$ of T and is such that if $y = Tx \in R$ then $T^{-1} y = x$. Consider any complex number λ. The number λ is an element of the *spectrum of the operator* T if $(T - \lambda I)^{-1}$ is not well-defined as bounded linear operator on M. The complimentary set of points is called the *resolvent set of* T. The *point spectrum of* T is a subset of the spectrum consisting of the complex numbers λ for which there is an element $x \in M$, $x \neq 0$, such that

$$Tx = \lambda x.$$

The element x is called the eigenelement corresponding to eigenvalue λ. If T is a bounded self-adjoint operator on a Hilbert space M then (Ty, y) is real for all $y \in M$ since $(Ty, y) = (y, T^* y) = (y, Ty) = \overline{(Ty, y)}$. The point spectrum of a bounded self-adjoint operator T must then be real for if $x \neq 0$ is an eigenelement with eigenvalue λ, then $(Tx, x) - \lambda(x, x) = 0$ and this is not possible unless λ is real. Actually the entire spectrum of a bounded self-adjoint operator T must be real. First of all, $(T - \lambda I)^{-1}$ is well-defined as a linear operator on M if λ is not real. This is so because every element $y \in M$ is in the range of $(T - \lambda I)$. If there were a $y \in M$ not in the range of $(T - \lambda I)$, then one could find an element $z \in M$ orthogonal to the range of $(T - \lambda I)$, that is,

$$((T - \lambda I)x, z) = 0$$

for all $x \in M$. However, since

$$((T - \lambda I)x, z) = (x, (T - \bar{\lambda} I)z) = 0$$

for all $x \in M$, this implies that $(T - \bar{\lambda} I)z = 0$ and we know that this is impossible since $\bar{\lambda}$ cannot belong to the point spectrum of T if it is not real. Since $(T - \lambda I)^{-1}$ is well-defined one only has to show that it is a bounded operator if λ is not real. Now

$$\begin{aligned}
\|(T - \lambda I)x\|^2 &= \|Tx\|^2 - 2 \operatorname{Re} \lambda (Tx, x) + |\lambda|^2 \, \|x\|^2 \\
&\geqslant \|Tx\|^2 - 2 \, |\operatorname{Re} \lambda| \, \|Tx\| \, \|x\| + |\lambda|^2 \, \|x\|^2 \\
&\geqslant (|\lambda|^2 - |\operatorname{Re} \lambda|^2) \, \|x\|^2
\end{aligned}$$

so that

$$\|(T - \lambda I)^{-1}\| \leqslant (|\lambda|^2 - |\operatorname{Re} \lambda|^2)^{-\frac{1}{2}}$$

if λ is not real.

We shall now make a few brief remarks about the spectral theorem for bounded self-adjoint operators T. This theorem is a generalization of the result in matrix theory that represents a Hermitian $n \times n$ (n finite) matrix in terms of its eigenvectors and real eigenvalues (see [20]). Let us call an operator-valued function $S(\lambda)$ on the reals *a resolution of the identity* operator I relative to T if the operators $S(\lambda)$ are self-adjoint projection operators on the Hilbert space M that commute with each other and with T and if

1. there are finite real numbers a, b with $a<b$ such that $S(a)=0$, the null operator mapping everything into the zero element 0 of M and $S(b)=I$ the identity operator;

2. $S(\cdot)$ is a non-decreasing function of λ in the sense that

$$S(\lambda)S(\lambda')=S(\lambda)$$

if $\lambda\leqslant\lambda'$.

Given any continuous function $f(\lambda)$ one can define an operator integral

$$\int f(\lambda)S(d\lambda)$$

by showing that the sequence of operator Riemann-Stieltjes sums

$$\sum_k f\left(\frac{k}{2^n}\right)\left[S\left(\frac{k+1}{2^n}\right)-S\left(\frac{k}{2^n}\right)\right]$$

converges in operator norm as $n\to\infty$. The spectral theorem says that a bounded self-adjoint operator T has a natural integral representation in terms of a resolution of the identity.

Theorem 5. *Let T be a bounded self-adjoint linear operator on a Hilbert space M. Then there is a resolution of the identity relative to T such that*

$$T=\int\lambda S(d\lambda). \tag{13}$$

Notice that (13) implies that

$$(Tx,y)=\int\lambda(S(d\lambda)x,y)$$

for any two elements $x,y\in M$ with $(S(\lambda)x,y)$ a function of bounded variation as a function of λ. That $(S(\lambda)x,x)$ is a non-decreasing function of λ follows from the fact that $S(\cdot)$ is non-decreasing as an operator-valued function of λ. A representation of the type (13) given in Theorem 5 is called a spectral representation of the operator since the integration can be thought of as being carried out over the spectral set of the operator. Such representations are of interest for operators generally. There is a similar spectral representation for *unitary operators*, that is, operators T whose adjoint T^* is the inverse

$$TT^*=T^*T=I.$$

Such operators preserve the norm of elements of M since

$$\|x\|^2=(x,x)=(T^*Tx,x)=(Tx,Tx)=\|Tx\|^2=\|T^*x\|^2.$$

The shift transformation on the double infinite sequence space representation of a stationary Markov process generates such a unitary operator.

Let τ be the shift transformation and $L^2(dP)$ the space of square integrable random variables on the probability space of the Markov process. Then

$$(Tf)(\omega) = f(\tau\omega)$$

for $f \in L^2(dP)$ is a unitary operator since τ and τ^{-1} are well-defined and measure-preserving. A unitary operator T has a spectral representation

$$T = \int_{-\pi}^{\pi} e^{i\lambda} S(d\lambda)$$

in terms of a resolution of the identity relative to T, which is $S(\lambda)$ defined on $[-\pi, \pi]$. The integrand is now $e^{i\lambda}$ because the spectrum of a unitary operator can be shown to be a subset of the complex numbers of absolute value one.

Appendix 5

Topological Groups

A group G is a nonvacuous set of elements with a binary operation ("multiplication") $g_1 g_2$ defined for pairs of elements $g_1, g_2 \in G$ and such that

1. $g_1 g_2 \in G$ if $g_1, g_2 \in G$;
2. the multiplication is associative so that

$$g_1(g_2 g_3) = (g_1 g_2) g_3$$

for $g_1, g_2, g_3 \in G$;

3. there is an identity element $e \in G$ such that

$$eg = ge = g$$

for every element $g \in G$;

4. every element $g \in G$ has an inverse g^{-1}

$$gg^{-1} = g^{-1}g = e.$$

If the group is commutative it is usual to use addition instead of multiplication to represent the binary group operation.

If there is a topology on the group such that gg'^{-1} is a continuous binary operation on G for $g, g' \in G$ then G with this topology is called a *topological group.* If G is a locally compact Hausdorff space with respect to this topology we call G a *locally compact* (Hausdorff) *topological group.* Let H be a closed subgroup of G. Let G/H be the set of left cosets gH of H. The left cosets constitute a partition of G, that is, for $g, g' \in H$ either $gH = g'H$ or else $(gH) \cap (g'H)$ is the empty set and $\bigcup_g gH = G$. The canonical map π of G onto G/H is given by

$$\pi(g) = gH.$$

One would like to set up a topology on G/H such that π is continuous. Furthermore, if α is a continuous map of G into a topological space J that is constant on each left coset of H in G, then the map β of G/H into J satisfying $\beta(\pi(g)) = \alpha(g)$ ought to be continuous. The topology on

G/H that insures this is the topology with closed sets A those subsets of G/H such that $\pi^{-1}(A)$ is closed on G. With this topology G/H is locally compact Hausdorff if G is locally compact Hausdorff. If H is a closed normal subgroup of G, that is, if $gH=Hg$ for all $g\in G$, then G/H is itself a topological group.

Let us consider from this point on a *compact topological* (Hausdorff) group G. There is then a regular probability measure η on the Borel field $\mathscr{B}$ generated by the topology invariant under group multiplication, this is,

$$\eta(Bg)=\eta(gB)=\eta(B)$$

for all $B\in\mathscr{B}$ and $g\in G$. This measure η is called the normalized *Haar measure or uniform measure* of the group G. Its existence and uniqueness was demonstrated for a large class of compact groups in section 5.3. We shall take the existence of the Haar measure for granted. Our object is to indicate that harmonic analysis can be carried out on a compact group G. This will be done by sketching the background of the Peter-Weyl theorem. Let M be a continuous homomorphic mapping of the group G in the group of nonsingular $r\times r$ (r finite) matrices with complex-valued elements, that is, for each $g\in G$ $M(g)$ is a nonsingular matrix of order r and for $g,g'\in G$

$$M(gg')=M(g)M(g').$$

M is called a *representation of G of degree r*. It is natural to call two representations M_1, M_2 (of degree r) *equivalent* if there is a nonsingular matrix L such that

$$M_1(g)=LM_2(g)L^{-1}$$

for all $g\in G$. Given the representation M, one can show that there is always an equivalent matrix representation M' with $M'(g)$ a *unitary matrix* for every $g\in G$. We shall show that this is so by constructing a positive Hermitian quadratic form invariant with respect to the representation M. Let $Q(u)=\sum_{i=1}^{r}|u_i|^2$ with u an r-vector. The desired quadratic form is

$$Q'(u)=\int Q(M(g)u)\eta(dg)$$

obtained by integrating $Q(M(g)u)$ with respect to the Haar measure η of G. From the invariance of η and the fact that M is a representation, it follows that

$$Q'(M(g')u)=\int Q(M(gg')u)\eta(dg)=\int Q(M(g)u)\eta(dg)=Q'(u)$$

is invariant for all $g' \in G$. However, this implies that there is a nonsingular $r \times r$ matrix L such that $M'(g) = LM(g)L^{-1}$ is unitary for all $g \in G$, the matrix that reduces $Q'(u)$ to the standard form $\sum_{i=1}^{r} |u_i|^2$ by a change of coordinates. Because one can always find an equivalent unitary representation, it is reasonable to limit oneself to considering unitary representations of the group G.

Let Δ be a set of $r \times r$ matrices. The set Δ is called *reducible* if a fixed s-dimensional linear subspace with $0 < s < r$ of the r-dimensional space of vectors with complex components is mapped into itself by all the matrices of Δ. By an s-dimensional space we mean a space spanned by s linearly independent vectors. If Δ is not reducible, it is said to be *irreducible*. The following important Lemma is due to Schur and will be stated without proof.

Lemma 1. *Let Δ and Σ be two irreducible sets of square matrices of order r and s respectively. Assume that there is an $r \times s$ matrix L such that*

$$\Delta L = L\Sigma,$$

that is, for each $M \in \Delta$ there is an $M' \in \Sigma$ such that $ML = LM'$ and conversely for each $M' \in \Sigma$ there is an $M \in \Delta$ with $ML = LM'$. Then all the elements of L are zero or else $r = s$ and L is a square matrix that is nonsingular.

A unitary matrix representation M of the group G is said to be irreducible if the set of unitary matrices $M(g)$, $g \in G$, is irreducible. If M is not irreducible, it is said to be reducible. Suppose M is a reducible unitary representation of G of degree r with an invariant subspace of dimension s, $0 < s < r$. Then there is a unitary matrix U that reduces M to block diagonal form

$$UM(g)U' = \begin{pmatrix} M_1(g) & 0 \\ 0 & M_2(g) \end{pmatrix}, \qquad g \in G,$$

with M_1 and M_2 unitary representations of G of degree s and $r - s$, respectively. We can therefore limit ourselves to a discussion of irreducible unitary representations of the group G. Schur's Lemma implies that the irreducible representations of a commutative compact group are of degree one, that is, they are numerical valued. For suppose M is an irreducible representation of degree $r > 1$. Then the matrices $M(g)$, $g \in G$, commute with each other. Let $\Delta = \{M(g), g \in G\}$. Let U be one of the matrices of Δ. Set $V = U - bI$ where I is the $r \times r$ identity matrix and b a number chosen so that V has determinant zero. It then follows from Schur's Lemma and $\Delta V = V\Delta$ that V is the matrix with zero elements or equivalently that

$$U = bI.$$

But then every element of Δ is some scalar multiplied by I and therefore the representation is not irreducible.

Lemma 2. *If G is a commutative compact topological group, all irreducible representations are of degree one, that is, they are numerical valued.*

Basic orthogonality relationships for the elements of unitary representations of the compact topological group G can be established by making use of Schur's Lemma.

Lemma 3. *Let* $U(g) = \{u_{i,j}(g)\}$, $M(g) = \{m_{i,j}(g)\}$ *be two distinct irreducible non-equivalent unitary representations of the compact topological group G. Then*

$$\int u_{i,j}(g)\overline{m_{k,l}(g)}\eta(dg) = 0 \tag{1}$$

where η *is the normalized Haar measure of the group. Further, if* r *is the degree of* M, *then*

$$\int m_{i,j}(g)\overline{m}_{k,l}(g)\eta(dg) = \delta_{i-k}\delta_{j-l}\frac{1}{r}. \tag{2}$$

Let s and r be the degrees of the representations U and M respectively. Consider an $s \times s$ matrix R. Set $S(g) = U(g)RM(g)$ and $S = \int S(g)\eta(dg)$. Then

$$\begin{aligned} U(g)SM(g^{-1}) &= \int U(g)U(h)RM(h^{-1})M(g^{-1})\eta(dh) \\ &= \int U(gh)RM((gh)^{-1})\eta(dh) = S \end{aligned} \tag{3}$$

so that $U(g)S = SM(g)$. If $r = s$ and the matrix S is nonsingular, the representations U and M would be equivalent, contrary to the assumption of non-equivalence. By Schur's Lemma the elements of S must all be zero. Let R be the matrix with the element in the jth row and l^{th} column one and all other elements zero. The relations (1) follows immediately. Consider equation (3) with $U = M$ and R an $r \times r$ matrix. It is still valid. Since $M(g)S = SM(g)$, the argument used in obtaining Lemma 1 implies that $S = bI$ where b is a number and I the $r \times r$ identity matrix. Thus,

$$\int M(g)SM(g^{-1})\eta(dg) = bI. \tag{4}$$

Take the trace of both sides of (5) to obtain

$$\sum_{j=1}^{r} s_{jj} = br. \tag{5}$$

By letting the matrix R have the element in the jth row and l^{th} column one and the remaining ones zero we obtain the relations (2) from (4) and (5).

We shall give a very brief sketch of the theorem of Peter-Weyl. By Zorn's Lemma there is a maximal set of mutually non-equivalent unitary irreducible representations of the group G.

Theorem 1. *Consider a maximal set of mutually non-equivalent unitary irreducible representations of finite degree of the compact topological group G. Let Δ be the collection of elements of all these representations. The set of finite linear combinations of functions of Δ are dense in the set of continuous functions on G in the absolute supremum norm.*

Let f be any given continuous function on G. Given any fixed $\varepsilon>0$ there is a neighborhood $B=B^{-1}=\{g:g^{-1}\in B\}$ of the identity e of G such that

$$|f(g)-f(h)|<\varepsilon$$

if $g^{-1}h\in B$. Let A be a neighborhood of the identity with $\bar{A}\subseteq B$. There is by Urysohn's Lemma a continuous function q, $0\leqslant q\leqslant 1$, with $q(g)=1$ on $\bar{A}$ and $q(g)=0$ on B^c. Let $k(g)=b(q(g)+q(g^{-1}))$ where $b\geqslant 0$ is a number chosen so that $\int k(g)\eta(dg)=1$. Notice that $k(g)=k(g^{-1})$ and $k(g)=0$ on B^c. Let

$$f'(g)=\int k(g^{-1}h)\,f(h)\eta(dh).$$

Then

$$|f'(g)-f(g)|=\left|\int k(g^{-1}h)[f(h)-f(g)]\eta(dh)\right|<\varepsilon. \tag{6}$$

Consider the eigenfunctions $\varphi(g)$ of the integral equation

$$\varphi(g)=\lambda\int k(g^{-1}h)\varphi(h)\eta(dh). \tag{7}$$

By a basic result on integral equations with a continuous symmetric kernel (see [88]), the function f' can be uniformly approximated to any given fixed degree by proper finite linear combinations of eigenfunctions φ of (7). But by (6) this implies that f can be uniformly approximated by finite linear combinations of eigenfunctions of (7) to any fixed degree. The argument for the Peter-Weyl theorem will be complete if we can show that any given eigenfunction φ of an integral equation of the type (7) is a linear combination of elements of a unitary representation of the group G. Let φ be an eigenfunction of (7) with eigenvalue λ. Let $\varphi_1,\ldots,\varphi_n$ be a complete orthonormal system of eigenfunctions of (7) with eigenvalue λ. There can be at most a finite number of such eigenfunctions since

$$\int\left\{k(g^{-1}h)-\lambda^{-1}\sum_{j=1}^{n}\varphi_j(g)\varphi_j(h)\right\}^2\eta(dh)=\int(k(g^{-1}h))^2\eta(dh)-\lambda^{-2}n\geqslant 0$$

and $\int (k(g^{-1}h))^2 \eta(dh)$ is finite. If $\varphi(g)$ is an eigenfunction of (7) with eigenvalue λ it follows that $\varphi(hg)$ is also for each $h \in G$ since

$$\varphi(hg) = \lambda \int k(g^{-1}h^{-1}hl)\varphi(hl)\eta(dl) = \lambda \int k(g^{-1}l)\varphi(hl)\eta(dl).$$

Thus $\varphi_1(hg), \ldots, \varphi_n(hg)$ are eigenfunctions of (7) with eigenvalue λ and so can be expressed as linear combinations of $\varphi_1(g), \ldots, \varphi_n(g)$

$$\varphi_i(hg) = \sum_{j=1}^{n} m_{ij}(h)\varphi_j(g). \tag{8}$$

The functions $\varphi_i(hg)$, $i = 1, \ldots, n$, are linearly independent since they are orthonormal

$$\int \varphi_i(hg)\varphi_j(hg)\eta(dg) = \int \varphi_i(g)\varphi_j(g)\eta(dg) = \delta_{i-j}.$$

The functions $\varphi_i(g)$, $i = 1, \ldots, n$, consequently can be expressed as linear combinations of the functions $\varphi_i(hg)$, $i = 1, \ldots, n$. However, this implies that the matrix $M(g) = (m_{ij}(g))$ has an inverse. The functions $m_{ij}(g)$ are continuous in g since $\varphi_i(hg)$ is jointly continuous in h, g and

$$m_{ij}(h) = \int \varphi_i(hg)\varphi_j(g)\eta(dg). \tag{9}$$

Equation (9) follows from (8) on multiplying by $\varphi_j(g)$ and integrating. We now wish to show that $M(g)$ is a homomorphism. Now

$$\varphi_i(ghl) = \sum_j m_{ij}(gh)\varphi_j(l) \tag{10}$$

and

$$\varphi_i(ghl) = \sum_j m_{ij}(g)\varphi_j(hl) = \sum_{k,l} m_{ik}(g)m_{kl}(h)\varphi_j(l) \tag{11}$$

by (8). It follows from (10) and (11) that

$$m_{ij}(gh) = \sum_{k=1}^{n} m_{ik}(g)m_{kl}(h)$$

so that M is a homomorphism. Setting $g = e$ in (8) leads to

$$\varphi_i(h) = \sum_j m_{ij}(h)\varphi_j(e).$$

The eigenfunctions φ of integral equations of the type given in (7) can be expressed as linear combinations of elements of group representations and therefore, in particular, of elements of irreducible unitary representations.

Bibliography

1. Arbib, M.: Realization of stochastic systems. Ann. Math. Statist. **38**, 927—933 (1967).
2. Arnold, V.I., Avez, A.: Ergodic Problems of Classical Mechanics. New York: Benjamin 1968.
3. Bernstein, S.: Equations differentielles stochastiques. Actualités Sci. Indust. **738**, 5—31 (1938).
4. Blackwell, D., Koopmans, L.: On the identifiability problem for functions of finite Markov chains. Ann. Math. Statist. **28**, 1011—1015 (1957).
5. Bush, R.R., Mosteller, F.: Stochastic Models for Learning. New York: John Wiley 1955.
6. Canavan, G.H.: Some properties of a Lagrangian Wiener-Hermite expansion. J. Fluid Mech. **41**, 405—412 (1970).
7. Chacon, R.V., Ornstein, D.: A general ergodic theorem. Illinois J. Math. **4**, 153—160 (1960).
8. Chung, K.L., Fuchs, W.H.J.: On the distribution of values of sums of random variables. Mem. Amer. Math. Soc. **6**, 12 (1951).
9. — Markov Chains with Stationary Transition Probabilities. Berlin-Göttingen-Heidelberg: Springer 1960.
10. Coddington, E.A., Levinson, N.: Theory of Differential Equations. New York: McGraw-Hill 1955.
11. Cogburn, R.: Asymptotic properties of stationary sequences. University of California Publications in Statistics **3**, 99—146 (1960).
12. — Conditional probability operators. Ann. Math. Statist. **33**, 634—658 (1962).
13. Collins, H.S.: Convergence of convolution iterates of measures. Duke Math. J. **29**, 259—264 (1962).
14. Crow, S.C., Canavan, G.H.: Relationship between a Wiener-Hermite expansion and an energy cascade. J. Fluid Mech. **41**, 387—403 (1970).
15. Dharmadhikari, S.W.: Sufficient conditions for a stationary process to be a function of a finite Markov chain. Ann. Math. Statist. **34**, 1033—1041 (1963).
16. Dobrushin, R.L.: The description of a random field by means of conditional probabilities and conditions of its regularity. Theor. Probability Appl. **13**, 197—224 (1968).
17. Doeblin, W.: Sur les propriétés asymptotiques de mouvement régis par certains types de chaines simples. Bull. Math. Soc. Roumanie Sci. **39**, No. 1, 57—115; No. 2, 3—61 (1937).
18. Doob, J.L.: Stochastic Processes. New York: John Wiley 1953.
19. Dudley, R.M.: Random walks on abelian groups. Proc. Amer. Math. Soc. **13**, 447—450 (1962).
20. Dunford, N., Schwartz, J.T.: Linear Operators, Part I. New York: Interscience 1958.
21. — — Linear Operators, Part II. New York: Interscience 1963.

22. Dvoretsky, A.: Central limit theorems for dependent random variables, to be published in Proc. 6th Berkeley Symp. Math. Statist. Prob.
23. Dynkin, E.B.: Markov Processes, vol. 1. Berlin-Heidelberg-New York: Springer 1965.
24. Erickson, R.V.: Functions of Markov chains. Ann. Math. Statist. **41**, 843—850 (1970).
25. Feller, W.: Non-Markovian processes with the semigroup property. Ann. Math. Statist. **30**, 1252—1253 (1959).
26. — An Introduction to Probability Theory and its Applications, vol. 1. New York: Wiley 1960.
27. — An Introduction to Probability Theory and its Applications, vol. 2. New York: Wiley 1966.
28. Foguel, S.R.: Existence of a σ finite invariant measure for a Markov process on a locally compact space. Israel J. Math. **6**, 1—5 (1968).
29. Gilbert, E.J.: On the identifiability problem for functions of finite Markov chains. Ann. Math. Statist. **30**, 688—697 (1959).
30. Goldstein, H.: Classical Mechanics. Cambridge, Mass.: Addison-Wesley 1950.
31. Gnedenko, B.V., Kolmogorov, A.N.: Limit Distributions for Sums of Independent Random Variables, Cambridge, Mass.: Addison-Wesley 1954.
32. de Groot, S.R., Mazur, P.: Non-equilibrium Thermodynamics. Amsterdam: North Holland 1962.
33. Hachigian, J., Rosenblatt, M.: Functions of reversible Markov processes that are Markovian. J. Math. Mech. **11**, 951—960 (1962).
34. Hanson, D.L.: On the representation problem for stationary stochastic processes with trivial tail field. J. Math. Mech. **12**, 293—301 (1963).
35. — A representation theorem for stationary Markov chains. J. Math. Mech. **12**, 731—736 (1963).
36. Harris, T.E.: The existence of stationary measures for certain Markov processes. Proc. 3rd Berkeley Symp. Math. Statist. Prob. **2**, 113—124 (1956).
37. Heble, M., Rosenblatt, M.: Idempotent measures on a compact topological semigroup. Proc. Amer. Math. Soc. **14**, 177—184 (1963).
38. Heller, A.: On stochastic processes derived from Markov chains. Ann. Math. Statist. **36**, 1286—1291 (1965).
39. Helson, H., Sarason, D.: Past and future. Math. Scand. **21**, 5—16 (1967).
40. Hopf, E.: The general temporally discrete Markov process. J. Math. Mech. **3**, 13—46 (1954).
41. — On the ergodic theorem for positive linear operators. J. Reine Angew. Math. **205**, 101—106 (1960/61).
42. — An inequality for positive linear integral operators. J. Math. Mech. **12**, 683—692 (1963).
43. Hunt, G.A.: Markov chains and Martin boundaries. Illinois J. Math. **4**, 313—340 (1960).
44. Ibragimov, I.A.: Some limit theorems for stationary processes. Theor. Probability Appl. **7**, 349—382 (1962).
45. — On the spectrum of stationary Gaussian sequences satisfying the strong mixing condition. I. Necessary conditions. Theor. Probability Appl. **10**, 85—106 (1965).
46. Iosifescu, M., Theodorescu, R.: Random Processes and Learning. Berlin-Heidelberg-New York: Springer 1969.
47. Ito, K.: On stochastic differential equations. Mem. Amer. Math. Soc. **4**, 51 (1951).

48. — Nisio, M.: On stationary solutions of a stochastic differential equation. J. Math. Kyoto Univ. **4**, 1—75 (1964).
49. — McKean, Jr., H.P.: Diffusion Processes and Their Sample Paths. Berlin-Heidelberg-New York: Springer 1965.
50. Jain, N., Jamison, B.: Contributions to Doeblin's theory of Markov processes. Z. Wahrscheinlichkeitstheorie und Verw. Gebiete **8**, 19—40 (1967).
51. Jirina, M.: On regular conditional probabilities. Czechoslovak Math. J. 445—450 (1959).
52. Kac, M.: Probability and Related Topics in Physical Sciences. London: Interscience 1959.
53. Kawada, Y., Ito, K.: On the probability distribution on a compact group I. Proc. Phys.-Math. Soc. Japan **22**, 226—278 (1940).
54. Kemeny, J.G., Snell, J.L., Knapp, A.W.: Denumerable Markov Chains. Princeton: Van Nostrand 1966.
55. Kendall, D.G.: Unitary dilations of Markov transition operators, and corresponding integral representations for transition probability matrices. Probability and Statistics (ed. U. Grenander), John Wiley 1959.
56. Kesten, H., Spitzer, F.: Random walk on countably infinite abelian groups. Acta Math. **114**, 237—265 (1965).
57. — The Martin boundary of recurrent random walks on countable groups. Proc. 5th Berkeley Symp. Math. Statist. Prob. **2**, 51—74 (1967).
58. Kloss, B.M.: Probability distributions on bicompact topological groups. Theor. Probability Appl. **4**, 237—270 (1959).
59. Kolmogorov, A.N., Rozanov, Yu.A.: On strong mixing conditions for stationary Gaussian processes. Theor. Probability Appl. **5**, 204—208 (1960).
60. — On the approximation of distributions of sums of independent summands by infinitely divisible distributions. Sankhya Ser. A. **25**, 159—174 (1963).
61. — Fomin, S.V.: Introductory Real Analysis. Prentice-Hall 1970.
62. Kurosh, A.G.: The Theory of Groups. New York: Chelsea 1956.
63. Le Cam, L.: On the distribution of sums of independent random variables. Bernoulli, Bayes, Laplace Anniversary Volume. Springer, 179—202 (1965).
64. de Leeuw, K., Glicksberg, I.: Applications of almost periodic compactifications. Acta Math. **105**, 63—97 (1961).
65. Levinson, N.: An elementary proof of the stationary distribution for an irreducible Markov chain. Amer. Math. Monthly **72**, 366—369 (1965).
66. Lévy, P.: Théorie de L'addition des Variables Aléatoires. Paris: Gauthier-Villars 1937.
67. — L'addition des variables aléatoires définis sur une circonférence. Bull. Soc. Math. France **67**, 1—41 (1939).
68. — Examples de processus pseudo-markoviens. C. R. Acad. Sci. Paris Sér. A, **228**, 2004—2006 (1949).
69. Ljapin, E.S.: Semigroups. Amer. Math. Soc. Transl. (1968).
70. Loéve, M.: Probability Theory. Princeton: Van Nostrand 1963.
71. Lorch, E.R.: Spectral Theory. New York: Oxford 1962.
72. MacLane, S., Birkhoff, G.: Algebra. Macmillan 1967.
73. Martin-Löf, P.: Probability theory on discrete semigroups. Z. Wahrscheinlichkeitstheorie und Verw. Gebiete **4**, 78—102 (1965).
74. McShane, E.J.: Stochastic integrals and non-linear processes. J. Math. Mech. **11**, 235—284 (1962).
75. Minlos, R.A.: Lectures on statistical physics. Russian Math. Surveys **23**, 137—196 (1968).
76. Meshalkin, L.D.: On the approximation of polynomial distributions by infinitely divisible laws. Theor. Probability Appl. **5**, 106—114 (1960).

77. Moore, III, B., Page, L. B.: The class ω of operator-valued weight functions. J. Math. Mech. **19**, 1011—1017 (1970).
78. Neveu, J.: Existence of bounded invariant measures in ergodic theory. Proc. 5th Berkeley Symp. Math. Statist. Prob. **2**, 461—472 (1967).
79. Nisio, M.: Remark on the canonical representation of strictly stationary processes. J. Math. Kyoto Univ. **1**, 129—146 (1961).
80. Norman, M. F.: Mathematical learning theory. Mathematics of the Decision Sciences, Part 2. (Ed. Dantzig, G. B., Veinott, A. F. J.) Amer. Math. Soc. 283—313 (1968).
81. — Some convergence theorems for stochastic learning models with distance diminishing operators. J. Mathematical Psychology **5**, 61—101 (1968).
82. Orey, S.: Recurrent Markov chains. Pacific J. Math. **9**, 805—827 (1959).
83. Ornstein, D. S.: Bernoulli shifts with the same entropy are isomorphic, to be published in Advances in Math.
84. Pfanzagl, J.: On the existence of regular conditional probabilities. Z. Wahrscheinlichkeitstheorie und Verw. Gebiete **11**, 244—256 (1968).
85. Port, S. C., Stone, C. J.: Potential theory of random walks on abelian groups. Acta Math. **122**, 19—114 (1969).
86. Prohorov, Yu. V.: On a uniform limit theorem of A. N. Kolmogorov. Theor. Probability Appl. **5**, 98—106 (1960).
87. Pym, J. S.: Idempotent measures on semigroups. Pacific J. Math. **12**, 685—698 (1962).
88. Riesz, F., Sz.-Nagy, B.: Functional Analysis. Ungar 1955.
89. Rosenblatt, D.: Aggregation in matrix models of resource flows. Amer. Statist. 36—39 (1965).
90. — Aggregation in matrix models of resource flows II. Amer. Statist. **21**, 32—37 (1967).
91. Rosenblatt, M.: A central limit theorem and a strong mixing condition. Proc. Nat. Acad. Sci. U. S. A. **42**, 43—47 (1956).
92. — Stationary processes as shifts of functions of independent random variables. J. Math. Mech. **8**, 665—682 (1959).
93. — Random Processes. New York: Oxford 1962.
94. — Equicontinuous Markov operators. Theor. Probability Appl. **9**, 205—222 (1964).
95. — Products of independent, identically distributed stochastic matrices. J. Math. Anal. Appl. **11**, 1—10 (1965).
96. — Central limit theorem for stationary processes, to be published in Proc. 6th Berkeley Symp. Math. Statist. Prob.
97. Rozanov, Yu. A., Volkonskii, V. A.: Some limit theorems for random functions I. Theor. Probability Appl. **4**, 178—197 (1959).
98. — Stationary Random Processes. San Francisco: Holden-Day 1967.
99. — On Gaussian fields with given conditional distributions. Theor. Probability Appl. **12**, 381—391 (1967).
100. Royden, H. L.: Real Analysis. New York: Macmillan 1968.
101. Rudin, W.: Fourier Analysis on Groups. New York: John Wiley 1962.
102. Schwartz, J. T.: Lectures on the Mathematical Method in Analytical Economics. New York: Gordon and Breach 1961.
103. Sinai, Ja. G.: On the foundations of the ergodic hypothesis for a dynamical system of statistical mechanics. Soviet Math. Dokl. 1818—1821 (1963).
104. Sperner, E.: Ein Satz über Untermengen einer endlichen Menge. Math. Z. **27**, 544—548 (1928).
105. Spitzer, F.: Principles of Random Walk. Princeton: Van Nostrand 1964.

106. Smorodinsky, M.: On Ornstein's isomorphism theorem for Bernoulli shifts, to be published in Advances in Math.
107. Sz. Nagy: Extensions of linear transformations in Hilbert space which extend beyond the space. New York: F. Ungar. Appendix to F. Riesz and B. Sz.-Nagy Functional Analysis 1960.
108. Suppes, P.: A linear model for a continuum of responses. Studies in Mathematical Learning Theory. (Ed. Bush, R. R. and Estes, W. K.) 400—414, Stanford 1959.
109. Tortrat, A.: Lois tendues μ sur un demi-groupe topologique complètement simple X. Z. Wahrscheinlichkeitstheorie und Verw. Gebiete **6**, 145—160 (1966).
110. Uhlenbeck, G.E., Ford, G.W.: Lectures in Statistical Mechanics. Amer. Math. Soc. 1963.
111. Wiener, N.: Nonlinear Problems in Random Theory. New York: John Wiley 1958.
112. Williamson, J.H.: Harmonic analysis on semigroups. J. London Math. Soc. **42**, 1—41 (1967).
113. Zygmund, A.: Trigonometric Series, **2**, Cambridge 1959.

Postscript

There are a number of recent publications that should be mentioned briefly since they relate to some of the topics that have been considered in this volume. The first of these is *S. R. Foguel, The ergodic theory of Markov processes, 102 pp., Van Nostrand 1969*. Foguel's little monograph has an excellent development of salient aspects of ergodic theory for iterates of the transition probability function of a Markov process.

Recently the paper of *A. Mukherjea and N. A. Tserpes*, "*Idempotent measures on locally compact semigroups*", *Proc. Amer. Math. Soc. 29, 143–150 (1971)* appeared and extended the representation obtained in section 5.4 for idempotent measures to idempotent measures on locally compact semigroups.

A paper of *C. Stein*, "*A bound for the error in the normal approximation to the distribution of a sum of dependent random variables*" will be published in the Proceedings of the 6th Berkeley Symposium on Probability and Mathematical Statistics. The most interesting result deals with an error term of the Berry-Esseen type for a class of stationary sequences. Let $\{X_i\}$ be a stationary sequence with $E X_i = 0$, $E X_i^2 = 1$, $\beta = E X_i^8 < \infty$, and

$$0 < c = \lim_{n \to \infty} \frac{1}{n} \operatorname{var}\left(\sum_1^n X_i\right) < \infty .$$

Under the assumption the maximal correlation $c(n)$ between $\mathscr{B}_0$ and $\mathscr{F}_n$ measurable square integrable random variables tends to zero exponentially as $n \to \infty$, it is shown that there is a constant A independent of n such that

$$\left| P\left\{ \frac{\sum_1^n X_i}{\sigma\left(\sum_1^n X_i\right)} \le a \right\} - \Phi(a) \right| \le A n^{-\frac{1}{2}} (\log n)^2 .$$

As a last remark we mention a manuscript of D. Sarason, "*An addendum to 'Past and Future'*", in which he effectively characterizes the spectral density w of a strongly mixing Gaussian stationary process as having the form

$$w = |P|^2 \exp(u + \tilde{v})$$

where $P(e^{i\lambda})$ is a trigonometric polynomial and $u(e^{i\lambda})$, $v(e^{i\lambda})$ are continuous functions with $\tilde{v}$ the conjugate function of v.

Author Index

Subject Index

Notation

1.1 Ω
$P(x, A)$, $x \in \Omega$, $A \in \mathscr{A}$, B
μ, ν
$A_0 \times A_1 \times \cdots \times A_n$
$\mathscr{A}_\infty$, Ω_∞, $\omega = (x_0, x_1, \ldots)$
P_μ
$X_n(\omega) = x_n$
$P_n(\cdot, \cdot)$
τ
$\mathscr{A}_m^n$, $\mathscr{B}_n$, $\mathscr{F}_m$
$P(C|\mathscr{B})(\omega)$
${}_kP(\cdot, \cdot)$, T
$L^p(d\mu)$
$\|f\|_{\mu, p}$, $\delta(x, A)$

1.2 $p_{j,k}$, $p_{j,k}^{(n)}$, P
ρ, $f_{j,k}^{(n)}$, $F_{j,k}(s)$, $G_{j,k}(s)$, ${}_ip_{i,j}^*$
$q_{i,j}^{(n)}$, $e_{h,i}$, E

1.3 λ, Φ, σ^2, m, $F * G$, $\hat{m}$, α, φ, d, ψ, π

1.4 N

1.5 θ, Re, L, $\oplus$

2.1 H, T, V, U, $\bar{p}$, $\bar{r}$, $\dfrac{dq_i}{dt}$, $\dfrac{\partial H}{\partial p_i}$

3.1 Y, cos, $\mathscr{A}'$, $(f, g)_\mu = \int f(x) g(x) \mu(dx) = \mu f g$
$E_\mu(f|\mathscr{C})$, S_μ, $T^{(n)}$, M
$E(\lambda)$, $N_\lambda^{\nu(\lambda)}$, $e(\lambda)$, U, $A_n \wedge A$

3.2 $\mathscr{I}$, ϕ, μ_A, B^c, η, N_n

3.3 $p(x_1, x_2, \ldots, x_n)$, A_S, ξ, α, β, (L, q, l_0), sS'-module

4.2 $P_A(x, B)$, Q_A

4.3 L_+^1, $|T|$, $Q_n(f, p)$, $S_n(f)$, f^+, f^-, ω, χ_n
$\hat{f}$, I_A, $\wedge$, f_ε

4.4 $\mathscr{M}$, I_f, $L(f)$, c_f, $\|\cdot\|$, $L_n(f;x)$, T_N

4.5 $\|f\|_2$, $\mathscr{U}_j$, $\mathscr{C}_j$, $\overline{\mathscr{U}}$, $\overline{\mathscr{C}}$, $\mathscr{D}$, $\mathscr{D}^*$

5.1 $\hat{v}$, $L(p)$, $\sigma(v)$

5.2 G, e, $s^{-1}A$, As^{-1}

5.3 $\overline{T}$, $S(v)$

5.4 J, K

5.5 $\Sigma(v)$, η_e

6.1 r, $\mathscr{M}_n$, $\mathscr{P}(X_n; \mathscr{M}_{-\infty})$

7.1 $f_{k,n}$, $Q_F(l)$, $F^{-1}(y)$, $\alpha(\theta)$, $\beta(\theta)$

Die Grundlehren der mathematischen Wissenschaften in Einzeldarstellungen mit besonderer Berücksichtigung der Anwendungsgebiete

---

Neu seit 1967

14. Klein: Elementarmathematik vom höheren Standpunkte aus. 1. Band: Arithmetik, Algebra, Analysis. DM 24,—
15. Klein: Elementarmathematik vom höheren Standpunkte aus. 2. Band: Geometrie. DM 24,—
16. Klein: Elementarmathematik vom höheren Standpunkte aus. 3. Band: Präzisions- und Approximationsmathematik. DM 19,80
22. Klein: Vorlesungen über höhere Geometrie. DM 28,—
26. Klein: Vorlesungen über nicht-euklidische Geometrie. DM 24,—
27. Hilbert/Ackermann: Grundzüge der theoretischen Logik. DM 38,—
30. Lichtenstein: Grundlagen der Hydromechanik. DM 38,—
32. Reidemeister: Vorlesungen über Grundlagen der Geometrie. DM 18,—
38. Neumann: Mathematische Grundlagen der Quantenmechanik. DM 28,—
40. Hilbert/Bernays: Grundlagen der Mathematik I. DM 68,—
43. Neugebauer: Vorlesungen über Geschichte der antiken mathematischen Wissenschaften. 1. Band: Vorgriechische Mathematik. DM 48,—
50. Hilbert/Bernays: Grundlagen der Mathematik II. DM 84,—
61. Maak: Fastperiodische Funktionen. DM 38,—
67. Byrd/Friedman: Handbook of Elliptic Integrals for Engineers and Scientists. 2nd edition, revised. DM 64,—
68. Aumann: Reelle Funktionen. DM 68,—
76. Tricomi: Vorlesungen über Orthogonalreihen. DM 68,—
78. Lorenzen: Einführung in die operative Logik und Mathematik. DM 54,—
87. van der Waerden: Mathematische Statistik. DM 68,—
94. Funk: Variationsrechnung und ihre Anwendung in Physik und Technik. DM 120,—
114. MacLane: Homology. DM 62,—
116. Hörmander: Linear Partial Differential Operators. DM 42,—
117. O'Meara: Introduction to Quadratic Forms. DM 68,—
120. Collatz: Funktionalanalysis und numerische Mathematik. DM 58,—
123. Yosida: Functional Analysis. DM 66,—
124. Morgenstern: Einführung in die Wahrscheinlichkeitsrechnung und mathematische Statistik. DM 38,—
127. Hermes: Enumerability. Decidability. Computability. DM 39,—
133. Haupt/Künneth: Geometrische Ordnungen. DM 68,—
134. Huppert: Endliche Gruppen I. DM 156,—
135. Handbook for Automatic Computation. Vol 1/Part a: Rutishauser: Description of ALGOL 60. DM 58,—
136. Greub: Multilinear Algebra. DM 32,—
137. Handbook for Automatic Computation. Vol. 1/Part b: Grau/Hill/Langmaack: Translation of ALGOL 60. DM 64,—
138. Hahn: Stability of Motion. DM 72,—
139. Mathematische Hilfsmittel des Ingenieurs. Herausgeber: Sauer/Szabó. 1. Teil. DM 88,—

140. Mathematische Hilfsmittel des Ingenieurs. Herausgeber: Sauer/Szabó. 2. Teil. DM 136,—
141. Mathematische Hilfsmittel des Ingenieurs. Herausgeber: Sauer/Szabó. 3. Teil. DM 98,—
142. Mathematische Hilfsmittel des Ingenieurs. Herausgeber: Sauer/Szabó. 4. Teil. DM 124,—
143. Schur/Grunsky: Vorlesungen über Invariantentheorie. DM 32,—
144. Weil: Basic Number Theory. DM 48,——
145. Butzer/Berens: Semi-Groups of Operators and Approximation. DM 56,—
146. Treves: Locally Convex Spaces and Linear Differential Equations. DM 36,—
147. Lamotke: Semisimpliziale algebraische Topologie. DM 48,—
148. Chandrasekharan: Introduction to Analytic Number Theory. DM 28,—
149. Sario/Oikawa: Capacity Functions. DM 96,—
150. Iosifescu/Theodorescu: Random Processes and Learning. DM 68,—
151. Mandl: Analytical Treatment of One-dimensional Markov Processes. DM 36,—
152. Hewitt/Ross: Abstract Harmonic Analysis. Vol. II. Structure and Analysis for Compact Groups. Analysis on Locally Compact Abelian Groups. DM 140,—
153. Federer: Geometric Measure Theory. DM 118,—
154. Singer: Bases in Banach Spaces I. DM 112,—
155. Müller: Foundations of the Mathematical Theory of Electromagnetic Waves. DM 58,—
156. van der Waerden: Mathematical Statistics. DM 68,—
157. Prohorov/Rozanov: Probability Theory. DM 68,—
159. Köthe: Topological Vector Spaces I. DM 78,—
160. Agrest/Maksimov: Theory of Incomplete Cylindrical Functions and their Applications. DM 84,60
161. Bhatia/Szegö: Stability Theory of Dynamical Systems. DM 58,—
162. Nevanlinna: Analytic Functions. DM 76,—
163. Stoer/Witzgall: Convexity and Optimization in Finite Dimensions I. DM 54,—
164. Sario/Nakai: Classification Theory of Riemann Surfaces. DM 98,—
165. Mitrinović: Analytic Inequalities. DM 88,—
166. Grothendieck/Dieudonné: Eléments de Géométrie Algébrique I. DM 84,—
167. Chandrasekharan: Arithmetical Functions. DM 58,—
168. Palamodov: Linear Differential Operators with Constant Coefficients. DM 98,—
170. Lions: Optimal Control of Systems Coverned by Partial Differential Equations. DM 78,—
171. Singer: Best Approximation in Normed Linear Spaces by Elements of Linear Subspaces. DM 60,—
172. Bühlmann: Mathematical Methods in Risk Theory. DM 52,—
173. F. Maeda/S. Maeda: Theory of Symmetric Lattices. DM 48,—
174. Stiefel/Scheifele: Linear and Regular Celestial Mechanics. DM 68,—
175. Larsen: An Introduction to the Theory of Multipliers. DM 84,—
176. Grauert/Remmert: Analytische Stellenalgebren. DM 64,—
177. Flügge: Practical Quantum Mechanics I. DM 70,—
178. Flügge: Practical Quantum Mechanics II. DM 60,—
179. Giraud: Cohomologie non abélienne. DM 109,—
186. Wilkinson/Reinsch: Handbook for Automatic Computation, Vol. II. Linear Algebra. DM 72,—